AF742107

Fish Chemosenses

Fish Chemosenses

Editors

Klaus Reutter

Anatomisches Institut
Universität Tübingen
Tübingen, Germany

B.G. Kapoor

Formerly Professor of Zoology
Gwalior University
Gwalior, India

Science Publishers, Inc.

Enfield (NH), USA Plymouth, UK

SCIENCE PUBLISHERS, INC.
Post Office Box 699
Enfield, New Hampshire 03784
United States of America

Internet site: *http://www.scipub.net*

sales@scipub.net (marketing department)
editor@scipub.net (editorial department)
info@scipub.net (for all other enquiries)

ISBN 1-57808-319-2

© 2005, Copyright reserved

All rights reserved. No part of this publication may be reproduced, stored in a retrieval system, or transmitted in any form or by any means, electronic, mechanical, photocopying or otherwise, without the prior permission.

This book is sold subject to the condition that it shall not. By way of trade or otherwise be lent, re-sold, hired out, or otherwise circulated without the publisher's prior consent in any form of binding or cover other than that in which it is published and without a similar condition including this condition being imposed on the subsequent purchaser.

Published by Science Publishers, Inc., Enfield, NH, USA

Preface

FISH CHEMOSENSES is the third book that exclusively comprises information on this subject; it follows Toshiaki Hara's "Chemoreception in Fishes" (1982) and "Fish Chemoreception" (1992). Besides these two works several comprehensive chapters concerning this or related fields were written, but these are scattered over books with more general titles such as "Mechanisms of Taste Transduction" (Simon and Roper, 1993), "The Neurobiology of Taste and Smell" (Finger *et al.*, 2nd. ed. 2000), "Sensory Biology of Jawed Fishes" (Kapoor and Hara, 2001) and "The Senses of Fish" (G. von der Emde *et al.*, 2004), etc. Due to the overwhelming growth of our knowledge about fish chemosenses during the last 15 years or so, it now seems to be time to edit a further chemoreception book that is reserved again solely to fish. Accordingly, this book focusses specially on the physiology of olfaction and taste, but also covers micromorphological, behavioural, ecological, ontogenetic and phylogenetic topics. It deals as well with the solitary chemosensory cells, the third chemosensory system of fishes. Not included in the book is the molecular biology of fish chemoreception: this matter is represented, in a more general manner in the volumes mentioned above and in the "Handbook of Olfaction and Gustation" (Doty ed., 2003). After all, it was not our intention to repeat the data that has recently been revised and reviewed; we favored such fields of fish chemosenses that during the last years gained importance and came into actual use, as for instance, chemosensory systems in fish ecology. As with all the other disciplines of science, the small field of fish chemosenses is becoming more and more specialized. As a consequence, it is difficult today, for example, for a morphologist to easily follow the data of an electrophysiologist, even when they both work on the same chemosensory organ. This dilemma presents an opportunity to a book such as this one which encompasses the various aspects of fish chemosenses. It introduces the reader to the main groupings within this field and guides him to widely spread literature. To the ambitious student/researcher it offers a chance to go deeper into a problem or even detect one!

It is evident that the chapters included in the book are authored by colleagues of varying backgrounds. Some of the chapters are more in the nature of review papers, whilst others are near to an original paper. This led to some variability in the chapters but we did not want to mold them into a distinct format: to us, the result looks quite lively!

After all the stress and frustrations we had in waiting for, and working with, the manuscripts we now really happy that Fish Chemosenses is finished! We dedicate the book to all persons who are working with this fascinating group of animals, the fishes, and even laymen who find fish fascinating. And lastly, many thanks to all the authors and friends for their contributions.

Tübingen and Gwalior, 2005 Klaus Reutter and B.G. Kapoor

Contents

List of Contributors

B. Albrecht
Physiologisches Institut der Universität, Humboldtallee 23, 37073 Göttingen, Germany

C. Conze
Physiologisches Institut der Universität, Humboldtallee 23, 37073 Göttingen, Germany

H. Dolle
Physiologisches Institut der Universität, Humboldtallee 23, 37073 Göttingen, Germany

Kjell B. Døving
Department of Biology, Division of General Physiology, University of Oslo, Box 1051 Blindern, N-0316 Oslo, Norway.
E-mail: k.b.doving@bio.uio.no

Lev Fishelson
Department of Zoology, George S. Wise Faculty of Life Sciences, Tel Aviv University, Tel Aviv 29978, Israel. Tel: 972 3 640 8655; Fax: 972 3 640 9403; E-mail: fishelv@post.tau.ac.il

M. Foitzik
Physiologisches Institut der Universität, Humboldtallee 23, 37073 Göttingen, Germany

A. Hamadeh
Physiologisches Institut der Universität, Humboldtallee 23, 37073 Göttingen, Germany

El Hassan Hamdani

Department of Biology, Division of General Physiology, University of Oslo, Box 1051 Blindern, N-0316 Oslo, Norway.
E-mail: e.h.hamdani@bio.uio.no

Anne Hansen

University of Colorado Health Sciences Center, Cell and Developmental Biology, Denver, CO, USA.
E-mail: AnneHansen@UCHSC.edu

Toshiaki J. Hara

Canada Department of Fisheries and Oceans, Freshwater Institute, Winnipeg, Manitoba, Canada R3T 2N6.
E-mail: thara@cc.umanitoba.ca

N. Hessenius

Physiologisches Institut der Universität, Humboldtallee 23, 37073 Göttingen, Germany

Erik Höglund

Department of Biology, Division of General Physiology, University of Oslo, Box 1051 Blindern, N-0316 Oslo, Norway.
E-mail: erik.hoglund@bio.uio.no

U. Jakob

Physiologisches Institut der Universität, Humboldtallee 23, 37073 Göttingen, Germany

Alexander O. Kasumyan

Department of Ichthyology, Faculty of Biology, Moscow State University, 119899 Moscow, Russia. E-mail: alex_kasumyan@mail.ru

Sadao Kiyohara

Department of Chemistry and BioScience, Faculty of Science, Kagoshima University, Kagoshima 890-0065, Japan.
E-mail: kiyohara@sci.kagoshima-u.ac.jp

R. Köhnke

Physiologisches Institut der Universität, Humboldtallee 23, 37073 Göttingen, Germany

A. Kokemüller

Physiologisches Institut der Universität, Humboldtallee 23, 37073 Göttingen, Germany

L.G.C. Lüthje
Physiologisches Institut der Universität, Humboldtallee 23, 37073 Göttingen, Germany

Elena S. Mikhailova
Department of Ichthyology, Faculty of Biology, Moscow State University, 119899 Russia

A.M. Möller-de Beer
Physiologisches Institut der Universität, Humboldtallee 23, 37073 Göttingen, Germany

Klaus Reutter
Anatomisches Institut, Universität Tübingen, Tübingen, Germany. E-mail: klaus.reutter@web.de

K. Rindermann
Physiologisches Institut der Universität, Humboldtallee 23, 37073 Göttingen, Germany

Junzo Tsukahara
Department of Chemistry and BioScience, Faculty of Science, Kagoshima University, Kagoshima 890-0065, Japan.
E-mail: junzo@sci.kagoshima-u.ac.jp

Arvo O. Tuvikene
Limnological Station, Institute of Zoology and Botany, Estonian Agricultural University, 61101 Rannu, Tartu County, Estonia.
E-mail: arvot@zbi.ee

Tine Valentinčič
Department of Biology, University of Ljubljana, Veèna pot 111, 1000 Ljubljana. E-mail: tine.valentincic@uni-lj.si

H.-J. Wagner
Graduate School of Neural & Behavioural Sciences and Max Planck Research School, Anatomisches Institut, Universität Tübingen, Österbergstr. 3, D-72074 Tübingen, Germany. Tel: x49 (0)7071 297 3019; Fax: x49 (0) 7071 29 4014; E-mail: hjwagner@anatu.uni-tuebingen.de

H.-G. Willms
Physiologisches Institut der Universität, Humboldtallee 23, 37073 Göttingen, Germany

Martin Witt

Anatomisches Institut, Technische Universität Dresden, Dresden, Germany. E-mail: martin.witt@mailbox.tu.dresden.de

Eckart Zeiske

Zoological Institute and Zoological Museum, University of Hamburg, 20146 Hamburg, Germany. E-mail: zeiske@zoologie.uni-hamburg.de

H.P. Zippel

Physiologisches Institut der Universität, Humboldtallee 23, 37073 Göttingen, Germany. Tel: ++49 551 395918; Fax: ++49 551 395923; E-mail: hpz@neuro-physiol.med.uni-goettingen.de

CHAPTER

1

Development and Evolution of the Olfactory Organ in Gnathostome Fish

Eckart Zeiske[1] and Anne Hansen[2]

ABSTRACT

Ample information is available on the morphological structures of the fully developed olfactory organs in gnathostome fish, but little is known about their morphogenesis from the olfactory placodes. At the ultrastructural level, remarkable differences were found when sturgeons (ancestral actinopterygians) and the teleost *Danio rerio* (zebrafish; a more advanced actinopterygian species) were compared. To facilitate generalizations on the morphogenesis of the olfactory organs in teleosts, we extended our investigations to atherinomorph fishes, since representatives of this group show conspicuous differences in the morphological structures of their fully grown olfactory organs. The results of our investigations revealed differences in the mechanisms by which the olfactory placode opens to the exterior and

Address for Correspondence: Eckart Zeiske[1] Zoological Institute and Zoological Museum, University of Hamburg, 20146 Hamburg, Germany. E-mail: zeiske@zoologie.uni-hamburg.de
[2]Department of Cell and Developmental Biology, University of Colorado Health Sciences Center at Fitzsimons, Aurora, CO 80045, USA. E-mail: Anne.Hansen@uchsc.edu

in the mode of formation of the incurrent and excurrent nostrils. Formation of the olfactory sensory epithelium follows the same pattern in atherinomorph fishes: the main cell types (basal, receptor and supporting cells, and in some cases ciliated non-sensory cells also) originate from a subepidermal layer, as in the zebrafish. This pattern of formation is probably the common type for teleosts. It clearly differs from the scenario seen in sturgeons (*Acipenser* spp.) and lower tetrapods (*Xenopus*) in which the olfactory sensory epithelia develop from both the subepidermal layer and the covering epidermal layer. Development of the olfactory organs in gnathostome fish follows several distinct patterns. These patterns of formation are discussed with respect to our understanding of the evolution of the olfactory organ in lower vertebrates.

Key Words: Olfaction; Development; Evolution; Fish; Teleosts.

1. GENERAL CONSIDERATIONS

Olfactory organs in gnathostome fish are paired structures located in the ethmoidal region of the skull. Each olfactory organ consists of a cavity which contains the olfactory epithelium and usually opens to the exterior through two openings (ditrematous type). Among vertebrates, fish show the greatest variability in the morphological characteristics of the olfactory organ, as described in comprehensive reviews (Kleerekoper, 1969, 1982; Yamamoto, 1982, Zeiske et al., 1992). Position and shape of the olfactory openings together with specific olfactory ventilation mechanisms are understood as adaptive traits of hydrodynamic properties (including boundary layer effects).

As in tetrapods, the olfactory epithelium of lower vertebrates develops from olfactory placodes (von Kupffer, 1894). Neurogenic placodes are ectodermal thickenings that give rise to the sensory neurons of various sensory systems of vertebrates including olfactory receptor neurons (ORNs) (Webb and Noden, 1993; Miyake et al., 1997; Whitlock and Westerfield, 2000; Baker and Bronner-Fraser, 2001; with further references). Development of the olfactory organs in gnathostome fish has been investigated at the light microscopic level: selachians (Balfour, 1877; Holm, 1895; Berliner, 1902); chondrosteans (Salensky, 1881); holosteans (Balfour and Parker, 1882; Dean, 1897; Brookover, 1914); lungfish (Fullarton, 1933; Bertmar, 1965); teleosts (Holm, 1894; Gawrilenko, 1910; Laibach, 1937; Reinke, 1937; Teichmann, 1964; Jahn, 1972; Verraes, 1976; Zeiske et al., 1976a) and at the ultrastructural level: elasmobranchs (Fishelson and Baranes, 1997); sturgeons (Pyatkina, 1976; Devitsina and Kazhlayev, 1993a, b; Zeiske et al., 2003); bichirs (Bartsch et al., 1997) and teleosts (Breucker et al., 1979; Elston et al., 1981; Evans et al., 1982; Appelbaum et al., 1983, Zielinski and Hara, 1988; Hara and Zielinski, 1989; Hansen and Zeiske, 1993; Olsén, 1993; Fishelson 1997).

Investigations of the early developmental process in gnathostome fish and its genetic control are rare. Strehlow and Gilbert (1993) observed cell lineages from the 8- to 16-cell stages in zebrafish and found that only probability statements were possible with respect to the fate of the blastomeres. Whitlock and Westerfield (2000) determined the origins of the olfactory placodes in zebrafish by labelling cells along the antero-lateral edge of the neural plate and examining their later fate. They found that the olfactory placode arises from a separate field of ectodermal cells, not from a discrete region of the anterior neural plate.

In this study we describe the formation of the olfactory epithelium in early ontogeny from the initial ectodermal thickening of the olfactory placode to the fully differentiated structure in different groups of actinopterygian fish. Formation of the sensory epithelium, the olfactory cavity and its openings to the exterior, are assessed phylogenetically. Basal groups are compared to find symplesiomorphies (shared primitive structures), synapomorphies (shared derived structures), or convergent modifications (independently evolved structures of similar functional significance in different groups).

The pseudostratified olfactory epithelium consists of basal cells, supporting cells and at least three different types of olfactory receptor neurons (ORNs): ciliated, microvillous, and crypt-type cells (Bannister, 1965; Morita and Finger, 1996; Eisthen, 1997; Hansen and Zeiske, 1998; Hansen and Finger, 2000; Zeiske et al. 2003). Ciliated (ventilatory) non-sensory cells as well as goblet cells may be scattered throughout the sensory and non-sensory epithelia which line the olfactory cavity. Semi-thin and ultrathin sections through the head of early embryos revealed a line of cells between the epidermis and the anterior nerve cord in zebrafish (Hansen and Zeiske, 1993, 1994, 1995a, b) as well as in sturgeons (*Acipenser* spp.) (Zeiske et al., 2003). We termed this cell layer the subepidermal layer. Initially, the subepidermal layer as well as the covering epidermal layer consist of just one layer of cells. Then both the epidermal and the subepidermal layers thicken on either side of the developing forebrain by adding cells to the respective layer. These thickenings represent the earliest visible sign of the olfactory placodes. Their further development varies considerably in different groups of osteichthyans.

In zebrafish, receptor neurons and their supporting cells together with the ciliated non-sensory cells originate only from the subepidermal layer (Hansen and Zeiske, 1993, 1994, 1995a, b). In sturgeons, the subepidermal layer gives rise to ORNs only, whereas supporting cells as well as goblet cells and ciliated non-sensory cells derive from epithelial cells of the epidermis (Zeiske et al., 2003). The latter type is similar to the formation of the olfactory epithelium seen in the African clawed frog, *Xenopus laevis* (Klein and Graziadei, 1983; Costanzo and Graziadei, 1987; Reiss and Burd, 1997).

During early phylogeny the olfactory receptor neurons in ancestral vertebrates presumably had a more superficial position since it is characteristic of early ontogenetic stages of the olfactory organ in all recent vertebrates (Bertmar, 1967, 1969; Bjerring, 1989). It is quite probable, indeed, that the complex and specialized chemosensory organ of recent vertebrates derives from a simple, superficially situated cluster of sensory cells. During evolution the delicate sensory epithelium presumably enlarged and shifted into the depth of a roofed cavity, thereby becoming protected against mechanical abrasion. Simultaneously, additional equipment became necessary to guarantee sufficient ventilation of the olfactory cavity.

In bony fish, the olfactory apparatus is ditrematous basically. The olfactory cavity communicates with the exterior through two separate openings, an anterior incurrent and a posterior excurrent nostril (or naris). Though not completely separate, functionally this is true even for chondrichthyans (e.g. Theisen et al., 1986; Zeiske et al., 1986, 1987). The openings may be closely situated (as in sturgeons and zebrafish) or rather far apart (as in rainbowfishes, silversides, topminnows, killifishes, and live-bearers). In general, the two olfactory openings arise simultaneously from a single primordial opening of the olfactory pit, which elongates and then is partly closed by dermal extensions that bridge the middle region of the groove. This type of formation is traditionally called the salmonid type, since it was first described for *Salmo salar*. *S. salar* was the first teleostean species to be investigated in greater detail with respect to development of the olfactory organ (Holm, 1884; Gawrilenko, 1910). This type is also found in rainbowfishes (Breucker et al., 1979) and sailfin silversides (Zeiske et al., 1992); for further details see below.

Development of the olfactory organs in the striped panchax, an Old World rivuline, and in New World live bearers (guppy and swordtail) differs significantly from that of the salmonid type: the two nostrils develop successively. The primary opening gives rise only to the incurrent nostril; the excurrent nostril is formed separately. At the light microscopic level, this second type was first described for the guppy *Lebistes reticulatus* (Poeciliidae) (Reinke, 1937) and was then confirmed for two other cyprinodontiform species (Zeiske et al., 1976a). The mechanism that forms the incurrent and excurrent nostrils remained unclear, however, until EM techniques were utilized.

Systematically, rainbowfishes, rivulines, and live-bearers belong to taxon Atherinomorpha *sensu* (Greenwood et al., 1966, or Nelson, 1994) (Table 1.1).

Table 1.1

Series **Atherinomorpha** *sensu* Nelson (1994)*		
Order Atheriniformes		
	Family	Melanotaeniidae—rainbowfishes
		[*Melanotaenia* (= *Nematocentris*) *maccullochi*; *Glossolepis incisus*]
	Family	Atherinidae—silversides
	Family	Telmatherinidae—sailfin silversides
		[*Marosatherina* (= *Telmatherina*) *ladigesi*]
Order Beloniformes		
Suborder Adrianichthyoidei		
		Subfamily Oryziinae—medakas or ricefishes
		Subfamily Adrianichthyinae
		Subfamily Horaichthyinae
Suborder Belonoidei (= Exocoetoidei)		
	Family	Belonidae—garfishes/needlefishes
		(*Belone belone*)
	Family	Scomberesocidae—sauries
	Family	Exocoetidae—flyingfishes
	Family	Hemiramphidae—halfbeaks
Order Cyprinodontiformes		
	Family	Aplocheilidae—rivulines
		(*Aplocheilus lineatus*)
	Family	Fundulidae—topminnows and killifishes
	Family	Anablepidae—four-eyed fishes
	Family	Poeciliidae—poeciliids/live bearers
		(*Xiphophorus hellerii*)
	Family	Goodeidae—goodeids/splitfins
	Family	Cyprinodontidae—pupfishes

*Scientific names of species included in the present study given in parentheses.

Two fundamentally distinct constructional types of olfactory organs can be distinguished in atherinomorph fish. The most widespread type is present in rainbowfishes, silversides, sailfin silversides, rivulines, topminnows and killifishes, four-eyed fishes, live-bearers, splitfins, pupfishes and some others. These fishes are grouped in orders Atheriniformes and Cyprinodontiformes (Table 1.1). The morphology of the olfactory organs in representatives of these two taxa is characterized by a cavity that opens exteriorly by two openings relatively wide set. The excurrent nostril can be opened and closed by a valve. The pumping action of an accessory ventilation sac ensures a pulsatile flow through the olfactory cavity (Kux et al., 1988). The topological arrangement of the olfactory epithelium is simplified. Olfactory rosettes are absent. More than 90 species of Atherinomorpha have been investigated with respect to the morphological characteristics of their olfactory organs. All exhibited the type just described (Zeiske, 1973, 1974; Zeiske et al., 1976b; Melinkat and Zeiske, 1979; Melinkat, 1982).

Representatives of suborder Belonoidei (=Exocoetoidei), i.e. garfishes (= needlefishes), sauries, flyingfishes and halfbeaks, possess a morphologically different type of olfactory organ. These fishes have wide-open grooves instead of olfactory cavities. The sensory epithelium covers the groove and a papilla-like protrusion extending above the surface level of the head. This type is unique and considered highly derived within teleosts (Burne, 1909; Gupta and Shrivastava, 1973; Collette, 1974, 1976; Theisen et al., 1980).

Investigation of the morphogenesis of the olfactory organs of atherinomorph fishes intrigued us since the remarkable differences described above occur in species grouped together in a systematic entity of close relationship on the basis of mostly osteological evidence. Atherinomorpha thus present a good model system for studying divergent functional and developmental specializations of the teleostean olfactory organ in rather closely related groups. Based on light (LM) and electron microscopic techniques (scanning EM and transmission EM) we investigated selected species of garfishes, rainbowfishes, sailfin silversides, rivulines, and live-bearers from early embryonic stages to hatched larvae and juveniles to establish adequate descriptions of the developing olfactory epithelium proper and accessory hydrodynamic structures.

2. ATHERINIFORMES

2.1 Melanotaeniidae—rainbowfishes

—*Melanotaenia* (= *Nematocentris*) *maccullochi* Ogilby, 1915—Macculloch's rainbowfish

—*Glossolepis incisus* Weber, 1907—red rainbowfish (= salmon-red rainbowfish, Comscale rainbowfish)

The gross morphology and ultrastructure of the adult olfactory organ of *Melanotaenia* were described by Zeiske et al., (1979). Although some details of the development of this organ were known earlier (Breucker et al., 1979), a re-examination was deemed necessary for a better understanding of the complete ontogenetic process.

The anterior and posterior nostrils of *Melanotaenia* form according to the so-called salmonid type as outlined above. Semithin sections through the head of embryos show that olfactory placodes are already present at 2.8 mm total length (TL). In a 3 mm embryo, each placode has a lumen situated adjacent to the epidermis which still covers this lumen. In *Melanotaenia*, the lumen opens to the exterior when the embryo hatches at 4 mm TL. In *Glossolepis* embryos the olfactory pit opens at a TL of 3.5 mm. SEM pictures at higher magnification reveal that the epidermal cells above the olfactory cavity lose contact with each other. When the pit opens, olfactory knobs of ORNs are evident at the surface of the olfactory epithelium. They possess ciliary buds or a single cilium radiating from the knob. A second type of cilia appears in 4 mm embryos of *Glossolepis* and in freshly hatched larvae of *Melanotaenia*. These cilia (kinocilia, motile cilia) grow from the free surface of non-sensory cells. Kinocilia are longer than cilia of ORNs and do not radiate from olfactory knobs. In the two species investigated, microvillus-like protrusions at the free surface of the sensory epithelium belong to supporting cells which surround the ORNs. Apical cell endings obviously grow into the lumen that separates the placode from the still intact overlying epidermis. At this ontogenetic stage ORNs have already developed axons. Axon bundles penetrate the basal lamina and enter the *lamina propria* which separates the olfactory placode from the neighbouring forebrain. Horizontal ultrathin sections through the apical surface of the olfactory epithelium of a 6 mm *Melanotaenia* larva (Fig. 1.1A, B) show dendrites of the ORNs characterized by neurotubuli. In contrast to ORNs, supporting cells contain tonofilaments and a terminal web. Cross sections through cilia which extend into the free

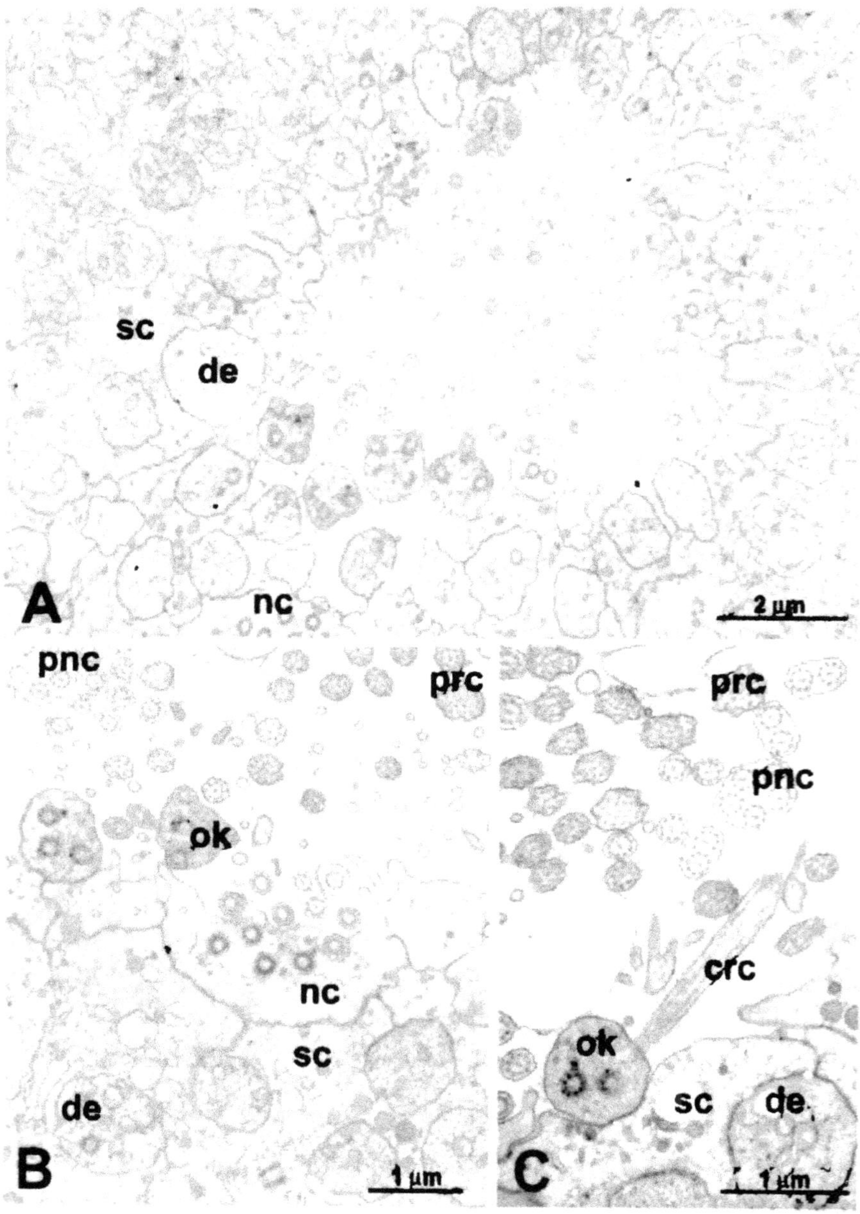

Fig. 1.1 A-C TEM: *Melanotaenia maccullochi*, larva, 6 mm TL. Sensory epithelium in horizontal section surrounding the lumen of the olfactory pit. *cnc*, cilia of non-sensory cell; *crc*, cilia of ORN; *de*, dendrite; *nc*, ciliated non-sensory cell; *ok*, olfactory knob; *pnc*, protrusion of non-sensory cell; *prc*, protrusion of ORN; *sc*, supporting cell. **A.** Overview. **B.** and **C.** Details in higher magnification.

lumen of the olfactory pit, show two different types of the axonemal pattern: the common 9 + 2 type of kinocilia and the 9 + 0 type of the ORNs. Microtubules of kinocilia have dynein arms; microtubules of ORN cilia have none. Some ORNs as well as some ciliated non-sensory cells have rod- or cone-shaped protrusions at their free surface. The pattern of their axonemal complexes conspicuously differs from the types described above. Rod- or cone-shaped protrusions of ciliated non-sensory cells contain multiple sets of microtubular complexes with the 9 + 2 complement enclosed in a single ciliary membrane. Axes through the central pairs of all microtubular sets in a single protrusion are aligned in the same direction (Fig. 1.1C). Rod- or cone-shaped protrusions of the ORNs show various numbers and irregular arrangements of microtubules (Fig. 1.1B, C). Protrusions of the ORNs as well as those of the ciliated non-sensory cells are only sporadically present in the olfactory epithelium of *Melanotaenia* and disappear early in ontogeny. Ciliated non-sensory cells are also true transitory structures in the developing olfactory epithelium of *Melanotaenia* (Breucker et al., 1979). Though ciliated non-sensory cells are totally absent in the adult olfactory epithelium (Zeiske et al., 1979), it is reasonable to assume that this cell type is genetically conserved and belongs to the basic pattern of the olfactory epithelium of rainbow-fish. Whether these cells maintain water transport through the olfactory organ during the early ontogenetic period or not is not known. Ciliated non-sensory cells decrease in number from juvenile to adult and are finally replaced by efficient accessory ventilation sacs (Breucker et al., 1979). Rapid developmental progress during the early larval stages involves cell turnover and cell death, as indicated by cell debris in the olfactory pit (Fig. 1.1A-C).

2.2 Telmatherinidae—sailfin silversides

Marosatherina (= *Telmatherina*) (Ahl, 1936)—Celebes rainbowfish

Formation of the olfactory openings is similar to that found in *Melanotaenia* and thus parallels the salmonid type (Fig. 1.2D) (see Zeiske et al., 1992). First signs of opening of the olfactory pit are seen in 4 mm embryos. SEM pictures reveal that the pit opens due to the demise of single epidermal surface cells covering the pit. Using SEM and TEM techniques, we were able to study this process in *Marosatherina* in more detail: SEM micrographs showed small holes in the epidermal cells covering the olfactory placode, indicating degenerative processes inside

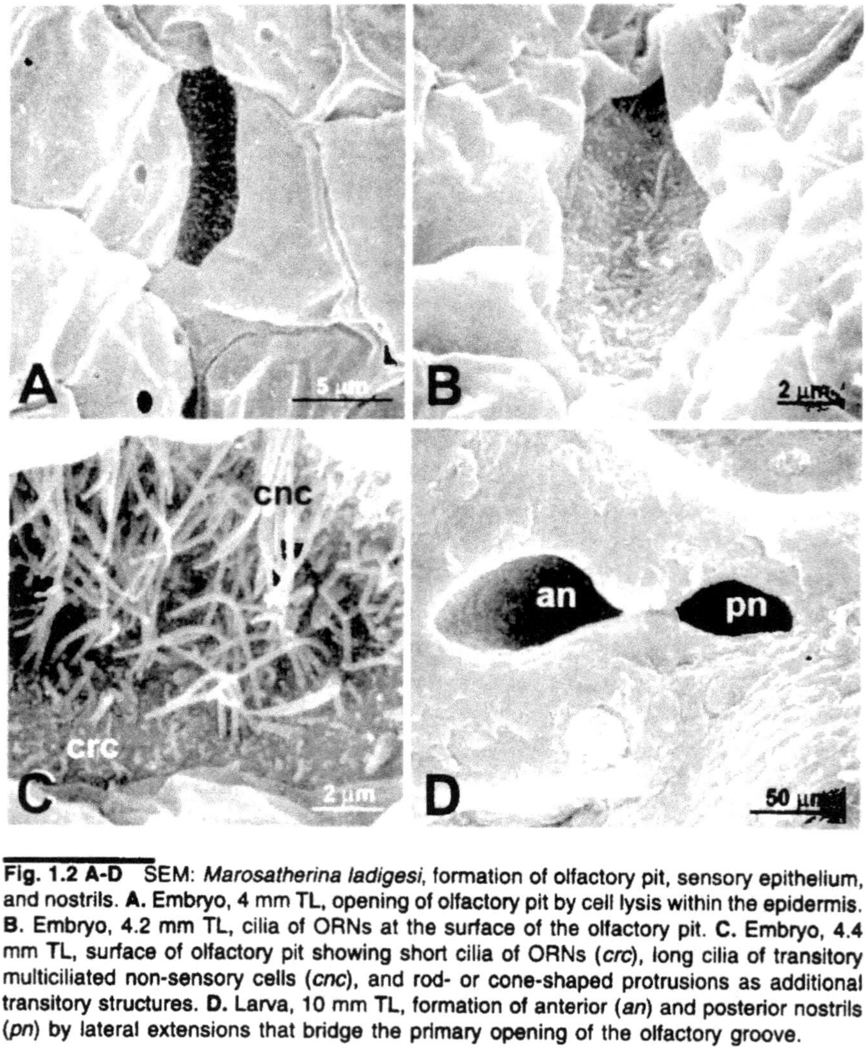

Fig. 1.2 A-D SEM: *Marosatherina ladigesi*, formation of olfactory pit, sensory epithelium, and nostrils. **A.** Embryo, 4 mm TL, opening of olfactory pit by cell lysis within the epidermis. **B.** Embryo, 4.2 mm TL, cilia of ORNs at the surface of the olfactory pit. **C.** Embryo, 4.4 mm TL, surface of olfactory pit showing short cilia of ORNs (*crc*), long cilia of transitory multiciliated non-sensory cells (*cnc*), and rod- or cone-shaped protrusions as additional transitory structures. **D.** Larva, 10 mm TL, formation of anterior (*an*) and posterior nostrils (*pn*) by lateral extensions that bridge the primary opening of the olfactory groove.

these cells (Fig. 1.2A). Ultrathin sections revealed more details of these processes. Initially the solid olfactory placode is closely attached to the epithelium that covers the placode (Fig. 1.3A). Cell lysis occurs in the cells located between the distal periphery of the placode and the innermost cell layer of the epidermis (Fig. 1.3A, inset), which leads to the formation of an internal lumen between the placode and the covering epidermis. This lumen is the primordium of the future olfactory cavity. Simultaneously the placodal cells change their shape, elongating

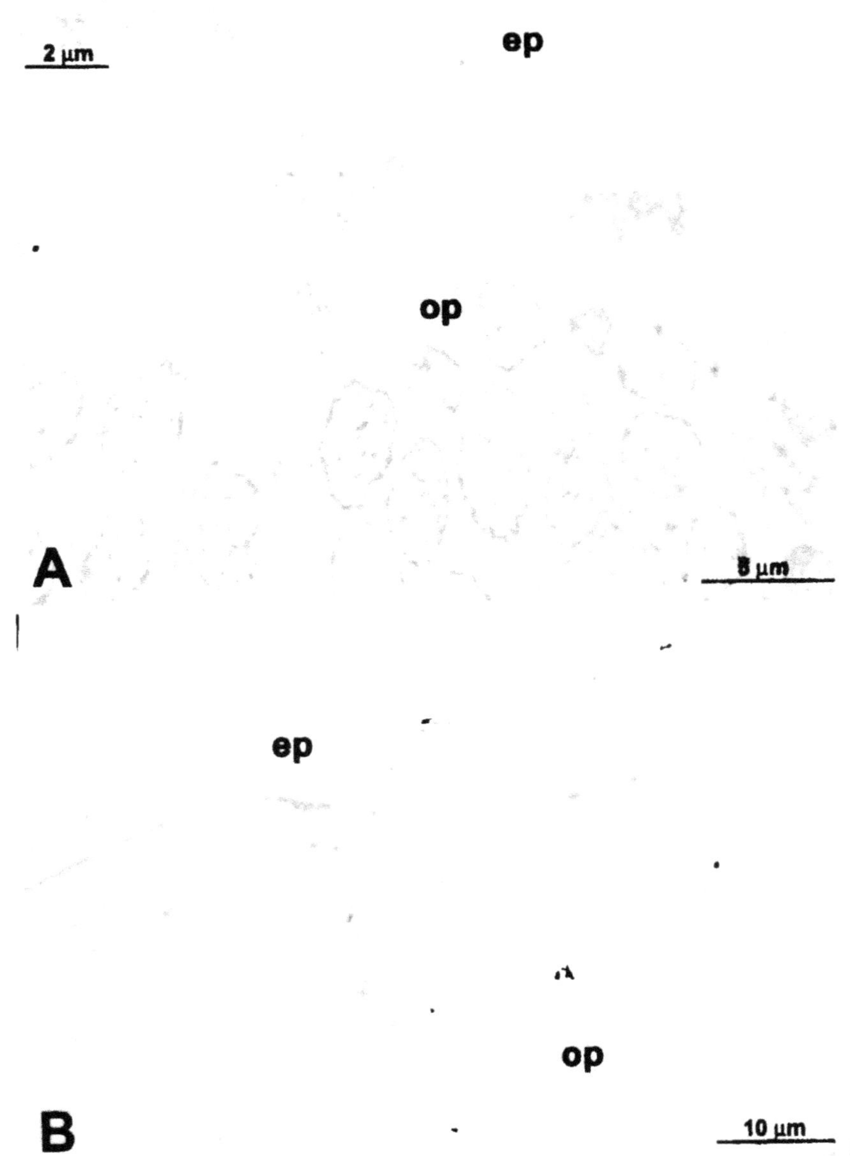

Fig. 1.3 A, B TEM: *Marosatherina ladigesi*, olfactory placode (*op*) and covering epidermis (*ep*). **A** and **Inset**. Embryo, 4.2 mm TL. Cell lysis in the inner epidermal layer leading to an internal lumen between olfactory placode and epidermis (see B). *mi*, cell mitosis. **B.** Embryo 4.4 mm TL. Internal lumen which gives rise to future olfactory cavity. Dendritic endings send cilia into the lumen.

disto-proximally. Dendrites of the ORNs taper toward the lumen (Figs. 1.2B, 1.3B). Their endings (olfactory knobs) send cilia into the lumen. Both TEM and SEM micrographs showed the two types of cilia—short sensory cilia of the ORNs and longer kinocilia of ciliated non-sensory cells (Fig. 1.2C). The latter are characterized by the 9 + 2 pattern of their axonemal complex. Rod- or cone-shaped protrusions may occur as described for *Melanotaenia* (see above). While the olfactory epithelium continues to differentiate and grow, cell death is apparent in the epidermal cell layer directly above the olfactory placode where later the opening of the olfactory pit will occur (Fig. 1.2A, B).

3. CYPRINODONTIFORMES

3.1 Aplocheilidae—rivulines

Aplocheilus lineatus (Valenciennes, 1846)—striped panchax (= lined panchax, sparkling panchax, or Malabar killie)

Semithin sections through the head of a 3 mm embryo show a spherical olfactory placode closely attached to the covering epidermis. The opening of the olfactory placode is visible in embryos 3.5 to 5 mm TL. When the epidermal cells above the placode separate, the free surface of the placode already shows dendritic endings (olfactory knobs). Some of the knobs bear cilia that protrude into the open pit. Based on SEM and TEM findings, the opening mechanism is as follows: (1) Apical placodal cells lose contact with the overlying epidermis so that the apical surface of the placode becomes successively free of the adjacent tissue (Fig. 1.4A, B). (2) Surface cell protrusions of the placode (cilia and microvilli) extend into the growing lumen (Fig. 1.5A, B). (3) The olfactory placode opens exteriorly when the overlying epidermal cells retract (Figs. 1.5A and 1.6A). The initial opening is slit-like. Goblet cells develop within the epidermal layer and release their mucus at the free surface (Fig. 1.5C).

At the time of hatching (6 mm TL) the olfactory pit is oval and about 15 by 30 μm in size. The posterior region of the olfactory pit invaginates (Fig. 1.6B). Increase in depth of the organ results from increase in surface of the growing sensory epithelium and growth of the bordering non-sensory epithelium. The posterior portion of the deepening pit remains covered by the surface tissue of the head, thus building a cavity. This cavity elongates caudally, then widens into an accessory ventilation sac that extends ventrocaudally to the primary olfactory cavity. The posterior

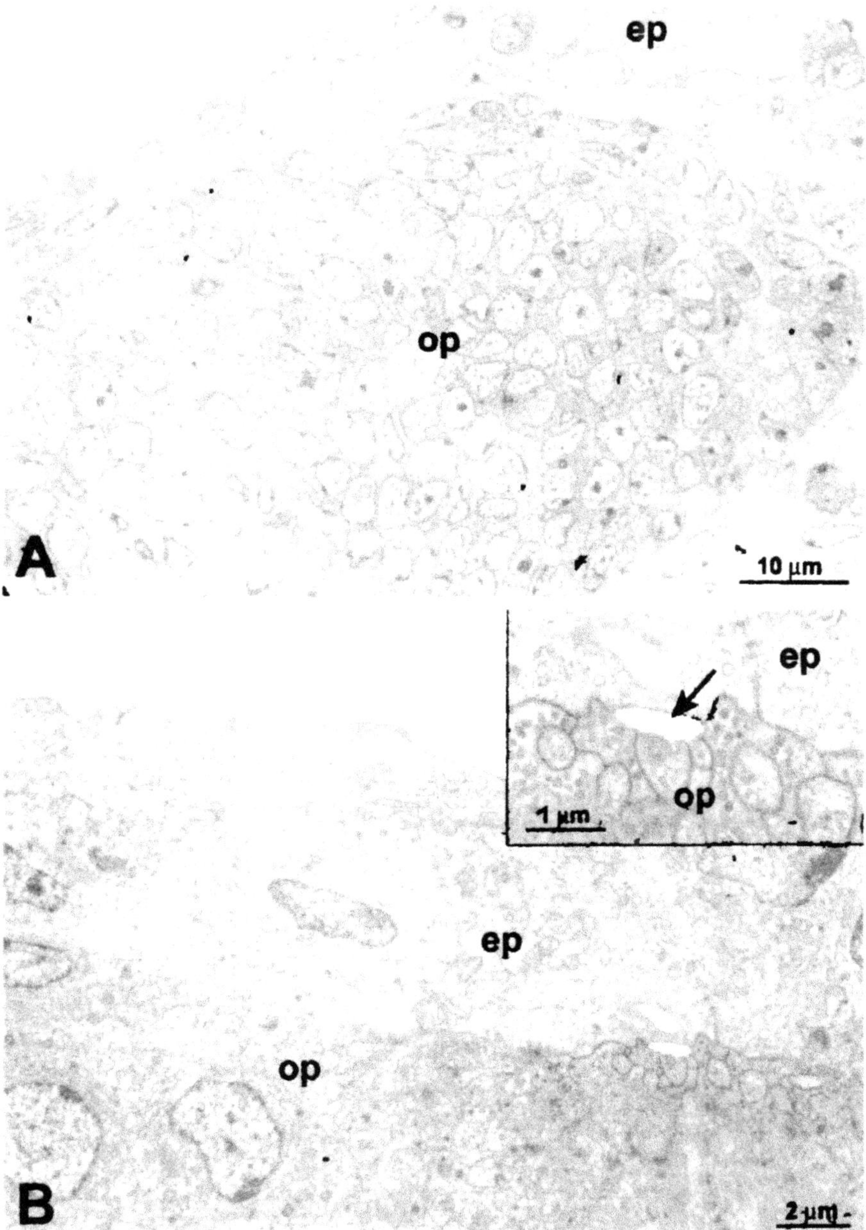

Fig. 1.4 A, B TEM: *Aplocheilus lineatus*, olfactory placode (*op*) and covering epidermis (*ep*). **A.** Embryo, 4.6 mm TL, thickened olfactory placode containing developing ORNs and their supporting cells. **B.** Embryo, 4.8 mm TL, epidermis and apical portion of the olfactory placode. **Inset** Refraction of epidermal and placodal cells followed by formation of an internal lumen (arrow) which gives rise to the future olfactory cavity.

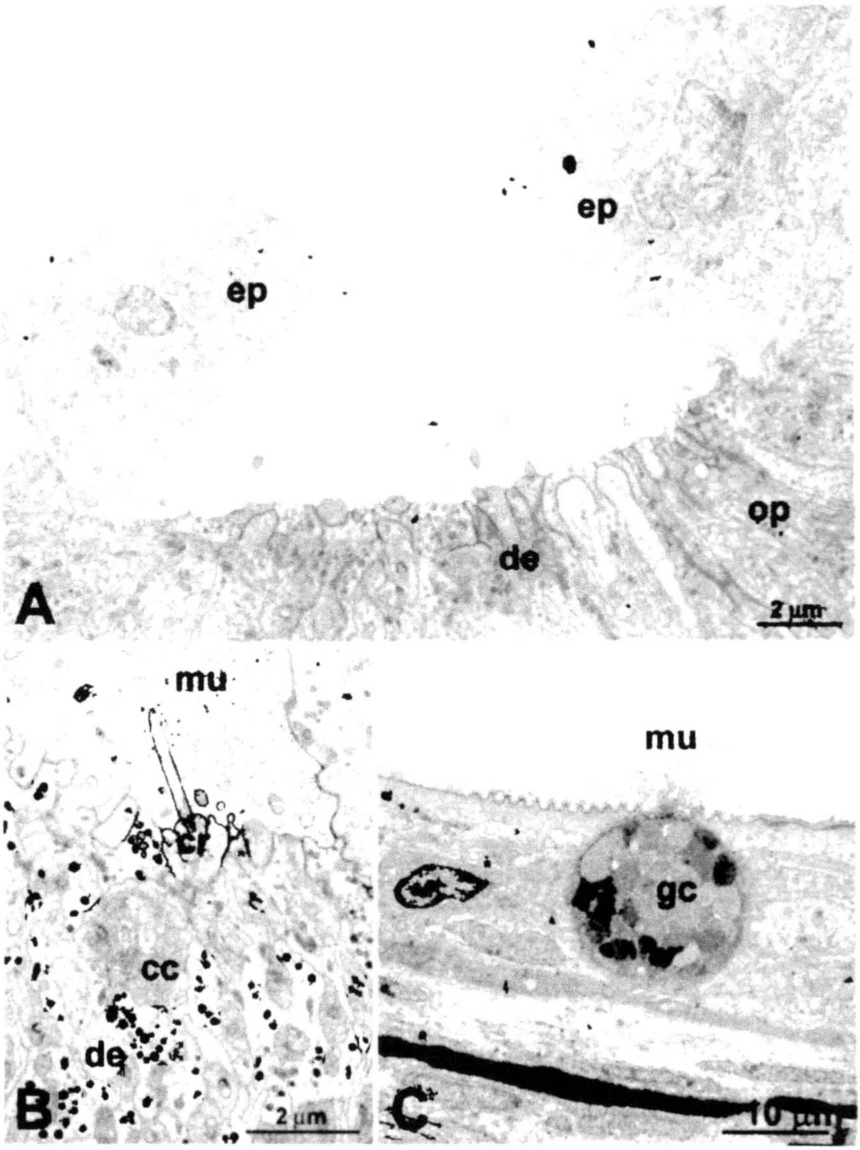

Fig. 1.5 A-C TEM: *Aplocheilus lineatus*. *cc*, crypt cell; *cr*, ciliated ORN; *de*, dendrite; *ep*, epidermis; *gc*, goblet cell; *mu*, mucus; *op*. olfactory placode. **A.** Embryo 4.8 mm TL, opening of olfactory pit by cell refraction and developing sensory epithelium. **B.** Larva, 7 mm TL, sensory epithelium showing ciliated and crypt-type ORNs. **C.** Embryo, 4 mm TL, epidermis containing a goblet cell which releases its mucus at the free surface.

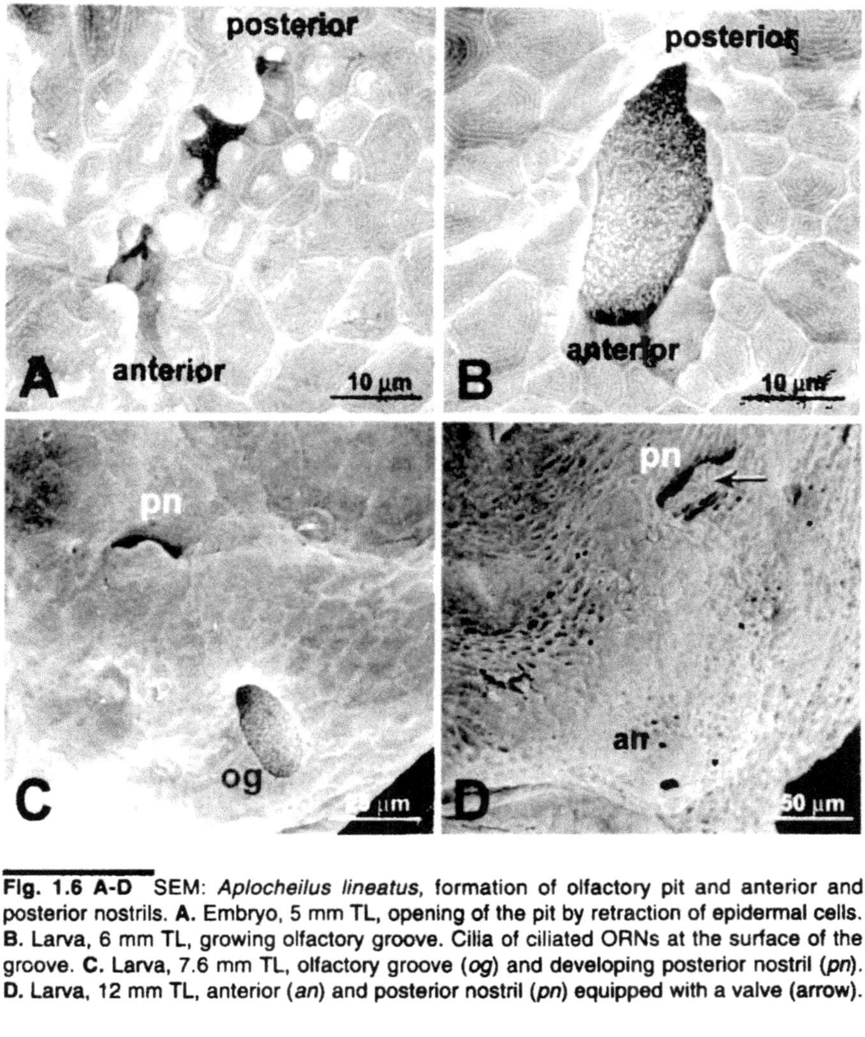

Fig. 1.6 A-D SEM: *Aplocheilus lineatus*, formation of olfactory pit and anterior and posterior nostrils. **A.** Embryo, 5 mm TL, opening of the pit by retraction of epidermal cells. **B.** Larva, 6 mm TL, growing olfactory groove. Cilia of ciliated ORNs at the surface of the groove. **C.** Larva, 7.6 mm TL, olfactory groove (*og*) and developing posterior nostril (*pn*). **D.** Larva, 12 mm TL, anterior (*an*) and posterior nostril (*pn*) equipped with a valve (arrow).

opening appears in hatched larvae of about 7 mm TL. The formation begins with a shallow superficial groove located dorsolaterally on the head some distance from the primary opening (Fig. 1.6C). The distance between the two openings is about 45 μm in a 6.6 mm larva. When the larva is about 8 mm in TL, the posterior opening is equipped with a valve arising from the posterolateral margin of the opening. The final shape of the openings has formed in larvae of 12 mm TL (Fig. 1.6D). The incurrent nostril is situated on a small rounded papilla. The excurrent nostril is slit-like and equipped with a valve.

The first few dendritic endings (olfactory knobs) are observed in 4 mm embryos. The knobs already bear the first few developing olfactory cilia. In 6 mm embryos microvillous ORNs appear in addition to ciliated ORNs. This order of appearance has been observed in all teleosts examined to date (Evans et al., 1982; Zielinski and Hara, 1988; Hansen and Zeiske, 1993). At hatching time (6 mm), the olfactory epithelium shows the typical ultrastructure of the fully developed epithelium of adults (see Zeiske et al., 1976a). In a 7 mm larva, a crypt-type receptor cell, the third type of ORN, was found (Fig. 1.5B). Due to their limited number it is not known when crypt-type ORNs first appear. In larvae of 7 mm, cilia of multiciliated non-sensory cells are also present. Sensory cilia show a 9 + 0 pattern in their axonemal complex, while the cilia of the ciliated non-sensory cells show the usual 9 + 2 microtubular pattern. As in *Melanotaenia* (order Atheriniformes), ciliated non-sensory cells are only transitory structures within the developing olfactory epithelium and have never been found in the adult olfactory organ of panchax (see also *Melanotaenia*).

3.2 Poeciliidae–poeciliids/live-bearers

Xiphophorus hellerii Haeckel, 1848—swordtail

Nutrition of the embryos of the viviparous swordtail is maintained via a follicular placenta and an extraembryonic pericardial sac (pericardial amniochorion) which envelops the head with its double-layered walls during embryonic stages (Turner, 1940; Tavolga and Rugh, 1947; Wourms, 1981). In order to facilitate controlled selection of various developmental stages, we raised embryos *in vitro* according to the method of Haas-Andela (1976).

Perforation of the head sac starts in embryos of about 5 mm TL (Fig. 1.7A). Concurrently, initial signs of the opening of the olfactory pit are visible in SEM micrographs (Fig. 1.7B). The opening process is more obvious in stages of 5.8–6.4 mm TL (Fig. 1.7C). The olfactory pit opens when epidermal surface cells simply part above the olfactory placode. The primary opening is slit-like. The surface of the olfactory epithelium in an advanced state of differentiation becomes exposed when the pit opens. Dendritic endings (olfactory knobs) of ORNs are visible at its surface. The knobs bear ciliary buds, a single cilium or even a few cilia (Fig. 1.7D, inset). Some olfactory knobs have microvilli in addition to cilia, which has also been reported for other fish species (Zeiske et al., 2003). Cells which surround the ORNs display microvillus-like protrusions. The posterior

ERRATA

PAGE / PARA / LINE	WRITTEN AS	TO BE READ AS
Title page / affiliation of B.G. Kapoor	Gwalior University , Gwalior	Jodhpur University, Jodhpur
vi / last para / 2nd line	with, the manuscripts we now really happy that Fish Chemosenses is	with, the manuscripts we now are really happy that Fish Chemosenses is
325 / footnote	*ff—The same item or expression of further cited in at least four susequent pages of this chapter	*ff—The same item or expression is further cited in at least four susequent pages of this chapter
Backcover / last para / last line	For the Reception of Natural Stimuli, 2004 (with G. Vonder Emde and J. Mogdans);	For the Reception of Natural Stimuli, 2004 (with G. von der Emde and J. Mogdans);

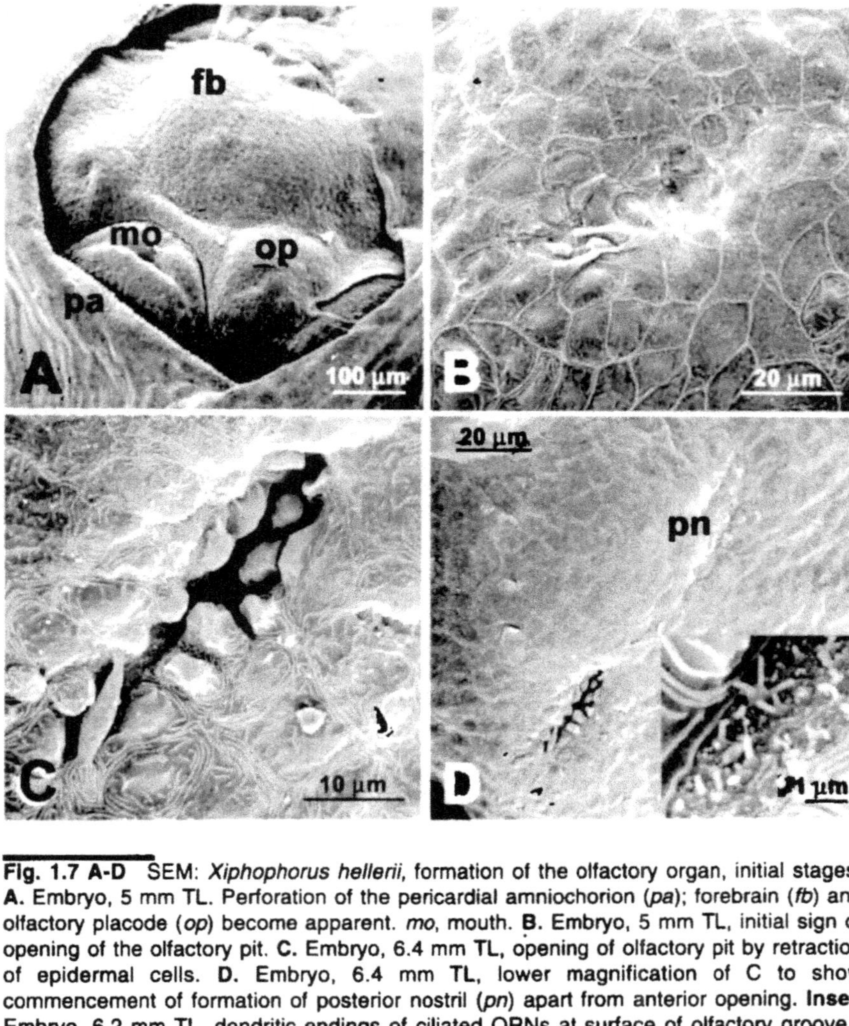

Fig. 1.7 A-D SEM: *Xiphophorus hellerii*, formation of the olfactory organ, initial stages. **A.** Embryo, 5 mm TL. Perforation of the pericardial amniochorion (*pa*); forebrain (*fb*) and olfactory placode (*op*) become apparent. *mo*, mouth. **B.** Embryo, 5 mm TL, initial sign of opening of the olfactory pit. **C.** Embryo, 6.4 mm TL, opening of olfactory pit by retraction of epidermal cells. **D.** Embryo, 6.4 mm TL, lower magnification of C to show commencement of formation of posterior nostril (*pn*) apart from anterior opening. **Inset.** Embryo, 6.2 mm TL, dendritic endings of ciliated ORNs at surface of olfactory groove.

part of the oval pit remains covered by epidermal cells (Fig. 1.7C). This part deepens by epithelial growth similar to that described for *Aplocheilus* (see above).

Formation of the excurrent nostril commences in embryos at 5.3 mm TL (Fig. 1.7D). The posterior opening becomes connected to a lumen that develops into the future accessory ventilation sac. This lumen may arise as an outpocketing of the primary olfactory pit or alternatively as an internal lumen initially isolated from the lumen of the primary olfactory pit

(Zeiske et al., 1976a). Semithin sections show that the primordium of the accessory ventilation sac is present at the lateral margin of the olfactory placode. The primary olfactory pit and simultaneously developing accessory ventilation sac fuse at a later stage. This might explain why at least in some cases the anterior and posterior olfactory openings in *Xiphophorus* embryos originate almost concomitantly, while in *Aplocheilus* the two openings seem to always form in sequential order, corresponding to the successive formation of the olfactory pit and ventilation sac.

Compared to *Aplocheilus*, development of the olfactory organ in *Xiphophorus* seems to be somewhat delayed. When *Aplocheilus* hatches, the olfactory organ consists of an open pit and the accessory ventilation sac is still missing. In the corresponding early larval stage of *Xiphophorus* (still within the ovary), part of the head is still covered by the extraembryonic pericardium, and only after this nutritional organ has receded from the nasal region, does development of the external nostrils begin. However, at birth the olfactory organ is fully developed, including the olfactory ventilation sac, in *Xiphophorus*.

4. BELONIFORMES

4.1 Belonidae–garfishes or needlefishes

Belone belone (L.)—garfish (= needlefish)

The conspicuous and unique morphological features of the olfactory organ in garfish have repeatedly attracted scientists (Blaue, 1884; Burne, 1909; Sewertzoff, 1931; Theisen et al., 1980). As mentioned above, the olfactory organ has the shape of an open groove with a protruding papilla. Development of the olfactory epithelium was described in an earlier study. However, how the ORNs of the primary pit gained access to the exterior remained unclear.

The various developmental stages of *Belone belone* investigated by SEM range from 5 mm embryos to fully grown adults (600-800 mm TL). The following sizes (mm TL) were included in our study: 5, 5.5, 7, 8, 8.5, 9, 9.5, 11, 14, 30, 33, 35, 38, 54. First signs of the opening process were found in embryos of 5 mm TL. Opening of the primary olfactory pit is achieved by retraction of surface cells (Fig. 1.8A, B). Retraction happens when cells that are initially attached, separate completely by losing cell contact. Sometimes single cells are even torn into two parts (Fig. 1.8B). As soon as the olfactory cavity is open to the exterior, sensory cilia of ORNs become visible. Kinocilia of non-sensory cells appear later. Ciliated

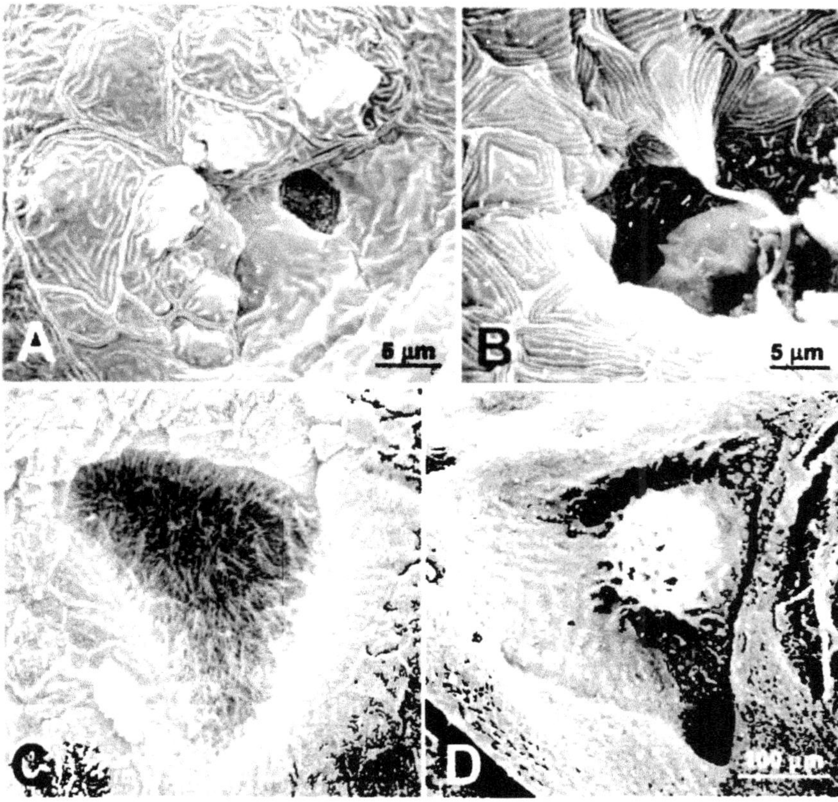

Fig. 1.8 A-D SEM: *Belone belone*, formation of olfactory groove. **A.** Embryo, 5.5 mm TL, initial stage of pit. **B.** Embryo, 9.5 mm TL, opening by epidermal cell retraction. Single cell torn into two parts still bridges the primary opening. Cilia of ORNs visible at surface of the groove. **C.** Larva, 14 mm TL, olfactory groove triangular with cilia of transitory multicilated non-sensory cells at periphery. **D.** Larva, 33 mm TL, olfactory papilla grows from bottom of the groove.

non-sensory cells are only transitory components of the olfactory epithelium and disappear during juvenile stages. They have never been found in adult olfactory epithelia of *Belone* (Theisen et al., 1980).

During the subsequent growth period the primary pit elongates and by the time of hatching (at about 13 mm TL) the opening of the olfactory organ has attained its persisting triangular shape (Fig. 1.8C). When the juvenile has reached a length of about 20 mm, a small elevation appears at the floor of the groove (Fig. 1.8D) and eventually develops into a large projection or papilla located in the middle of the groove (Sewertzoff, 1931; Theisen et al., 1980).

5. CONCLUSIONS

Development of the olfactory organ in fishes can be divided into two major phases: the formation of the olfactory epithelium proper and formation of the olfactory cavity and its accessory components. Formation of the cavity, including incurrent and excurrent nostrils, and–if present–ventilation sacs, shows a striking diversity leading to various morphologies of the mature olfactory organ. The same is true for the various mechanisms with which the olfactory epithelium gains access to the outside world. Contrarily, formation of the olfactory epithelium proper and, more precisely, the origin of the ORNs seems to follow the same pattern in all 'modern teleosts'.

The olfactory epithelium of rainbowfishes, sailfin silversides, rivulines, live-bearers, and garfishes/needlefishes differentiates from a subepidermal layer. Formation of an olfactory epithelium from a subepidermal layer has also been reported for the zebrafish *Danio rerio* (Hansen and Zeiske, 1993). Earlier observations of the development of the olfactory organ in the salmon *Salmo salar* (Gawrilenko, 1910; Reinke, 1937; see also Kleerekoper, 1969) seem to correspond to the formation described for zebrafish and atherinomorph fishes. Hence, salmon, zebrafish, and atherinomorph fishes seem to represent the model type of formation of the olfactory epithelium which might be ubiquitous for all 'modern teleosts'. More basal lineages of teleosts such as elopomorphs, osteoglossomorphs, and clupeomorphs remain to be studied.

The subepidermal layer may correspond to the so called nervous layer of the African clawed frog *Xenopus laevis*. However, in contrast to the subepidermal layer of teleosts, the nervous layer of *Xenopus* leads only to the ORNs, whereas supporting cells originate from cells of the superficial non-nervous layer (Klein and Graziadei, 1983; Costanzo and Graziadei, 1987; Reiss and Burd, 1997). In *Xenopus*, the non-nervous layer, like the epidermis, does not disappear. The olfactory pit is formed by invagination of both the nervous and the non-nervous layer. We recently found the same type of development in sturgeons (*Acipenser* species), which are considered a basic group of actinopterygian (ray-finned) fish and thus rather close to the common origin of actinopterygians and sarcopterygians, including tetrapods (Zeiske et al., 2003). Formation of the olfactory epithelium and the primary olfactory pit as found in *Acipenser* and *Xenopus* is phylogenetically considered a plesiomorphic (primitive) type, as opposed to the derived (apomorphic) mode found in teleosts

(Zeiske et al., 2003). Similarities may be based on symplesiomorphies (shared primitive structures), synapomorphies (shared derived structures), or convergent modifications (independently evolved structures). Provided that convergences did not occur, it may be speculated that the plesiomorphic developmental type has been conserved in the branch that leads from fish to tetrapods.

Despite similarities in development of the olfactory epithelium described here for atherinomorph fishes and as seen in salmonids and zebrafish, differences in the formation of the olfactory cavity are striking (construction of an internal lumen and its primary opening to the exterior, and formation of the anterior and posterior nostrils).

In atherinomorph fishes, the epidermal epithelium which transiently covers the olfactory placode disappears when the first dendritic endings of the ORNs and other surface structures appear on the surface of the olfactory epithelium. In Atheriniformes and Cyprinodontiformes, an internal lumen or at least a small gap between epidermis and subepidermal layer occurs before the epidermis disappears. Dendritic endings of the ORNs differentiating in the olfactory placode grow into this lumen. This internal lumen is the primordium of the future olfactory cavity and forms either by lysis of cells of the innermost cell layer of the epidermis (e.g. in *Marosatherina*), or by retraction of cells of the placode and the epidermis respectively (e.g. in *Aplocheilus*). The primary opening of the internal lumen to the exterior results from processes similar to those described for the origin of the internal lumen, i.e., cell death (e.g. in *Marosatherina* and *Glossolepis*) or cell retraction (e.g. in *Aplocheilus* and *Xiphophorus*). Retraction of cells is achieved by separation of cells initially attached to each other. Sometimes a cell may be torn apart during this process.

The most striking differences in the development of the olfactory organ occur in the formation of the two openings which become the anterior (incurrent) and the posterior (excurrent) nostrils. The two openings arise either simultaneously from a single primordial opening or consecutively by two processes which are topographically separated from each other from the beginning. The former 'salmonid' type is widespread in gnathostome fish, e.g. in paddlefishes, sturgeons, bichirs, and teleosts (Breucker et al., 1979; Bemis and Grande, 1992; Hansen and Zeiske, 1993; Olsén, 1993; Bartsch et al., 1997; Zeiske et al., 2003) and is also present in some species of Atheriniformes (*Melanotaenia* and *Marosatherina*; this study). It is considered the primitive or plesiomorphic character of osteichthyans (Kleerekoper, 1969; Zeiske et al., 1992).

Conversely, in some fishes the primary opening gives rise to the anterior nostril only, while the posterior nostril originates separately at a more caudal position. This type of formation, found in Cyprinodontiformes (*Aplocheilus* and *Xiphophorus*), is assumed to be derived (Reinke, 1937; Zeiske et al., 1976a, 1992). It is not known why the atheriniform and cyprinodontiform species included in the present study have a similar structure and shape with regard to their olfactory cavities and nostrils, but nonetheless differ in the formation of their olfactory nostrils (*Melanotaenia* and *Marosatherina* versus *Aplocheilus* and *Xiphophorus*). Perhaps this observation has to be viewed in the context of systematic relationships, since monophyly is accepted for Cyprinodontiformes (*Aplocheilus* and *Xiphophorus*), while Atheriniformes are suspected to form a paraphyletic taxon (Nelson, 1994).

The olfactory organ of *Belone* (and also the closely related sauries, flyingfishes, and halfbeaks) is unique in shape and differs greatly from the olfactory organs of all other atherinomorph fish species. Its pattern is highly derived and considered a synapomorphy (shared derived structure) of the suborder Belonoidei (= Exocoetoidei) *sensu* Nelson (1994; see Table 1.1). Interestingly, representatives of Adrianichthyoidei (development of the olfactory organ not investigated here), the sister group of Belonoidei, possess ditrematous olfactory organs of the same type as described for atheriniform and cyprinodontiform species. As shown in Table 1.1, both the taxa Belonoidei and Adrianichthyoidei (comprising three subfamilies) are grouped into the higher taxon Beloniformes. Consequently, the overall character of the olfactory organs of Adrianichthyoidei as far as known has to be considered plesiomorphic compared to the highly derived (synapomorphic) features of the sister group Belonoidei.

By means of dye filaments and suspensions of fine metal particles as tracers for a cinematographic recording in a laminar flow channel, we found that elongation of nasal papillae in *Belone belone* is an adaptation to the hydrodynamic situation, including boundary layer effects of the longirostrate fish (unpubl. results). Since direct driving forces for ventilation of the olfactory organ are lacking, water flow around the papillae is passively achieved by motion of the fish. The olfactory groove of the adult fish is wide open. Consequently, the primary olfactory pit has to widen during development and growth. The mechanism, however, by which the primordial pit opens to the exterior through a small opening is similar to that found in atheriniform and cyprinodontiform species investigated in the present study.

The anatomy of the olfactory organ of *Belone belone* differs clearly from so-called monotrematous olfactory organs. Such organs have a single narrow opening causing water to flow in and out through the same nostril. This type is known e.g. for *Gasterosteus aculeatus* (three-spined stickleback) (Solger, 1894), *Spinachia spinachia* (sea stickleback, fifteen-spined stickleback) (Theisen, 1982), *Aphanopus carbo* (black scabbard fish, black espada) (Holl and Meinel, 1968), *Zoarces viviparus* (viviparous blenny, eelpout) (Pipping, 1926), *Pholis gunnellus* (gunnel, rock gunnel, butterfish), and *Liparis montagui* (Liermann, 1933). Monotrematous olfactory organs are considered derived and have evolved convergently several times. Formation of the opening of the olfactory organ in *Belone belone* differs clearly from the formation of monotrematous olfactory organs. In monotrematous olfactory organs, one of the nostrils (mostly the posterior one) is either secondarily closed or the primary opening remains almost unchanged in size and finally becomes the single nostril (e.g. in *Gasterosteus*) (Liermann, 1933).

In summary, development of the olfactory organs in gnathostome fish varies considerably. Nevertheless, the modifications can be related to distinct types:

1. Epidermal and subepidermal cell layers contribute to formation of the olfactory sensory epithelium. The olfactory pit forms by invagination of these primordial layers, e.g. in sturgeons (ancestral actinopterygians) and in *Xenopus* (lower tetrapods). This is considered a primitive (plesiomorphic) type of formation.
2. Only the subepidermal layer gives rise to the olfactory placode and its future cell types. This mode has been found in all teleosts studied so far and is considered a derived (apomorphic) mode of formation.
3. In the plesiomorphic type, the olfactory pit is open to the exterior from the beginning of its formation by invagination. In teleosts the opening mechanism is group specific and differs even in closely related systematic groups. The opening results from local cell death in the epidermal layer just above the placode or by retraction of epidermal cells that lose their cell contacts.
4. In ditrematous olfactory organs, incurrent and excurrent nostrils may arise from a single primordial opening of the pit, or the primary opening gives rise only to the incurrent nostril and the excurrent nostril develops separately. The first type is widespread (in elasmobranchs, chondrosteans, 'holosteans', and many teleosts)

and is considered the primitive (plesiomorphic) type. The derived (apomorphic) type is found in only a few teleostean species.

5. Monotrematous olfactory organs are considered derived and have convergently evolved several times. A single opening causes water to flow in and out through the same nostril. This opening originates either from the primary opening of the pit or is formed by secondary closure of one of two transitory openings.
6. The olfactory organ of *Belone belone* and other representatives of the suborder Belonoidei (= Exocoetoidei) is unique in its anatomical structure and cannot be assigned to either the ditrematous or monotrematous type of olfactory organs in teleosts. Rather, it is characterized by reduction of the nostril formation and wide exposure of the sensory epithelium.

Acknowledgements

We thank Dr. Peter Bartsch for helpful comments on various versions of this manuscript and Dr. Haide Breucker, Ms. Kirsten Kruse, Dr. Dietmar Keyser, Dr. Reinhard Melinkat, Ms. Karin Meyer, Ms. Brigitte Segner, Mr. Günter Willis and Ms. Martina Wichmann for assistance with tissue preparation and imaging. This work was supported in part by the Deutsche Forschungsgemeinschaft Ze 141 (to Eckart Zeiske) and the National Institutes on Deafness and Other Communication Disorders Grant DC-03792 (to John Caprio, a co-researcher).

References

Appelbaum, S., J.W. Adorn, S.G. George, A.M. Mackie and B.J.S. Pirie. 1983. On the development of the olfactory and gustatory organs of the dover sole, *Solea solea*, during metamorphosis. *J. Mar. Biol. Assoc.* UK 63: 97–108.

Baker, C.V.H., and M. Bronner-Fraser. 2001. Vertebrate cranial placodes. I. Embryonic induction. *Devel. Biol.* 232: 1–61.

Balfour, F.M. 1877. On the development of elasmobranch fishes. *J. Anat. Physiol.* 11: 406–490.

Balfour, F.M. and W.N. Parker. 1882. On the structure and development of *Lepisosteus*. *Phil. Trans. Roy. Soc.* (Lond.) 2: 359–442.

Bannister, L.H. 1965. The fine structure of the olfactory surface of teleostean fishes. *Quart. J. micr. Sci.* 106: 333–342.

Bartsch, P., S. Gemballa and T. Piotrowski. 1997. The embryonic and larval development of *Polypterus senegalus* Cuvier, 1829: its staging with reference to external and skeletal features, behaviour and locomotory habits. *Acta Zool.* (Stockholm) 78: 309–328.

Bemis, W.E. and L. Grande. 1992. Early development of the actinopterygian head. I. External development and staging of the paddlefish *Polyodon spathula*. *J. Morphol.* 213: 47–83.

Berliner, K. 1902. Die Entwicklung des Geruchsorgans der Selachier. *Arch. Mikrosk. Anat. Entw.-Mech.* 60: 386–406.

Bertmar, G. 1965. The olfactory organ and upper lips in Dipnoi, an embryological study. *Acta Zool.* (Stockholm) 46: 1–40.

Bertmar, G. 1967. Lungfish phylogeny. In: *Current Problems of Lower Vertebrate Phylogeny*, T. Ørvig (Ed.). Nobel Symposium 4. Almqvist & Wiksell, Stockholm, pp. 259–283.

Bertmar, G. 1969. The vertebrate nose, remarks on its structural and functional adaptation and evolution. *Evolution* 23: 131–152.

Bjerring, H.C. 1989. Apertures of craniate olfactory organs. *Acta Zool.* (Stockholm) 70: 71–85.

Blaue, J. 1884. Untersuchungen über den Bau der Nasenschleimhaut bei Fischen und Amphibien, namentlich über Endknospen als Endapparate des Nervus olfactorius. *Arch. Anat. Physiol., Anat. Abt.* 1884: 231–309.

Breucker, H., E. Zeiske and R. Melinkat. 1979. Development of the olfactory organ in the rainbow fish *Nematocentris maccullochi* (Atheriniformes, Melanotaeniidae). *Cell Tissue Res.* 200: 53–68.

Brookover, C. 1914. The development of the olfactory nerve and its associated ganglion in *Lepidosteus*. *J. Comp. Neurol.* 24: 113–130.

Burne, R.H. 1909. The anatomy of the olfactory organ of teleostean fishes. *Proc. Zool. Soc.* (Lond.) 1909: 610–663.

Collette, B.B. 1974. South American freshwater needlefishes (Belonidae) of the genus *Pseudotylosurus*. *Zool. Meded.* 48: 169–186.

Colette, B.B. 1976. Indo-West Pacific halfbeaks (Hemiramphidae) of the genus *Rhynchorhamphus* with descriptions of two new species. *Bull. Mar. Sci.* 26: 72–98.

Costanzo, R.M. and P.P.C. Graziadei. 1987. Development and plasticity of the olfactory system. In: *Neurobiology of Taste and Smell*, T.E. Finger and W.L. Silver (Eds). John Wiley, New York, NY, pp. 233–250.

Dean, B. 1897. On the larval development of *Amia calva*. *Zool. Jb. Abt. Syst.* 9: 639–672.

Devitsina, G.V. and A.A. Kazhlayev. 1993a. Chemoreceptor organs in early juvenile paddlefish, *Polyodon spathula*. *J. Ichthyol.* 33: 143–149.

Devitsina, G.V. and A.A. Kazhlayev. 1993b. Development of chemosensory organs in Siberian sturgeon, *Acipenser baeri*, and stellate sturgeon, *A. stellatus*. *J. Ichthyol.* 33: 9–19

Eisthen, H.L. 1997. Evolution of vertebrate olfactory system. *Brain Behav. Evol.* 50: 222–233.

Elston, R., L. Corazza and J.G. Nickum. 1981. Morphology and development of the olfactory organ in larval walleye *Stizostedion vitreum*. *Copeia* 1981: 890–893.

Evans, R.E., B. Zielinski and T.J. Hara. 1982. Development and regeneration of the olfactory organ in rainbow trout. In: *Chemoreception in Fishes*, T.J. Hara (Ed.). Elsevier, Amsterdam, pp. 15–37.

Fishelson, L. 1997. Comparative ontogenesis and cytomorphology of the nasal organs in some species of cichlid fish (Cichlidae, Teleostei). *J. Zool.* (Lond.) 243: 281–294.

Fishelson, L. and A. Baranes. 1997. Ontogenesis and cytomorphology of the nasal olfactory organs in the Oman shark, *Lago omanensis* (Triakidae), in the Gulf of Aqaba, Red Sea. *Anat. Rec.* 249: 409–421.

Fullarton, M.H. 1933. On the development of the olfactory organ in *Protopterus*. *Proc. Roy. Soc.* (Edinburgh) 53: 1–6.

Gawrilenko, A. 1910. Die Entwicklung des Geruchsorgans bei *Salmo salar*. (Zur Stammesentwicklung des Jacobson'schen Organs). *Anat. Anz.* 36: 411–427.

Greenwood, P.H., D.E. Rosen, S.H. Weitzman and G.S. Myers. 1966. Phyletic studies of teleostean fishes, with a provisional classification of living forms. *Bull. Amer. Mus. Nat. Hist.* 131: 339–445.

Gupta, O.P. and R.K. Shrivastava. 1973. An interesting type olfactory organ in Indian gar-fish of the family Belonidae, *Xenentodon cancila* (Ham.). *Zool. Jb. Anat.* 90: 450–453.

Haas-Andela, H. 1976. *In vitro* Kultur und Aufzucht von Embryonen lebendgebärender Zahnkarpfen der Gattung *Xiphophorus*. *Zool. Anz.* (Jena) 197: 1–5.

Hansen, A. and E. Zeiske. 1993. Development of the olfactory organ in the zebrafish, *Brachydanio rerio*. *J. Comp. Neurol.* 333: 289–300.

Hansen, A. and E. Zeiske. 1994. Cell proliferation in the olfactory organ of the embryonic and larval zebrafish, *Brachydanio rerio*. In: *Olfaction and Taste*, K. Kurihara, N. Suzuki and H. Ogawa (Eds). Springer-Tokyo, Vol. 11, p. 755.

Hansen, A. and E. Zeiske. 1995a. Development of the olfactory organ in zebrafish: an electron microscopic and immunocytochemical study of early differentiation and growth. *Biofizika* 40: 151–162. (Russian with English summary)

Hansen, A. and E. Zeiske. 1995b. Development of the olfactory organ in zebrafish: an electron microscopic and immunocytochemical study of early differentiation and growth. *Biophysics* 40: 159–170.

Hansen, A. and E. Zeiske. 1998. The peripheral olfactory organ in the zebrafish, *Danio rerio*: an ultrastructural study. *Chem. Senses* 23: 39–48.

Hansen, A. and T.E. Finger. 2000. Phyletic distribution of crypt-type olfactory receptor neurons in fishes. *Brain Behav. Evol.* 55: 100–110.

Hara, T.J. and B. Zielinski. 1989. Structural and functional development of the olfactory organ in teleosts. *Trans. Amer. Fish. Soc.* 118: 183–194.

Holl, A. and W. Meinel. 1968. Das Geruchsorgan des Tiefseefisches *Aphanopus carbo* (Percomorphi, Trichiuridae). *Helgoländer wiss. Meeresunters.* 18: 404–423.

Holm, J.F. 1894 The development of the olfactory organ in the Teleostei. *Morph. Jb. Gegenbaur* 21: 620–624.

Holm, J.F. 1895. Some notes on the early development of the olfactory organ of *Torpedo*. *Anat. Anz.* 10: 201–207.

Jahn, L.A. 1972. Development of the olfactory apparatus of the clutthroat trout. *Trans. Amer. Fish. Soc.* 101: 284–289.

Kleerekoper, H. 1969. *Olfaction in fishes*. Indiana University Press, Bloomington

Kleerekoper, H. 1982. Research on olfaction in fishes: historical aspects. In: *Chemoreception in Fishes*, T.J. Hara (Ed.). Elsevier, Amsterdam, pp. 1–14.

Klein, S.L. and P.P.C. Graziadei. 1983. The differentiation of the olfactory placode in *Xenopus laevis*: A light and electron microscope study. *J. Comp. Neurol.* 217: 17–30.

Kupffer, C. von. 1894. Studien zur vergleichenden Entwicklungsgeschichte des Kopfes der Kranioten. 2. Heft. Die Entwicklung des Kopfes von *Ammocoetes planeri*. J.F. Lehmann, München, Leipzig.

Kux, J., Zeiske, E. and Y. Osawa. 1988. Laser Doppler velocimetry measurement in the model flow of a fish olfactory organ. *Chem. Senses* 13: 257–265.

Laibach, E. 1937. Das Geruchsorgan des Aals (*Anguilla vulgaris*) in seinen verschiedenen Entwicklungsstadien. *Zool. Jb., Abt. Anat.* 63: 37–72.

Liermann, K. 1933. Über den Bau des Geruchsorgans der Teleostier. *Z. Anat. Entw.-Gesch.* 100: 1–39.

Melinkat, R. 1982. Funktionsmorphologische Untersuchungen zur Ventilationsmechanik des Geruchsorgans bei ditremen Ährenfischartigen (Pisces, Atheriniformes). PhD Thesis, University of Hamburg, Germany.

Melinkat, R. and E. Zeiske. 1979. Functional morphology of ventilation of the olfactory organ in *Bedotia geayi* Pellegrin 1909 (Teleostei, Atherinidae). *Zool. Anz.* (Jena) 203: 354–386.

Miyake, T., I.H. von Herbing and B.K. Hall. 1997. Neural ectoderm, neural crest, and placodes: Contribution of the otic placode to the ectodermal lining of the embryonic opercular cavity in Atlantic cod (Teleostei). *J. Morphol.* 231: 231–252.

Morita, Y. and T.E. Finger. 1996. Olfactory receptor cell morphology correlates with site of axon termination in the olfactory bulb. *Soc. Neurosci. Abstr.* 22: 1072.

Nelson, J.S. 1994. Fishes of the World. 3rd edition. John Wiley, New York, NY.

Olsén, K.H. 1993. Development of the olfactory organ of the Arctic charr, *Salvelinus alpinus* (L.) (Teleostei, Salmonidae). *Can. J. Zool.* 71: 1973–1984.

Pipping, M. 1926. Der Geruchssinn der Fische mit besonderer Berücksichtigung seiner Bedeutung für das Aufsuchen des Futters. *Soc. Sci. Fenn., Commentat. Biol.*, II. 4: 1–28.

Pyatkina, G.A. 1976. Receptor cells of various types and their proportional interrelation in the olfactory organ of larvae and adults of acipenserid fishes. *Tsitologiya* 18: 1444–1449. (Russian with English summary)

Reinke, W. 1937. Zur Ontogenie und Anatomie des Geruchsorgans der Knochenfische. *Z. Anat. Entw.-Gesch.* 106: 600–624.

Reiss, J.O. and G.D. Burd. 1997. Cellular molecular interactions in the development of the *Xenopus* olfactory system. *Semin. Cell Devel. Biol.* 8: 171–179.

Salensky, W. 1881. Recherches sur le développement du sterlet (*Acipenser ruthenus*). *Arch. Biol.* 2: 223–341.

Sewertzoff, A.N. 1931. *Morphologische Gesetzmässigkeiten der Evolution.* Gustav Fischer, Jena, Germany.

Solger, B. 1894. Notiz über die Nebenhöhle des Geruchsorgans von *Gasterosteus aculeatus* L. *Z. wiss. Zool.* 57: 186.

Strehlow, D. and W. Gilbert. 1993. A fate map for the first cleavages of the zebrafish. *Nature* (Lond.) 361: 451–453.

Tavolga, W.N. and R. Rugh. 1947. Development of the platyfish, *Platypoecilus maculatus. Zoologica* (NY) 32: 1–15.

Teichmann, H. 1964. Experimente zur Nasenentwicklung der Regenbogenforelle (*Salmo irideus* W. Gibb.). *Wilhelm Roux Arch. Entw.-Mech. Org.* 155: 129–143.

Theisen, B. 1982. Functional morphology of the olfactory organ in *Spinachia spinachia* (L.) (Teleostei, Gasterosteidae). *Acta Zool.* (Stockholm) 63: 247–254.

Theisen, B., E. Zeiske and H. Breucker. 1986. Functional morphology of the olfactory organ in the spiny dogfish (*Squalus acanthias* L.) and the small-spotted catshark (*Scyliorhinus caniculus* (L.)). *Acta Zool.* (Stockholm) 67: 73–86.

Theisen, B., H. Breucker, E. Zeiske and R. Melinkat. 1980. Structure and development of the olfactory organ in the garfish *Belone belone* (L.) (Teleostei, Atheriniformes). *Acta Zool.* (Stockholm) 61: 161–170.

Turner, C.L. 1940. Pseudoamnion, pseudochorion, and follicular pseudoplacenta in poeciliid fishes. *J. Morphol.* 67: 59–89.

Verraes, W. 1976. Postembryonic development of the nasal organs, sacs and surrounding skeletal elements in *Salmo gairdneri* (Teleostei: Salmonidae), with some functional interpretations. *Copeia* 1976: 71–75.

Webb, J.F. and D.M. Noden. 1993. Ectodermal placodes: Contributions to the development of the vertebrate head. *Amer. Zool.* 33: 434–447.

Whitlock, K.E. and M. Westerfield. 2000. The olfactory placodes of the zebrafish form by convergence of cellular fields at the edge of the neural plate. *Development* 127: 3645–3653.

Wourms, J.P. 1981. Vivipary: the maternal-fetal relationship in fishes. *Amer. Zool.* 21: 473–515.

Yamamoto, M. 1982. Comparative morphology of the peripheral olfactory organ in teleosts. In: *Chemoreception in Fishes*, T.J. Hara (Ed.). Elsevier, Amsterdam, pp. 39–59.

Zeiske, E. 1973. Morphologische Untersuchungen am Geruchsorgan von Zahnkarpfen (Pisces, Cyprinodontoidea). *Z. Morph. Tiere* 74: 1–16.

Zeiske, E. 1974. Morphologische und morphometrische Untersuchungen am Geruchsorgan oviparer Zahnkarpfen (Pisces). *Z. Morph. Tiere* 77: 19–50.

Zeiske, E., J. Kux and R. Melinkat. 1976a. Development of the olfactory organ of oviparous and viviparous cyprinodonts (Teleostei). *Z. zool. Syst. Evolut-forsch.* 14: 34–40.

Zeiske, E., R. Melinkat, H. Breucker and J. Kux. 1976b. Ultrastructural studies on the epithelia of the olfactory organ of cyprinodonts (Teleostei, Cyprinodontoidea). *Cell Tissue Res.* 172: 245–267.

Zeiske, E., H. Breucker and R. Melinkat. 1979. Gross morphology and fine structure of the olfactory organ of rainbow fish (Atheriniformes, Melanotaeniidae). *Acta Zool.* (Stockholm) 60: 173–186.

Zeiske, E., J. Caprio and S.H. Gruber. 1986. Morphological and electrophysiological studies on the olfactory organ of the lemon shark, *Negaprion brevirostris* (Poey).

Indo-Pacific Fish Biology: Proceedings of the 2nd International Conference on Indo-Pacific Fishes. T. Uyeno, R. Arai, T. Taniuchi, and K. Matsuura (Eds).*Ichthyol. Soc. Japan* Tokyo, pp. 381–391.

Zeiske, E., B. Theisen and S.H. Gruber. 1987. Functional morphology of the olfactory organ of two carcharhinid shark species. *Can. J. Zool.* 65: 2406–2412.

Zeiske, E., B. Theisen and H. Breucker. 1992. Structure, development, and evolutionary aspects of the peripheral olfactory system. In: *Fish Chemoreception*, T.J. Hara (Ed.). Chapman & Hall, London, New York, pp. 13–39.

Zeiske, E., A. Kasumyan, P. Bartsch and A. Hansen. 2003. Early development of the olfactory organ in sturgeons of the genus *Acipenser*: a comparative and electron microscopic study. *Anat. Embryol.* 206: 357–372.

Zielinski, B. and T.J. Hara.1988. Morphological and physiological development of olfactory receptor cells in rainbow trout (*Salmo gairdneri*) embryos. *J. Comp. Neurol.* 271: 300–311.

CHAPTER

2

Olfactory Responses to Amino Acids in Rainbow Trout: Revisited

Toshiaki J. Hara

ABSTRACT

Fish, unlike terrestrial vertebrates, detect chemical compounds dissolved in the surrounding water and the entire process of olfaction takes place in water. Of the four main classes of chemical compounds (amino acids, bile acids, prostaglandins, and sex steroids) identified as specific olfactory stimuli (odorants) for fish to date, amino acids are by far the most widely studied chemicals for fish olfaction. The purpose of this study was to re-examine data on electro-olfactogram (EOG) experiments in rainbow trout in an attempt to determine receptor types through which naturally occurring amino acids might be detected and thereby information translated into fish behaviours. Three lines of experimental evidence–concentration-response relationship, cross-adaptation, and binary mixture–indicate that at least three receptors for

Address for Correspondence: Toshiaki J. Hara, Canada Department of Fisheries and Oceans, Freshwater Institute, Winnipeg, Manitoba, Canada R3T 2N6; and Department of Zoology, University of Manitoba, Winnipeg, Manitoba, Canada R3T 2N2. E-mail: thara@cc.umanitoba.ca cysteine, arginine, and glutamate exist, playing dominant roles in discrimination of amino acids. Evidence further suggests that each of these receptors, distinct from those for other odorant classes, is likely to be expressed in separate neurons and detects a set of amino acids. Each receptor recognizes the characteristic feature, or epitope, of an amino acid molecule including: 1) hydrocarbon chain length, 2) difference in functional groups, and 3) position and nature of functional groups. Because all amino acids have some of these features in common, they could all function as partial agonists, activating all available receptors depending on their affinities and concentrations. The possibility that fish olfactory receptors for amino acids are members of a multigene family of G protein-coupled receptors, evolving from a common ancestral gene characterized in mammals is discussed.

Key Words: Fish; Olfaction; Electro-olfactogram; Amino acids; Receptors.

1. INTRODUCTION

Olfaction begins with the binding of an odorant molecule to a receptor on the olfactory neuron (or olfactory sensory neuron or receptor neuron) surface, initiating a cascade of enzymatic reactions that results in the production of a second messenger and the eventual depolarization of the neuronal membrane, which leads to triggering of action potential (Shepherd, 1994; Buck, 1996). Olfactory receptors belonging to G-protein-coupled receptor superfamily encoded by a large multigene family were first identified in rat (Buck and Axel, 1991) and then in other species including fishes (Ngai et al., 1993b; Cao et al., 1998; Freitag et al., 1998; Naito et al., 1998). The size of the receptor repertoire is estimated to be 50-100 in fishes (Ngai et al., 1993a; Barth et al., 1996; Weth et al., 1996) and as many as 1000 in higher vertebrates (Buck and Axel, 1991; Parmentier et al., 1992; Raming et al., 1993). In the channel catfish (*Ictalurus punctatus*), each receptor gene expresses in approximately 1% of olfactory neurons (Ngai et al., 1993a), suggesting that each neuron may express only a single receptor gene. To understand how olfactory neurons transduce the information represented by the molecular structure of odorants, it is essential to identify definite pairing of receptors with odorants. To date, however, functional evidence that cloned olfactory receptors indeed mediate specific odorants has been obtained only in a few species (Zhao et al., 1998; Speca et al., 1999; Touhara et al., 1999; Gaillard et al., 2002).

Fish, unlike terrestrial vertebrates, detect chemical compounds dissolved in the surrounding water and the entire process of olfaction takes place in water. The chemical substances thus perceived by fish olfaction are primarily small molecules with aqueous solubility. Although the volatility of odorants is less relevant for fish than for those in air, volatile chemicals primarily odorous to humans have been widely utilized as chemical stimuli in the study of fish olfaction (for review see Hara, 1971, 1993). Sutterlin and Sutterlin (1971) and Suzuki and Tucker (1971) respectively demonstrated that white catfish (*Ictalurus catus*) and Atlantic salmon (*Salmo salar*), are highly sensitive to non-volatile amino acids and thereby revolutionalized the aforesaid practice. To date, four main classes of chemical compounds have been identified as specific olfactory stimuli (odorants) for fish and their stimulatory effectiveness characterized: amino acids, bile acids, sex steroids, and prostaglandins. These four odorant/pheromone classes, detected by separate receptors, are non-volatile chemicals that are typically non-odorous to humans. Of these, amino acids are by far the most widely studied chemicals for fish olfaction. To date, high sensitivity of the olfactory system to amino aids has been shown in more than 20 fish species (e.g. Hara, 1994a, b). Recently, specific olfactory sensitivities to nucleotides in channel catfish (*Ictalurus punctatus*; Nikonov and Caprio, 2001), catecholamines in goldfish (*Carassius auratus*; Hubbard et al., 2003), and polyamines in goldfish (*C. auratus*; Rolen et al., 2003) and zebrafish (*Danio rerio*; Michel et al., 2003) have been reported, but their stimulatory characteristics have yet to be thoroughly examined. Although odorants play different roles in fish behaviour involving feeding, reproduction, migration, kin recognition, and predator-prey interactions, amino acids have been shown to be involved in all these behaviours (Hara, 1994a, b).

Amino acids are small biomolecules with an average molecular weight of about 135. These organic acids exist naturally in a zwitterion state, with the carboxylic acid moiety ionized and the basic amino group protonated. The entire class of amino acids has a common backbone of an organic carboxylic acid group and an amino group attached to a saturated carbon atom. The simplest member of this group is glycine, in which the saturated carbon atom is unsubstituted, rendering it optically inactive. The rest of the 20 most common amino acids are optically active and exist as both D and L stereoisomers. Naturally occurring amino acids are, for the most part, a laevorotatory (L) isomer. Substituents on the alpha (or saturated) carbon atom vary from lower alkyl groups to aromatic amines and alcohols. There are also acidic and basic side chains as well as thiol chains that can be oxidized to dithiol linkages. Amino acid side chains can thus be polar, non-polar, or practically neutral. Concentrations of dissolved amino acids in natural waters are generally low, ranging 10^{-7}–10^{-8} M (Gardner and Lee, 1975), though well within the detective range as shown below. For instance, recent studies show that the migrating Pacific salmon (*Oncorhynchus masou*) are able to recognize home-stream waters by olfaction solely based on the composition of dissolved amino acids (Shoji et al., 2000). Earlier studies using electroencephalogram (EEG) recordings determined the specificity of olfactory stimulation of amino acids and analogues in rainbow trout (*O. mykiss*; formerly *Salmo gairdneri*) and established that: 1) only α-amino acids are stimulatory at biologically relevant concentrations, 2) the natural L-isomer of an amino acid is significantly more stimulatory than its D-isomer, 3) ionized α-amino and α-carboxyl groups are required, 4) the α-hydrogen of an amino acid must be free, and 5) the size and polar nature of the fourth α-moiety are important determinant factors for stimulatory effectiveness (Hara, 1973, 1976, 1977). The amino acids effective as odorants for rainbow trout are thus characterized by being simple, short, and straight chained, with only certain substituent groups. Glycine is the smallest amino acid molecule that fulfils all the requirements. Among homologous series of aliphatic amino acids, the response is maximal for those whose number of carbon atoms in the chain is three to four. Based on these findings, a hypothetical three-subsite amino acid receptor model was proposed involving two charged subsites, one anionic and one cationic, capable of interacting with ionized amino and carboxylic groups of the stimulant amino acid molecule (Hara, 1977). As just mentioned, because the size and polar nature of the fourth α-moiety of the amino acid are determinant factors of effectiveness, there must be another site in the receptor that recognizes and accommodates the general profile and charge of the molecular residue.

As detailed below, EEG responses to amino acids increase with stimulus concentrations of more than four to five log units. To explain this, a kinetic analysis of EEG responses in rainbow trout was done (Hara, 1982). The analysis revealed that three receptor components, or transduction mechanisms with different specificities, exist for amino acids in this species. Kinetic analysis is based on the assumption that the EEG response is linearly related to the primary events of olfactory transduction. The questions then raised were whether the EEG responses elicited in the secondary neurons of the olfactory bulb directly represent the activity of the primary olfactory neurons.

This study summarizes data on electro-olfactogram (EOG) experiments in rainbow trout so as to determine/confirm possible receptor types through which water-borne amino acids might be detected and thereby information translated into fish behaviours. Throughout the discussion, it is emphasized that each of the three experimental approaches, i.e. the concentration-response (C-R) relationship, cross-adaptation and mixture experiments for determining receptor specificity, is of potential value in assessing the specificity of a response to a particular receptor population. All these lines of experimental evidence are consistent with the contention that cysteine- (*CysR*), arginine- (*ArgR*), and glutamate- (*GluR*) receptors exist in the rainbow trout olfactory epithelium, playing dominant roles in detecting and discriminating naturally occurring amino acids. Evidence further suggests that each of these receptors is likely to be expressed in separate neurons and binds/detects a set of amino acids. In the following discussion, the term 'receptor (s)' is loosely used to accommodate the stimulus-binding sites and all subsequent events (transduction processes) leading to the production of generator potentials within the olfactory neuron.

2. ELECTRICAL RESPONSES TO AMINO ACIDS

2.1 Electro-olfactogram (EOG) Response

Most of the data presented in this chapter were obtained by EOG recordings. The EOG, a negative slow potential caused by summated current flow through the extracellular resistance of the olfactory epithelium, is a reliable indicator of olfactory neuronal activity to amino acids as well as other odorants, including pheromones, widely used in the study of fish olfaction (Ottoson, 1956; Getchell, 1974; Hara, 1992). The EOG response to amino acid stimulation is characterized by an initial phasic component followed by a rapid decline to a steady state tonic level that continues throughout the stimulus duration (Fig. 2.1A; Evans and Hara, 1985). This slow adaptation process is consistent with electrophysiological evidence obtained in the olfactory systems of other animal classes (Ottoson, 1956). Approximately 2 min is required for complete recovery when two identical amino acid stimuli are applied successively at increasing intervals (Evans and Hara, 1985).

Coinciding with the EOG, the phasic and tonic response patterns are recorded in olfactory nerve facicles (NTR) as well as in the olfactory bulb

(Hara, 1982; Sveinsson and Hara, 1990a, b). These data suggest the existence of two olfactory neuronal types based on electrophysiological characteristics, fast and slow adapting. Recent study shows that this feature is common across a wide range of vertebrates and that firing patterns of olfactory neurons can be determined by their individual passive electrical properties and not by membrane capacitance and voltage-dependent conductance (Madrid et al., 2003).

2.2 Electroencephalogram (EEG) Response

Electrical signals evoked in the primary olfactory neurons in response to chemical stimulation are transmitted through the olfactory nerve fibres to the olfactory bulb, where the spontaneous electrical activity (intrinsic wave) is immediately interrupted by rhythmic oscillatory potentials (induced waves) (Fig. 2.1B). The induced waves, commonly known as electroencephalographic (EEG) responses, are interpreted as produced in the dendritic network within the olfactory glomeruli as a result of synchronous activity of the olfactory bulbar neuron (Satou, 1990). Rhythmic waves of the fish olfactory bulb induced by odorants are slower (3-15 Hz) compared to those in other animals. A possible role for specific frequency patterns has been hypothesized (Hara et al., 1973; Kaji et al., 1975; Satou and Ueda, 1975). The idea stems from studies using

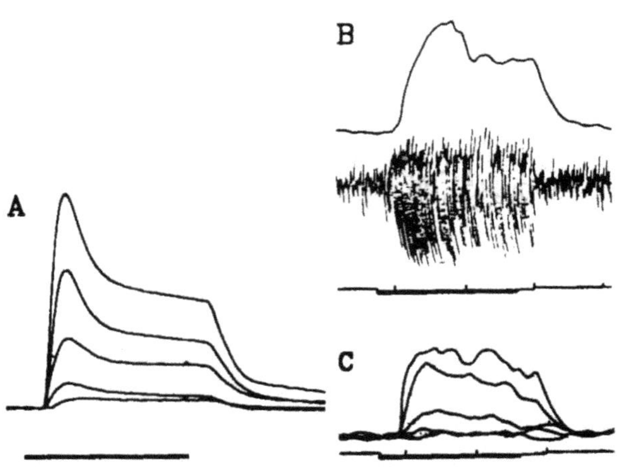

Fig. 2.1 Examples of electro-olfactogram (EOG) and electroencephalogram (EEG) responses to amino acids in rainbow trout. **A.** Superimposed EOG responses to 10^{-9}–10^{-5} M L-cysteine. **B.** Oscillatory EEG response to 10^{-5} M L-cysteine (lower trace) and its simultaneous integration (upper trace). **C.** Superimposed integrated EEG responses to 10^{-9}–10^{-5} M L-serine (from Hara and Zhang, 1998).

band-pass filters and spectral analysis to extract the frequency components of the EEG waveforms (Kudo et al., 1972). These analyses show that some stimuli, such as a single amino acid elicit a single dominant frequency in response, while others such as food extracts elicit a wider frequency range (Hara et al., 1973). However, a more recent study in goldfish using many odorant classes showed that the dominant frequencies of EEGs to various stimuli are similar, though their size and distribution are specific for odorant and pheromone types (Hanson et al., 1998).

In fish, the primary olfactory neurons specific for amino acids as well as other odorants are randomly distributed over the entire surface of the olfactory epithelium and yet project topographically to spatially segregated regions of the olfactory bulb (Fig. 2.2; Ngai et al., 1993a; Hara and Zhang, 1996). In rainbow trout, amino acid information is processed in a conical mass occupying a large portion of the lateroposterior olfactory bulb, while

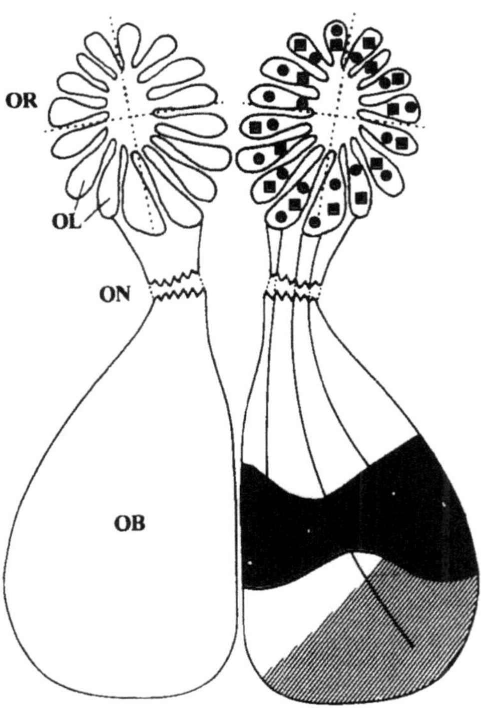

Fig. 2.2 Diagram showing olfactory neurons specific for amino acids (dots) and bile acids (squares) are distributed throughout the olfactory rosette and project to segregated olfactory bulbar regions, the lateroposterior (grey) and medial (black) respectively. OB, olfactory bulb; OL, olfactory lamellae; ON, olfactory nerve; and OR, olfactory rosettes (from Hara and Zhang, 1996).

that of bile acids is processed in the central region across the mid-bulb (Fig. 2.3; Hara and Zhang, 1998). These data are consistent with a model in which randomly distributed primary olfactory neurons with common receptor specificities project to common glomeruli in the olfactory bulb. Interestingly, though not the primary subject of the present discussion, none of the pheromones known to function in other fish species induces EEG responses in the rainbow trout olfactory bulb (see below).

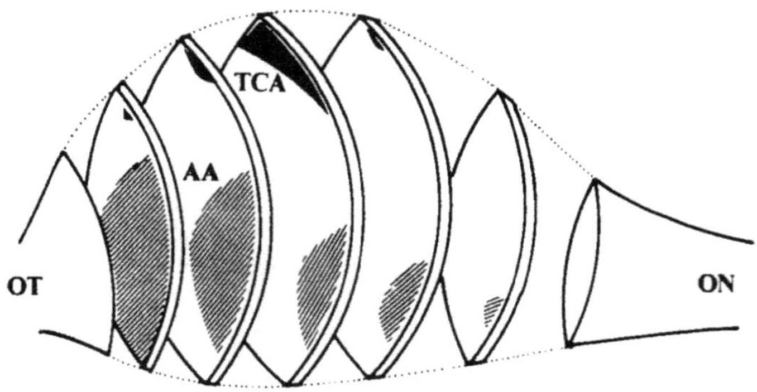

Fig. 2.3 Diagram illustrating bulbar projection patterns of the primary olfactory neurons specific for amino acids (AA; grey) and bile acids (TCA; black) in rainbow trout. ON, olfactory nerve; OT, olfactory tract (from Hara and Zhang, 1998).

3. DETERMINATION OF RECEPTOR SPECIFICITY

3.1 Concentration-response (C-R) Relationship

Existence of receptors is predicted from a large number of observations demonstrating the extraordinary specificity with which a response is elicited when a series of amino acid homologues is evaluated. The very specificity of an amino acid action persuades us that an observed effect is receptor mediated rather than a non-specific phenomenon independent of specific ligand-receptor interaction.

The C-R curve provides important clues about the efficacy and potency of stimulants and possible involvement of multiple receptor interactions. Curves relating receptor responses to the logarithmic concentration of stimuli with decreasing affinities will be progressively displaced to the right, but their pseudolinear central parts will remain parallel to each other (Boeynaems and Dumont, 1980). Complete C-R curves are useful not only to compare the relative potency of amino acids involved, but to determine whether all amino acids evoke the same maximal effect, evidence for the existence of a single receptor population. A wide range of amino acids is detected by the rainbow trout olfactory system, with threshold concentrations ranging from 10^{-9}–10^{-7}M, which approximates levels of free amino acids found in natural waters. As shown in Fig. 2.4, response magnitude increases almost exponentially and C-R curves of most amino acids exhibit characteristic broad dynamic ranges, covering over six to seven log units, which represent a 1 to 10 millionfold change in amino acid concentration. The broad dynamic range exhibited may be the result of binding of amino acid molecules to multiple receptor types with varying binding affinities or the result of negative co-operativity among receptor sites (Boeynaems and Dumont, 1980).

The quantitative feature of the C-R curve for each amino acid varies considerably. These analyses are useful in comparing relative affinities of amino acid stimulants (ligands) with receptors and in determining whether the stimulant of interest in a series of compounds can cause the same maximal response, indicative of a closely similar interaction with the

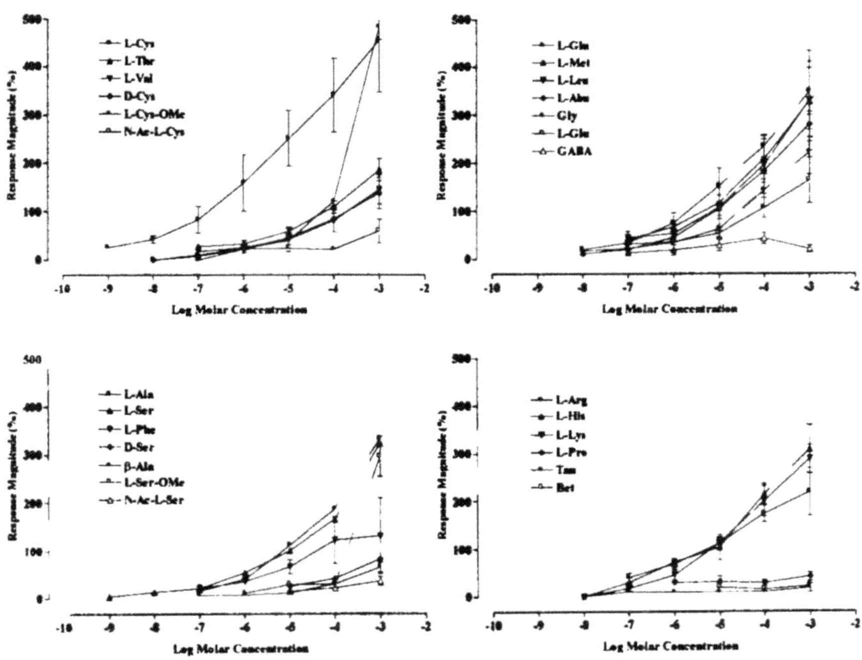

Fig. 2.4 Concentration-response (C-R) relationships of EOG responses of rainbow trout to amino acids and related chemicals (arbitrarily grouped). L-Cys-OMe, L-cysteine methyl ester; N-Ac-L-Cys, N-acetyl-L-cysteine; GABA, γ-aminobutylic acid; Tau, taurine; Bet, betaine.

recognition site, as other analogues. Once complete C-R curves are obtained for all amino aids, their relative potencies can be estimated (Fig. 2.4). However, the relative potency of amino acids is valid only when all the amino acids under study are capable of eliciting the same maximal response. Clearly, based on maximal responses, several different amino acid groups can be recognized in Figure 2.4, notably (1) cysteine (Cys), (2) arginine (Arg), (3) glutamic acid (Glu), and possibly (4) phenylalanine (Phe). The rest are either assigned to one of these groups or non-stimulatory. The following features, most of which have been established from EEG studies, are also observable: 1) responses to D-enantiomers are considerably smaller than those to L-enantiomers; 2) acetylation of the α-amino group diminishes EOGs; 3) esterification of the α-carboxyl group also diminishes EOGs up to 10^{-4} M, where they take off sharply, suggesting involvement of non-binding activation processes such as transport (see below); and 4) GABA, β-alanine, proline, taurine, and betaine are not stimulatory. A comparison of EOG and EEG responses to a set of amino acids shows that a high correlation exists (Fig. 2.5). This would disclaim the earlier question regarding the validity of use of EEG responses for the purpose of receptor evaluation. It can be seen in the Figure 2.5 that cysteine, arginine, and glutamic acid groups are slightly shifted from the rest of the amino acids, which may provide additional support for the grouping made based on C-R curves above.

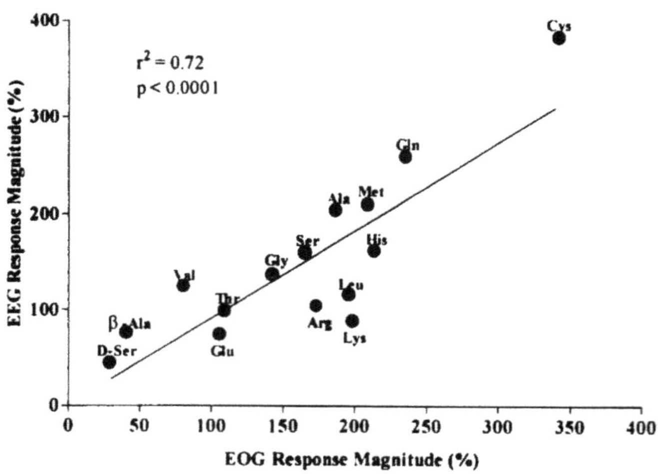

Fig. 2.5 Correlations between EOG and EEG responses to selected amino acids in rainbow trout.

Some of the other characteristic features of olfactory responses to amino acids deserve special attention here. In aqueous solution under normal physiological conditions the amino acid exists as a dipolar ion; the carboxylic group is dissociated and the amino group is associated. The pH of the solution determines the degree of ionization of the amino acids and of the charged groups in the receptor membrane, and consequently may affect the interaction of the stimulus molecule with the receptor site. In fact, the olfactory responses to most amino acids are highly pH-dependent, displaying the maximal activity near their isoelectric points where dipolar ions are maximal and bear no net charge (Hara, 1976). Binding studies with membrane fractions derived from the olfactory epithelium show good agreement between the rank order of binding and electrophysiological potency, supporting the conclusion that the binding data reflect the interaction of stimulus amino acids with physiologically relevant receptors (Cagan and Zeiger, 1978). However, experiments with transport inhibitors, low extracellular Na^+ levels, countertransport, and increasing osmotic strength show some involvement of amino acid transport, especially for neutral ones (Brown and Hara, 1981). It is likely that transport could be involved in the sharp increases in EOG responses to neutral amino acids occurring at or near 10^{-5}–10^{-4}M and above, which may be partly responsible for the broad dynamic C-R relationships seen in EOG responses (see Fig. 2.4).

3.2 Cross-adaptation

The important criterion of a specific receptor-mediated event is selectivity of the blockade by antagonist agents. For example, β-adrenergic effects are selectively blocked by propranolol whereas α-adrenergic effects are selectively blocked by phentolamine (Limbird, 1986). Unfortunately, except for a competitive inhibitor N_{ω}-nitro-L-arginine for L-arginine responses in sea lamprey (Zielinski et al., 1996), no specific antagonist is known to exist for any of amino acid-mediated olfactory responses. Similarly, lack of specific blocking agents for the amino acid receptor makes it almost impossible to conduct protection studies. An alternative is an experimental approach known as cross-adaptation or cross-desensitization. The general principle is as follows: if exposure to amino acid A results in a decline in the subsequent sensitivity to it and amino acid B, but not to amino acid C, then one interpretation of these findings is that A and B interact with a common receptor whereas C interacts with a distinct receptor population(s). This approach has been widely exploited to demonstrate a multiplicity of functional receptors for fish olfactory as well as gustatory systems. These studies assume that the desensitization process under study is homologous in nature, i.e. that a particular amino acid desensitizes the tissue only to effects mediated by its specific receptor and not to all receptor populations sharing a particular effector mechanism. In most cross-adaptation experiments with fish, plain water perfusing the olfactory organ is replaced by an adapting amino acid solution until the response declines and stabilizes at the tonic level, and test stimuli prepared with the adapting solution are tested. Percent adaptation is calculated by pre- and active adaptation responses. Two cross-adaptation protocols differing in principle have been practised in fish olfactory studies: 1) a set of amino acids is randomly selected and tested at equipotent or identical concentrations for both adapting and test stimuli (e.g. Caprio and Byrd Jr., 1984; Ohno et al., 1984; Michel and Derbidge, 1997), and 2) adapting amino acids and their concentrations are selected based on C-R curves and adaptation is made at high concentrations to achieve near-receptor saturation (e.g. Sveinsson and Hara, 1990a; Laberge and Hara, 2003; this study). The former addresses the similarity or dissimilarity between two competing stimuli, while the latter determines the capacity of inhibition/desensitization by the adapting stimuli. The competition paradigm can be used in the situation wherein chemicals from across groups are involved, e.g. amino acids *vs* bile acids (Michel and Derbidge, 1997), or often results in estimating as many receptor types as the number of adapting stimuli employed (Caprio and Byrd Jr., 1984; Ohno et al., 1984; Michel and Derbidge, 1997). The inhibition paradigm using a high concentration of adapting stimuli virtually eliminates all receptors interacting with the adapting stimuli without physical isolation (see Sveinsson and Hara, 1990b). The response induced by testing stimuli, if any, is thus the result of reaction with receptors other than adapting stimuli. It is prerequisite to obtain complete C-R curves for each amino acid before any cross-adaptation experiment is attempted.

Figure 2.6 shows a typical cross-adaptation protocol used in the present study. EOG responses to test amino acids are first recorded with plain water perfusing the olfactory organ (L-Ser and L-Arg at 0.1 mM). Plain water is then switched to an adapting amino acid (1 mM L-Cys) and the same test stimuli prepared in the adapting solution (1 mM L-Cys) are tested in the same fashion. Responses to the test stimuli are superimposed

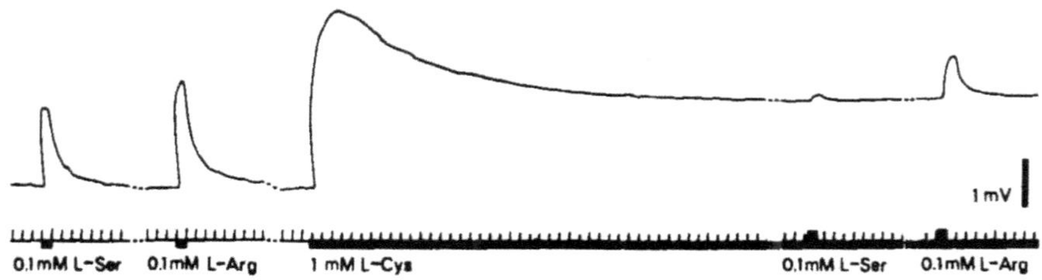

Fig. 2.6 Example of EOG tracings showing the cross-adaptation protocol employed in the study. L-Serine and L-arginine are tested before and during adaptation to L-cysteine.

on top of the tonic response to the adapting stimulus. In cross-adaptation, it is essential that interactions between two chemicals be competitive. Therefore, if one amino acid is tested under exposure to the same amino acid with consecutively higher concentrations (self-adaptation), the C-R curves thus obtained would be parallel and shift gradually to the right (Fig. 2.7A). These data also estimate that for complete suppression the adapting concentrations require approximately 100 times those of testing stimuli, e.g. a response to 10^{-5} M L-Ala is completely suppressed under adaptation with 10^{-3} M L-Ala solution. The fact that a response to a given amino acid is not suppressed completely can be taken as evidence that the amino acids stimulate more than one receptor type (Fig. 2.7B). In Figure 2.7, L-Ala responses to a fixed concentration (10^{-5}M) are suppressed by increasing concentrations of L-Ala, but L-Arg and possibly L-His remain unsuppressed by L-Ala, indicating that L-Arg and possibly L-His interact with separate receptors.

When the rainbow trout olfactory organ is adapted to the neutral amino acid L-Cys at 10^{-3} M, responses to all neutral amino acids are suppressed, while responses to basic amino acids Arg and Lys and acidic amino acid Glu remain unsuppressed (Fig. 2.8A). Similarly, only basic amino acids are suppressed when adapted to 10^{-3} M Arg (Figs. 2.8B and 2.9). This situation is also shown in a competition experiment in which EOG responses to a fixed concentration (10^{-5} M) of Ala, Arg, and His are recorded under adaptation to alanine at successively increasing concentrations (cf. Fig. 2.7B). These data are consistent with the view that at least three amino acid receptor types exist in the rainbow trout olfactory system or, more specifically, the cysteine- (*CysR*), arginine- (*ArgR*) and glutamic acid (*GluR*) olfactory receptors play major roles in detecting and discriminating naturally occurring amino acids. Cross-adaptation also shows that neither Cys nor Arg cross-adapts tyrosine (Tyr)

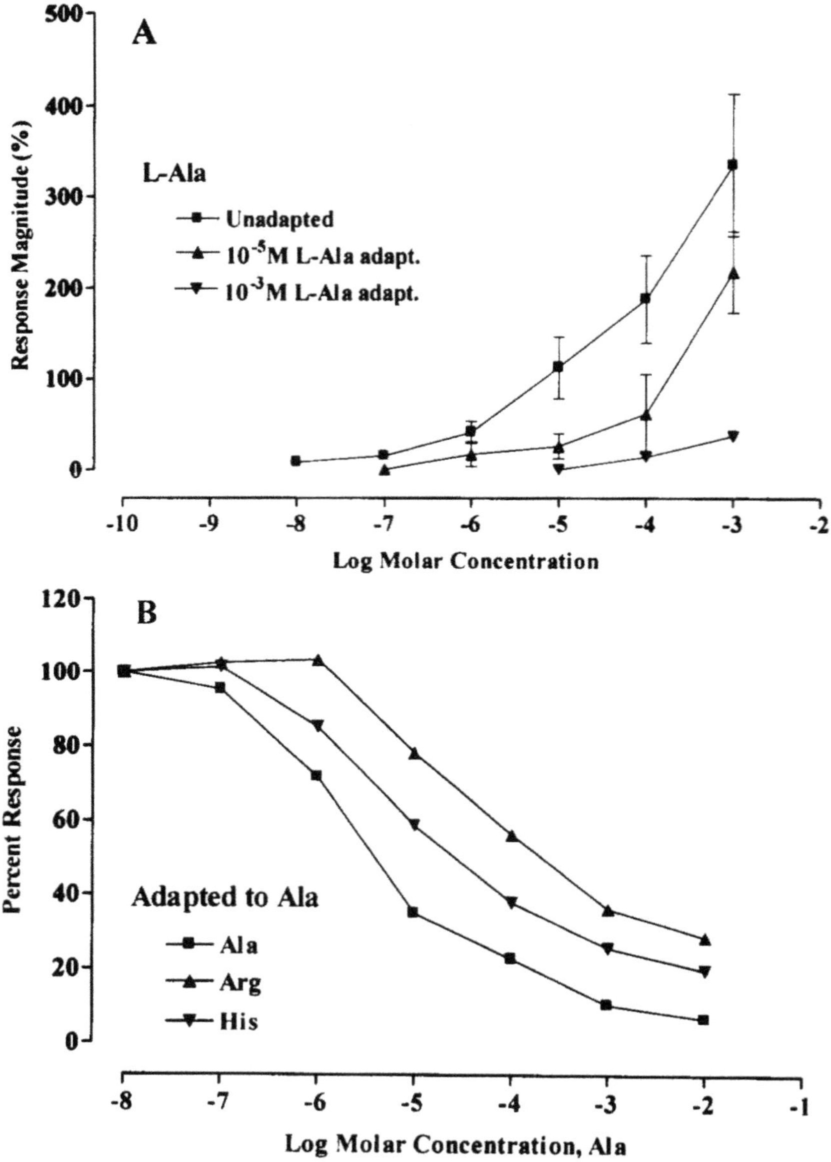

Fig. 2.7 Two of the characteristics used to evaluate cross-adaptation data. **A.** Self-adaptation data used to examine the behaviour of interactions between amino acids and the prospective receptors. Parallel shifting of the C-R curves is expected if a ligand interacts with a receptor in a truly competitive and fully reversible fashion as illustrated. **B.** Displacement data. EOGs to 10^{-5} M to Ala, Arg and His are plotted against adapting stimulus Ala with increasing concentrations. The full agonist or competitor (Ala) displaces/suppresses completely, whereas the partial agonists (Arg, His) suppress partially.

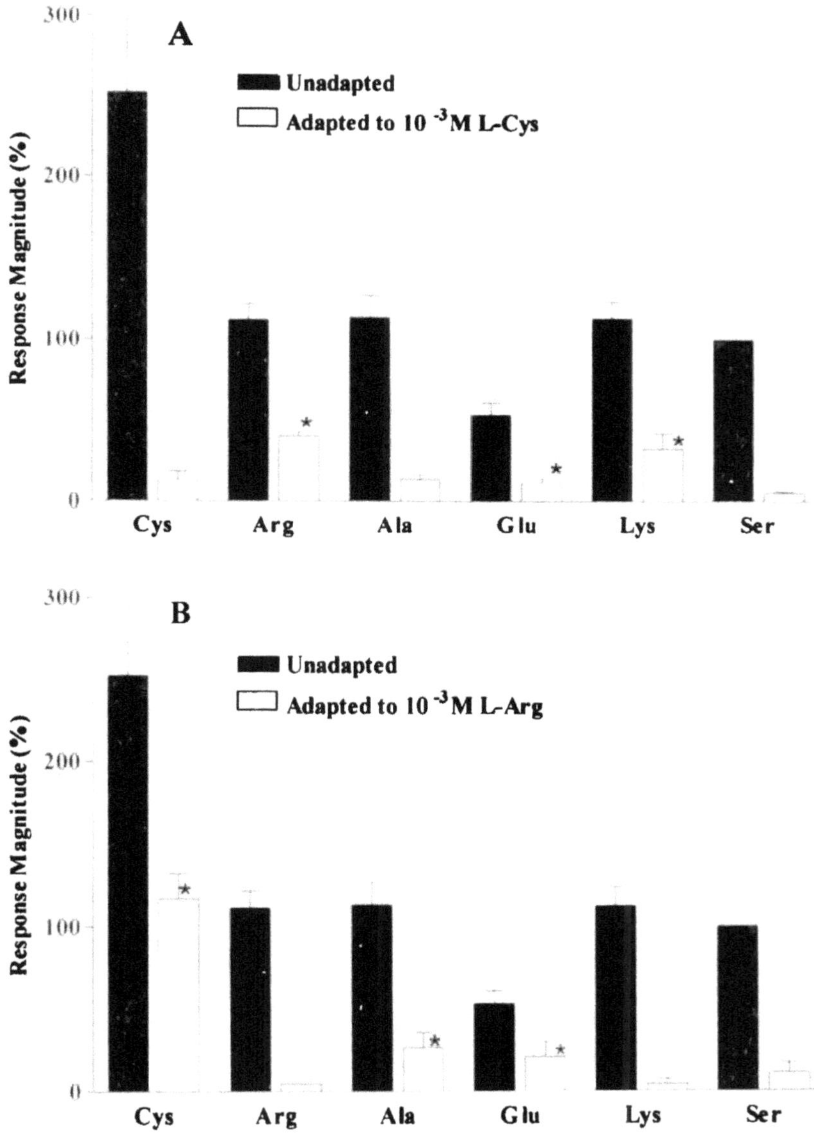

Fig. 2.8 Cross-adaptation data. EOG responses to selected amino acids tested at 10^{-5} M during adaptation to 10^{-3} M L-Cys (**A**) and L-Arg (**B**). Note that Arg, Glu and Lys are not suppressed by Cys significantly, while Cys, Ala and Glu are not suppressed by Arg significantly.

and tryptophan (Trp), suggesting that receptors exist for amino acids with aromatic rings (data not shown). These three amino acid receptors bind to other neutral, basic and acidic amino acids, all of which may stimulate respective receptors at varying but lower affinities than those of the primary ligands, Cys, Arg and Glu. In other words, none of these receptors is fully independent i.e., most if not all receptors can be activated by multiple amino acids depending on concentrations, and conversely most amino acids are able to activate more than one type of receptor.

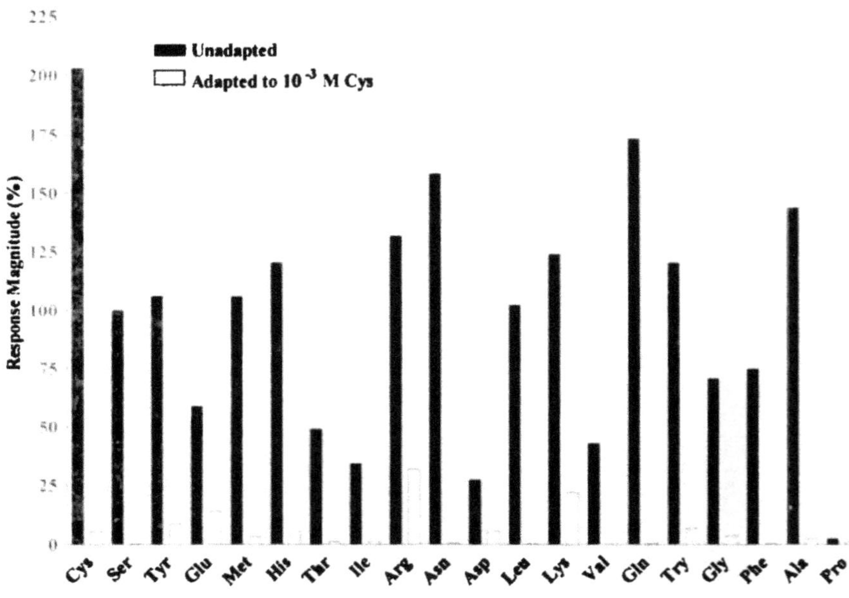

Fig. 2.9 Cross-adaptation data showing that Cys depresses all neutral amino acids.

3.3 Binary Mixture

Further supportive evidence for the existence of independent *CysR*, *ArgR* and *GluR* receptors comes from binary mixture studies. If two stimulants activate separate receptors, responses to a mixture of the two will be greater than that of twice the concentration of either of the mixture component. Here, the mixture discrimination index (MDI) is measured by dividing a response to a mixture by the larger of the responses to twice the concentration of each component tested individually, while the independent component index (ICI) is calculated by dividing the response to the mixture by the sum of the responses to its components (Hyman and Frank, 1980; Caprio et al., 1989). Theoretically, the MDI equals 1 unless mixture suppression (MDI ≤ 1) or mixture enhancement (MDI ≥ 1) occurs. The ICI equals 1 if the response to the mixture equals the sum of the responses to the components. For example EOGs to mixtures of Cys/Arg, Cys/Glu and Arg/Glu are significantly greater than those of the twice the concentration of either of the respective mixture components (Laberge and Hara, 2004). The MDI and ICI are 1.224 and 0.894, 1.220 and 0.887, and 1.336 and 0.856 respectively. The calculated MDI values for all, except arginine-lysine and serine-cysteine combinations are larger than 1, indicating the mixture enhancement. The result of binary mixture experiments presented here supports the cross-adaptation data, indicating that Cys, Arg and Glu activate the respective receptor mechanisms independent of each other.

4. NATURE OF THE AMINO ACID RECEPTORS

4.1 'Epitopic' Nature of Amino Acid Receptor Interactions

Three lines of evidence strongly suggest that approximately 20 naturally occurring amino acids are detected and discriminated by at least three olfactory receptor types, *CysR*, *ArgR* and *GluR* and that Cys, Arg and Glu are respectively the primary ligands for the three receptors. First, different patterns of C-R curves, typified by these three amino acid groups, are exhibited. Second, adapting the olfactory epithelium to any one of the three suppresses all amino acids within the group, but not others. Third, binary mixtures of either two of the three amino acids produce responses significantly greater than those elicited by double the concentrations of either of the mixture components. Although each of the three approaches employed has the inherent limitations, the result from combining all three lines of experimental evidence can be declared quite definitive.

The hypothetical amino acid receptor site proposed earlier involves two charged subsites capable of interacting with ionized amino and carboxyl groups of amino acid molecules (Hara, 1976). These two subsites are so arranged as to accommodate the α-hydrogen atom of an amino acid molecule, thus making L-isomers more accessible to the receptor site. This triple-subsite receptor model is a minimal prerequisite for an amino acid to be stimulatory. As emphasized earlier, the size and polar nature of the fourth α-moiety are important determinants for effective stimulation and discrimination of the amino acid quality could take place in this region. Let us take a brief look at the molecular architecture of the primary ligands

Cys, Arg and Glu, designated as the respective full agonists for the respective amino acid receptors (Fig. 2.10). Cys, a neutral amino acid with a 3-carbon chain, is one of the few amino acids that contains sulphur. This very reactive thiol side chain (mercaptan group, -SH) enables binding to nitric oxide, other thiols, and superoxides. Arg, the most basic amino acid with a 7-carbon chain, is a positively charged polar amino acid. It contains a complex guanidinium group on its side-chain that encompasses geometrical and charge distributions ideal for binding negatively charged groups on the receptors. The hydrogen of both NH1 and NH2 can be readily bonded. Although lysine is another positively charged amino acid, it contains only a single amino group, meaning it is more limited in the number of hydrogen bonds it can form. Glu (glutamate), an acidic amino acid with a 5-carbon chain on the other hand, is one of just two amino acids (the other being aspartic acid) that has a net negative charge at physiological pH, because it has a carboxylic acid moiety on the side chain. This negative charge makes Glu a very polar molecule. Thus, each of three amino acids is characterized by two specific structural features: primary (two functional groups, amino and carboxylic, specifically arranged around the α-carbon atom) and secondary (carbon chain length and terminal functional groups). Structural diversity of olfactory receptors might well recognize different structural features or *epitope* of an odorant molecule including: 1) length and/or structure of hydrocarbon chain, 2) difference in functional group and 3) position of the functional group within the molecule.

As already mentioned, direct evidence for the existence of binding sites for amino acids in the olfactory tissue was first provided by Cagan and Zeiger (1978), who studied the binding of radiolabelled amino acid to a sedimentable fraction derived from the olfactory rosettes of rainbow trout. Physical parameters such as saturability, reversibility, affinity, and quantity

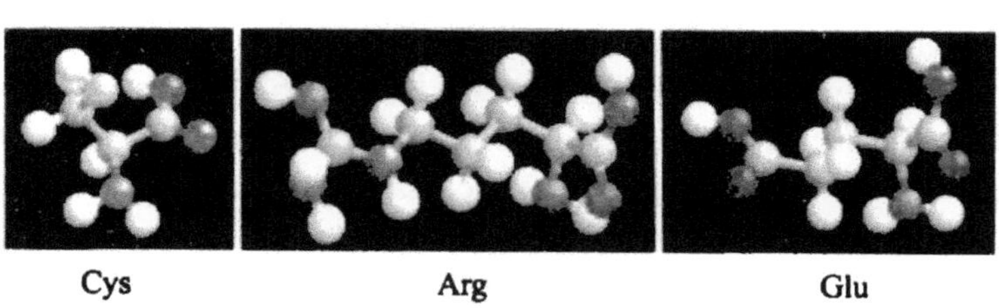

Fig. 2.10 Three-dimensional models of the Cys, Arg and Glu molecules representing respectively the primary agonists for *CysR*, *ArgR* and *GluR* olfactory receptors in rainbow trout.

of binding sites are reasonably consistent with receptor criteria. In addition, good agreement between the rank order of binding and electrophysiological potency further supports the conclusion that binding data reflect the interaction of amino acids with physiologically relevant receptors. Competition studies demonstrate multiple binding sites for amino acids: site TSA, which binds threonine, serine and alanine; site L, which binds lysine; site V, which binds valine; site A_B, which binds β-alanine; site H, which binds histidine; and site A_D, which binds D-alanine. However, the kinetics of association is slow and not in accord with values expected of sensory receptors. Inconsistencies between the various aspects of amino acid binding by the sedimentable fraction and electrophysiological response thus preclude unequivocal classification of the binding as representing olfactory receptors. Further, the Na-dependence of part of the binding, sensitivity to an amino acid transport inhibitor, accelerative exchange resulting from ligand preloading and reduction of binding by high osmotic strength suggest that the assay procedure is measuring, at least in part, a transport-related phenomenon (Brown and Hara, 1981, 1982). A recent study showed that decarboxylated arginine, agmatine (1-amino-4-guanidobutane), permeates cation channels involved in olfactory transduction in the zebrafish olfactory neurons (Michel et al., 1999). Although binding studies provided direct evidence for the existence of receptors, the receptor types determined differ considerably from those presented in the present EOG studies. This discrepancy may be due in part to the choice of unlikely ligands because of the paucity of physiological data at that time.

4.2 Specificity of Amino Acid Receptors

Then, how are the rest of the naturally occurring amino acids detected via the three receptors? All these amino acids share the primary structural features and some share some of the secondary features, which may allow them to bind to one or more of the three receptors. In other words, they could all function as a partial agonist for one or more of the receptors. A partial agonist is defined as a substance endowed with some intrinsic activity and thus behaving as an agonist. In the presence of an agonist of higher intrinsic activity, with which it shares a common binding site on the receptor, it could behave as a competitive antagonist. Also, when there are multiple receptor populations with differing affinities for the ligand, the ligand will occupy the high-affinity population first, so that binding to the lower affinity population(s) will only be detected as ligand concentrations increase. A ligand could stimulate receptor populations with lower affinities non-specifically at higher concentrations. This complex situation also makes interpretation of the competitive cross-adaptation data discussed earlier extremely difficult. In a cross-adaptation study with zebrafish, for instance, Michel and Derbridge (1997) explain that shared patterns of partial adaptation of the structurally related odorants such as amino acids are due to the result that either the odorant receptors or olfactory neurons involved have overlapping ligand-binding specificities. Under somewhat different situations, Sato and Suzuki's (2001) results of an inward current study using isolated rainbow trout neurons are a bit confusing. Some neurons responsive to a single amino acid do not respond to an amino acid mixture, and vice versa. These authors suggest that individual olfactory neurons of rainbow trout have a complex combination of multiple amino acid receptor sites.

In nature, fish generally encounter amino acids as complex mixtures rather than single amino acids or simple mixtures usually tested in laboratory studies (e.g. Shoji et al., 2000; Sato and Suzuki, 2001). Given that a fish encounters a mixture of 20 amino acids at equal concentrations, from the above discussion it is expected that it would first detect Cys, Arg and Glu from the mixture through their respective receptors, followed by other amino acids depending on their affinities for the remaining spare receptors. The fish would thus never individually perceive all amino acids; rather it would recognize an entire picture or image formed by the mixture. The constituents and their concentrations in the mixture determine the nature of images. Fish foods for example, emit a unique chemical image to which fish are conditioned (Atema et al., 1980; Valentinčič et al., 1994). Casual observations show that fish often reject a new food until they recognize/recondition to its new chemical image. However, a recent study demonstrated that each of three amino acids, Cys, Arg or Glu, do individually play dominant roles in fish behaviour (Hara, unpubl. data).

4.3 Distribution (Expression) of Amino Acid Receptors in Individual Neurons

How are these amino acid receptors expressed and distributed in individual olfactory neurons? *In situ* hybridization studies tell us that each olfactory receptor gene is expressed in only a small fraction of vertebrate olfactory neurons (for review see Buck, 1996). In the channel catfish, each olfactory neuron gene is expressed in ~1% olfactory neurons, suggesting that each neuron may express only a single receptor gene (Ngai 1993a). Let us take a close look at the cross-adaptation data again (Figs. 2.6 to 2.9). A small but significant portion of EOGs to Arg remain unaffected when the olfactory epithelium is adapted to a high concentration of Cys, or vice versa. This clearly indicates that extremely specific, independent *CysR*, *ArgR* or *GluR* receptors, though small in number, do exist, with each receptor type residing in a separate neuron. The majority of cross-adapts seen may be partly explained by sharing receptors or some levels of the receptive process in less differentiated neurons, or more likely by experimental artifacts in which unnaturally high concentrations ($\leq 10^{-5}$ M) of both adapting and test stimuli are used to unnecessarily overstate data. As mentioned before, the amino acid levels fish generally encounter in natural waters would rarely exceed 10^{-8}–10^{-7} M. Thus, depending on conditions, one odorant receptor type could detect multiple odorants with similar molecular profiles and chemical natures, and a single odorant could be detected by more than one receptor type, though it is extremely unlikely that odorants from two different chemical groups, e.g. amino acids and bile acids, are detected via the same receptor type under natural conditions. Speca et al. (1999) suggested the possibility that individual olfactory neurons may express multiple odorant receptors based on their observations demonstrating general expression of the 5.24 and 5.3 receptors in goldfish, combined with electrophysiological evidence shown in channel catfish (Kang and Caprio, 1995) and coho salmon, *Oncorhynchus kisutch* (Nevitt and Dittman, 1999). In channel catfish, 55% of the single olfactory neurons respond to more than one amino acid and ~ 40% respond to 10^{-4} M Arg. This is not surprising considering the fact that a higher concentration of an amino acid activates more receptor types, that is the identity of a olfactory signal could be lost with increasing concentration (Fuss and Korsching, 2001; Kajiya et al., 2001; Laberge and Hara, 2004). However, the demonstration that more than 60% of isolated coho salmon olfactory neurons respond to lower concentrations (10^{-5}–10^{-8} M) of L-Ser is hardly explained. In support of the present view that each receptor type resides in a separate neuron in rainbow trout, Serizawa et al. (2000, 2003) proposed that the functional expression of an olfactory receptor gene inhibits activation of other genes, which thus ensures maintenance of the one receptor-one neuron rule in the mammalian olfactory system.

The ciliated neuron-microvillar neuron controversy continues (see review, Zielinski and Hara, 2001). Whether these two morphologically and ontogenetically distinct olfactory neuronal types are functionally differentiated has been the subject of speculation based primarily on indirect studies in various fish species. A recent immunocytochemical study against olfactory receptor coupled G-proteins in the round goby (*Neogobius melanostomus*) shows that the ciliated neurons express the G protein G_{olf} and the microvillar neurons express $G_{\alpha o}$ (Belanger et al., 2003). The $G_{\alpha o}$-protein is expressed by vomeronasal receptor neurons in terrestrial vertebrates (Matsuoka et al., 2001) and is localized in microvillar neurons in catfish and goldfish (Hansen et al., 2001). Amino acid responses from rainbow trout embryos containing only ciliated neurons indicate that ciliated neurons respond physiologically to amino acid stimulation (Zielinski and Hara, 1988). Sato and Suzuki (2001) suggested that ciliated neurons are 'generalists' responding to a wide variety of odorants including amino acids, whereas microvillar neurons are 'specialists' responding only to amino acids. It is pointed out however, that odorant groups employed in their study are amino acids and urine samples. As their analyses show, the urine samples are basically amino acid mixtures containing a large amount of free amino acids. Most interesting is the observation that the ciliated vs microvillar neuron ratio is approximately 1:11 (Sato and Suzuki, 2001). This large numerical latitude seems favourable to specificity of ciliated neurons for amino acids, considering the fact that amino acid responsive neurons project to a huge area of the lateroposterior olfactory bulb (Hara and Zhang, 1998). Thommesen (1983) earlier reported that in rainbow trout, Arctic char (*Salvelinus alpinus*) and brook char (*S. fontinalis*), microvillar neurons respond specifically to amino acids, whereas ciliated neurons respond specifically to bile acid-like substances, which supports the Sato and Suzuki (2001) finding.

According to Zippel and co-authors (1997), in goldfish, as in rainbow trout, ciliated neurons regenerate before the microvillar neurons following olfactory nerve section, coinciding with differential functional restoration (Zippel et al., 1997). The authors suggested that microvillar neurons mediate responses to pheromones while ciliated neurons mediate responses to amino acids. In support of this contention, recent *in situ* hybridization studies in goldfish suggest that odorant receptors are expressed in ciliated neurons and pheromone receptors in the microvillar (Cao et al., 1998). Further, odorant receptors preferentially tuned to recognize Arg with significant similarities with some vertebrate vomeronasal organ receptors are expressed in microvillar neurons (Speca et al., 1999). Not directly related to the ciliated-microvillar neurons issue, Kashiwayanagi et al. (1988) showed that in the carp *Cyprinus carpio*, olfactory responses to amino acids remained unchanged after total deciliation treatment of the olfactory epithelium. They suggested that the olfactory cilia are not necessary for receptor function in the carp. Alternatively however, amino acid responses may have been mediated by microvillar neurons, unless the 'calcium-ethanol shock' treatment *demicrovillated* at the same time. Microvillar neurons are capable of detecting amino acids in zebrafish, but the possibility that ciliated neurons may also detect amino acids cannot be ruled out (Lipschitz and Michel, 2002). In other non-salmonid species, sharks with only microvillar neurons and sea lampreys with only ciliated neurons both respond to amino acids (Zeiske et al., 1986; Li et al., 1995). In sea lampreys, the olfactory sensibility is limited to the basic amino acid Arg; most of the ciliated neurons seem to be steroidal detectors (bile acids and gonadal steroids; Li and Sorensen, 1992; Zielinski et al., 1996).

In the studies mentioned above, neither of the neuron types is exclusive; identification of neuron types is mostly indirect, based on neuron height and nuclei location, with the ciliated part situated nearer the apical surface (Morita and Finger, 1998). However, olfactory neurons are continuously renewed in adults, whether ciliated or microvillar, and show variable, intermediate sizes and nuclear locations in the olfactory epithelium (Evans et al., 1982). The issue of the functional differentiation of the ciliated-microvillar neuron is intriguing, considering the fact that this most ancient sensory system is involved in mediating the functions most basic to the survival of the individual and the species: feeding and reproduction. Only a direct method can determine unequivocally the functional differentiation, if any, between the two neuronal types.

4.4 Amino Acid Receptors in Other Fishes

Competitive cross-adaptation experiments have identified multiple amino acid olfactory receptor groups in three other fish species. In the common carp (*Cyprinus carpio*), a few chemically similar amino acids (Thr/Ser; Asp/Glu; Try/Phe) compete for shared receptors, but most other amino acids, including some chemically similar ones such as the basic amino acids Lys and Arg, and the neutral amino acids Val and Leu, interact with relatively distinct receptors (Ohno et al., 1984). Truth to tell, the amino acid receptors almost equate the number of amino acids identified in this species. In channel catfish, four relatively independent receptors are identified for: (1) the acidic (including Glu and Asp), (2) basic (including Lys and Arg), (3) short side-chain hydrophilic neutral amino acids (including Gly, Ala, Ser, Gln and possibly Cys), and (4) long side-chain hydrophobic amino acids (including Met, Val, Leu, His, Phe and possibly Cys) (Caprio and Byrd, 1984). This basically follows the patterns presented for the rainbow trout. Most surprising though is the occurrence of separate receptors for Pro and D-Ala. Both chemicals are totally inactive in rainbow trout at biologically significant concentrations. Strikingly similar amino acid receptors are found in the zebrafish (Michel and Derbidge, 1997). Cluster analyses of competition cross-adaptation data have identified three amino acid subgroups—neutral, basic and acidic—but for reasons discussed earlier, failed to address specific receptors interacting with the specific ligand amino acids.

5. RECEPTORS FOR OTHER ODORANT GROUPS

Bile acids are the only other chemical group established to be specific odorants/pheromones in rainbow trout. Extreme olfactory sensibilities to bile acids, coupled with their wide distribution and chemical variations, have been implicated for their role in fish behaviour (e.g. Zhang et al., 2001). Bile acids are hydrophilic derivatives of cholesterol which are synthesized in the liver. They are then exported to the small intestine via the gall bladder where they aid in lipid digestion. In fish, metabolites and residues are excreted into water with faeces and urine. A bile acid molecule is composed of a cyclopentenophenanthrene nucleus, an alkyl side-chain at C-17, two methyl groups at C-10 and -13, and other substituents. In teleosts, the principal biliary bile acids are 1) sulphated bile alcohol, mainly 5-cyprinol and 5-chimaerol and 2) C_{24} bile acids, mainly cholic acid, deoxycholic acid, and chenodeoxycholic acid (Zhang et al., 2001). The C_{24} bile acids are taurine or glycine amidated and/or sulphated. Of more than 20 bile acids tested in rainbow trout, the six most potent are cholic acid (CA), deoxycholic acid (DC), chenodeoxycholic acid (CD), taurocholic acid (TCA), taurochenodeoxycholic acid (TCD), and taurolithocholic acid 3-sulphate (TLS). Based on their characteristic C-R relationships, they are grouped into 1) free bile acids (CA, DC and CD), 2) taurine-conjugated amino acids (TCA and TCD) and 3) sulphated bile acid (TLS). Cross-adaptation experiments further showed that at least three independent classes of receptors exist for bile acids in rainbow trout (Giaquinto and Hara, unpubl. data).

EOGs to bile acids are unchanged while adapting to amino acids and vice versa, indicating that two groups of chemicals are mediated by a separate receptor population (Fig. 2.11). Information on these two distinct odorant groups is transmitted by separate neural pathways and processed independently in separate regions of the olfactory bulb. Therefore, when amino acid L-serine and bile acid taurocholic acid are simultaneously introduced into the nares of rainbow trout, the resultant EOG to the mixture is nearly the additive of those elicited individually, whereas EEGs to the same mixture, recorded at the amino acid and bile acid regions of the bulb respectively, are indistinguishable from those induced individually (Hara and Zhang, 1998).

A goldfish preovulatory steroid pheromone, 17α, 20β-dihydroxy-4-pregnen-3-one, was not olfactory stimulatory for rainbow trout and other salmonids tested (Hara and Zhang, 1998). Of the other gonadal steroids tested, only etiocholanolone glucuronide had a limited stimulatory effect in rainbow trout. The only other steroid known for its olfactory

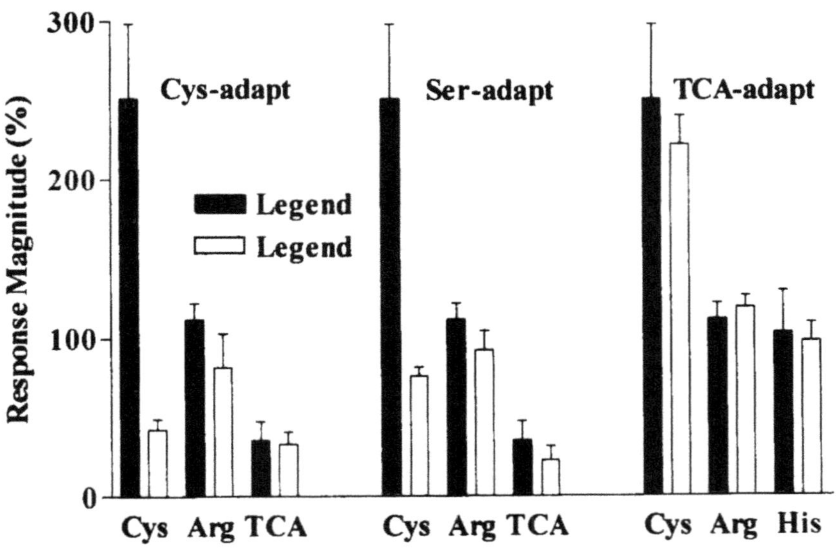

Fig. 2.11 Cross-adaptation data showing that independent amino acid and bile acid receptor groups exist in the rainbow trout olfactory system.

stimulation is testosterone; it stimulates the precocious male Atlantic salmon (*Salmo salar*) olfactory organ only a short period during spawning (Moore and Scott, 1992). Although not rigorously characterized in salmonids, their general EOG characteristics, combined with those obtained from goldfish, do suggest a multiple receptor population exists independent of those for amino acids and bile acids (Sorensen et al., 1987, 1988).

Most unexpected is that, unlike other salmonids rainbow trout lack the olfactory sensitivity to any of the goldfish postovulatory pheromones prostaglandins (Kitamura et al., 1994; Hara and Zhang, 1998; Laberge and Hara, 2003b). Whether this is a common feature throughout *Oncorhynchus* requires further investigation. Nevertheless, widespread olfactory sensitivity to prostaglandins among salmonids suggests their importance in behaviours. In fact, exposure of brown trout and lake whitefish (*Coregonus clupeaformis*) to prostaglandins increases locomotory activities and further, induces digging and nest building in female brown trout (Laberge and Hara, 2003a). No prostaglandin induces EEGs in salmonids (Hara and Zhang, 1998). Instead, in lake whitefish a neuron population responsive to prostaglandins is situated in the ventromedial brain tissue strip connecting the olfactory bulb to the telencephalon through which putative reproductive pheromone signals are integrated (Laberge and Hara, 2003a). Unlike the bulbar secondary neurons responsive to amino acids, this neuron population does not induce synchronized oscillatory waves upon stimulation with prostaglandins, despite massive discharges. In the vertebrate vomeronasal (VNO) system, the pattern of receptor neuron projection to glomeruli in the accessory olfactory bulb (AOB) is variable and less elaborate than that of the main olfactory bulb (Rodriguez et al., 1999; Brennan and Keverne, 2004). The pheromone responsive neuron population in lake whitefish could be a remnant of the mammalian VNO and the non-oscillatory discharges may be related to the less elaborated neural network in the pheromonal system.

Rainbow trout are unique among salmonids in that they totally lack olfactory sensitivity to prostaglandins. Although functional variability in gustatory responses to amino acids exists among various rainbow trout strains, lack of olfactory sensitivity to prostaglandins seems to be widespread (Kitamura et al., 1994; Hara and Zhang, 1998). These may be related to their life history factors which are extremely variable depending on location, type and habitat.

6. CONCLUDING REMARKS

Rainbow trout, like other fishes, show remarkable olfactory sensitivity to naturally occurring amino acids. Based on EOG studies, it is concluded that rainbow trout and possibly other fish species, detect amino acids through three main olfactory receptor types, *ArgR*, *CysR* and *GluR*. These receptors are specifically activated by their primary ligands, Arg, Cys and Glu. Each receptor recognizes the characteristic structural feature or epitope of an amino acid including 1) length and/or hydrocarbon chain, 2) difference in functional group and 3) position of the functional group within the molecule. In essence, this conforms to the concept of Stereochemical Theory of Odour originally proposed by Lucretius in 55 BC, and modernized by Moncrief (1967) and Amoore (1970). Because all amino acids have the primary, and in some cases secondary, features in common, they could all share the receptors depending on their affinities for them, i.e. as partial agonists they could activate these receptors at adequately high concentrations. Experimental evidence suggests that each receptor type is expressed in separate olfactory neurons. Shared patterns of partial adaptation often encountered may be due in part to experimental artifacts produced by the use of artificially high ligand concentrations, or to the existence of undifferentiated receptors at various developmental stages.

We humans detect Glu not by olfaction, but by gustation or taste. A flavour known as UMAMI, originally discovered by Kikunae Ikeda almost 100 years ago, is based on glutamate or the ionic form of Glu. Ikeda (1909) wrote '...UMAMI is attributed to the properties of the ionic glutamic acid....The UMAMI of the ionic glutamic acid is related first to the characteristics of an amino acid and second to the fact that its amino group occupies the alpha position for one carboxyl group and the gamma position for the other carboxyl group....' (translation by Ogiwara and Ninomiya, 2002). Recent sequencing and functional expression studies of rat and human taste receptors for Glu provide a molecular basis for Ikeda's pioneering work. Metabotropic G protein-coupled glutamate receptors T1R1 and T1R3 combine function as a broadly tuned L-amino acid receptor responding to most of the 20 amino acids in rat (Nelson et al., 2002). In contrast, human T1R1/T1R3 specifically responds to Glu (Li et al., 2002). The ligand specificities of the *Glu* receptors in the fish olfactory system seem to correspond to those of the mammalian gustatory system quite well. One major difference however, is their C-R relationship; that for fish in the order of µM and that for mammals in the order of mM. The presence of similar T1Rs receptor genes in pufferfish (*Fugu rubripes*) olfactory tissues (Naito et al., 1998) suggests that the fish olfactory *GluR* receptor and mammalian taste receptor may be evolved from a common ancestral gene. The goldfish odorant receptor in microvillar neurons preferentially tuned to recognize basic amino acids is also a member of a multigene family of G protein-coupled receptors, sharing significant sequence similarities with the calcium-sensing, metabotropic glutamate, and V2R class of mammalian VNO receptors (Speca et al., 1999). Although not all ligand specificities of these receptors have been demonstrated, the findings suggest that the family of olfactory receptors may in fact share sequence similarities and subunits.

Molecular biology has provided a pathway for identifying a novel multigene family that encodes proteins with seven transmembrane domains that bind odorant molecules and transduce chemical signals through interactions with G-proteins. To understand how olfactory receptors transduce the information represented by the molecular structure of odorant, it is essential to identify a specific odorant for a given olfactory receptor. As is often the case, it is not clear whether a cloned receptor recognizes the same set of odorants originally used to identify the olfactory neuron from which it was isolated. The reason for the failure in this pairing may be due in part to the difficulty in functionally expressing olfactory receptors (Katada et al., 2003), and in part to lack of information on natural odorants specific for animals. We should know that odorants such as isoamyl acetate, 2-heptanone, and eugenol are indeed detected via specific olfactory receptors through rigorous *in vivo* screening. The three conventional experimental approaches for determining receptor specificity described here are of potential value in assessing the specificity of a response to a particular receptor population.

Acknowledgements

The research discussed here was supported by the Natural Science and Engineering Research Council of Canada grant (OGP 0007576). I thank Drs. S. Brown, R. Evans, F. Laberge, T. Sveinsson and C. Zhang for their encouragement, ideas and assistance. I also thank Dr. B. Zielinski for critically reviewing the manuscript.

References

Amoore, J.E. 1970. *Molecular Basis of Odor.* Charles C. Thomas Publisher. Springfield, IL, USA, 200 pp.

Atema, J., K. Holland and W. Ikehara. 1980. Olfactory responses of yellowfin tuna (*Thonnus albacares*) to prey odors: chemical search image. *J. Chem. Ecol.* 6: 457–465.

Barth, A.L., N.J. Justice and J. Ngai. 1996. Asynchronous onset of odorant receptor expression in the developing zebrafish olfactory system. *Neuron* 16: 23–34.

Belanger, R.M., C.M. Smith, L.D. Corkum and B.S. Zielinski. 2003. Morphology and histochemistry of the peripheral olfactory organ in the round goby, *Neogobius melanostomus* (Teleostei: Gobiidae). *J. Morphol.* 257: 62–71.

Boeynaems, J.M. and J.E. Dumont. 1980. *Outline of Receptor Theory.* Elsevier/North-Holland, Amsterdam, 226 pp.

Brennan, P.A. and E.B. Keverne. 2004. Something in the air? New insights into mammalian pheromones. *Curr. Biol.* 14: R81–R89.

Brown, S.B. and T.J. Hara. 1981. Accumulation of chemostimulatory amino acids by a sedimentable fraction isolated from olfactory rosettes of rainbow trout (*Salmo gairdneri*). *Biochim. Biophys. Acta* 675: 149–162.

Brown, S.B. and T.J. Hara. 1982. Biochemical aspects of amino acid receptors in olfaction and taste. In: *Chemoreception in Fishes*, T.J. Hara (Ed.). Elsevier, Amsterdam, pp. 159–180.

Buck, L.B. 1996. Information coding in the vertebrate olfactory system. *Annu. Rev. Neurosci.* 19: 517–544.

Buck, L.B. and R. Axel. 1991. A novel multigene family may encode odorant receptors: a molecular basis for odor recognition. *Cell* 65: 175–187.

Cagan, R.H. and W.N. Zeiger. 1978. Biochemical studies of olfaction: binding specificity of radioactively labeled stimuli to an isolated olfactory preparation from rainbow trout (*Salmo gairdneri*). *Proc. Natl. Acad.* Sci. USA 75: 4679–4683.

Cao, Y., B.C. Oh and L. Stryer. 1998. Cloning and localization of two multigene receptor families in goldfish olfactory epithelium. *Proc. Natl. Acad. Sci.* USA 95: 11987–11992.

Caprio, J. and R.P. Byrd, Jr. 1984. Electrophysiological evidence for acidic, basic, and neutral amino acid olfactory receptor sites in the catfish. *J. Gen. Physiol.* 84: 403–422.

Caprio, J., J. Dudek and J.J. Robinson II. 1989. Electro-olfactogram and multiunit olfactory receptor responses to binary and trinary mixtures of amino acids in the channel catfish, *Ictalurus punctatus. J. Gen. Physiol.* 93: 245–262.

Evans, R.E. and T.J. Hara. 1985. The characteristics of the electro-olfactogram (EOG): its recovery following olfactory nerve section in rainbow trout (*Salmo gairdneri*). *Brain Res.* 330: 65–75.

Evans, R.E., B. Zielinski and T.J. Hara. 1982. Development and regeneration of the olfactory organ in rainbow trout. In: *Chemoreception in Fishes,* T. J. Hara (Ed.). Elsevier, Amsterdam, pp. 15–37.

Freitag, J., G. Ludwig, I. Andreini, P. Rössler and H. Greer. 1998. Olfactory receptors in aquatic and terrestrial vertebrates. *J. Comp. Physiol.* A 183: 635–650.

Fuss, S. H. and S. I. Korsching. 2001. Odorant feature detection: activity mapping of structure response relationships in the zebrafish olfactory bulb. J. Neurosci. 21: 8396–8407.

Gaillard, I., S. Rouquier, J.-P. Pin, P. Mollard, S. Richard, C. Barnabé, J. Demaille and D. Giorgi. 2002. A single olfactory receptor specifically binds a set of odorant molecules. *Eur. J. Neurosci.* 15: 409–418.

Gardner, W.S. and G.F. Lee. 1975. The role of amino acids in the nitrogen cycle of Lake Mendota. *Limnol. Oceanog.* 20: 379–388.

Getchell, T.V. 1974. Electrogenic sources of slow voltage transients recorded from frog olfactory epithelium. *J. Neurophysiol.* 37: 1115–1130.

Hansen, A., K.T. Anderson and T.E. Finger. 2001. Immunohistochemical and ultrastructural identification of G-proteins in the olfactory epithelium. *Chem. Senses* 26: 1128.

Hanson, L.R., P.W. Sorensen and Y. Cohen. 1998. Sex pheromones and amino acids evoke distinctly different spatial patterns of electrical activity in the goldfish olfactory bulb. *Ann. N. Y. Acad. Sci.* 85: 521–524.

Hara, T.J. 1971. Chemoreception. In: *Fish Physiology*, vol. 5. Sensory Systems and Electric Organs. W.S. Hoar and D.J. Randall (Eds). Academic Press, New York, NY, pp. 79–120.

Hara, T.J. 1973. Olfactory responses to amino acids in rainbow trout, *Salmo gairdneri*. *Comp. Biochem. Physiol.* 44A: 407–416.

Hara, T.J. 1976. Structure-activity relationships of amino acids in fish olfaction. *Comp. Biochem. Physiol.* 54A: 31–36.

Hara, T.J. 1977. Further studies on the structure-activity relationships of amino acids in fish olfaction. *Comp. Biochem. Physiol.* 56A: 559–565.

Hara, T.J. 1982. Structure-activity relationships of amino acids as olfactory stimuli. In: *Chemoreception in Fishes*, T.J. Hara (Ed.). Elsevier, Amsterdam, pp. 135–157.

Hara, T.J. 1992. Mechanisms of olfaction. In: *Fish Chemoreception*, T.J. Hara (Ed.). Chapman & Hall, London, UK, pp. 150–170.

Hara, T.J. 1993. Role of olfaction in fish behaviour. In: *Behaviour of Teleost Fishes*, T.J. Pitcher (Ed.). Chapman & Hall, London, UK, pp. 171–199.

Hara, T.J. 1994a. Olfaction and gustation in fish: an overview. *Acta Physiol. Scand.* 152: 207–217.

Hara, T.J. 1994b. The diversity of chemical stimulation in fish olfaction and gustation. *Rev. Fish Biol. Fish.* 4: 1–35.

Hara, T.J. and C. Zhang. 1996. Spatial projections to the olfactory bulb of functionally distinct and randomly distributed primary neurons in salmonid fishes. *Neurosci. Res.* 26: 65–74.

Hara, T.J. and C. Zhang. 1998. Topographic bulbar projections and dual neural pathways of the primary olfactory neurons in salmonid fishes. *Neuroscience* 82: 301–313.

Hara, T. J., M. Freese and K. R. Scott. 1973. Spectral analysis of olfactory bulbar responses in rainbow trout. *Jpn. J. Physiol.* 23: 325–333.

Hubbard, P. C., E. N. Barata and A. V. M. Canário. 2003. Olfactory sensitivity to catecholamines and their metabolites in the goldfish. *Chem. Senses* 28: 207–218.

Hyman, A. M. and M. E. Frank. 1980. Effects of binary taste stimuli on the neural activity of the hamster chorda tympani. *J. Gen. Physiol.* 76: 125–142.

Ikeda, K. 2002. New seasonings. *Chem. Senses* 27: 847–849.

Kaji, S., M. Satou, Y. Kudo, K. Ueda and A. Gorbman. 1975. Spectral analysis of olfactory responses of adult spawning chum salmon (*Oncorhynchus keta*) to stream waters. *Comp. Biochem. Physiol.* 51A: 711–716.

Kajiya, K., K. Inaki, M. Tanaka, T. Haga, H. Kataoka and K. Touhara. 2001. Molecular bases of odor discrimination: reconstitution of olfactory receptors that recognize overlapping sets of odorants. *J. Neurosci.* 21: 6018–6025.

Kang, J. and J. Caprio. 1995. *In vivo* responses of single olfactory receptor neurons in the channel catfish, *Ictalurus punctatus*. *J. Neurophysiol.* 73: 172–177.

Kashiwayanagi, M., T. Shoji and K. Kurihara. 1988. Large olfactory responses of the carp after complete removal of olfactory cilia. *Biochem. Biophys. Res. Commun.* 154: 437–442.

Katada, S., T. Nakagawa, H. Kataoka and K. Touhara. 2003. Odorant response assays for a heterologously expressed olfactory receptor. *Biochem. Biophys. Res. Commun.* 305: 964–969.

Kitamura, S., H. Ogata and F. Takashima. 1994. Activities of F-type prostaglandins as releaser sex pheromones in cobitid loach, *Misgurnus anguillicaudatus* *Comp. Biochem. Physiol.* 107A: 161–169.

Kudo, Y., M. Satou, S. Kaji and K. Ueda. 1972. Frequency analysis of olfactory response in fish by band-pass filters. *J. Fac. Sci., Univ. Tokyo*, Ser. IV 12: 3365–383.

Laberge, F. and T.J. Hara. 2003a. Behavioural and electrophysiological responses to F-prostaglandins, putative spawning pheromones, in three salmonid fishes. *J. Fish Biol.* 62: 206–221.

Laberge, F. and T.J. Hara. 2003b. Non-oscillatory discharges of an F-prostaglandin responsive neuron population in the olfactory bulb-telencephalon transition area in lake whitefish. *Neuroscience* 116: 1089–1095.

Laberge, F. and T. J. Hara. 2004. Electrophysiological demonstration of independent olfactory receptor types and associated neuronal responses in the trout olfactory bulb. *Comp. Biochem. Physiol.* 137A: 397–408.

Li, W. and P. W. Sorensen. 1992. The olfactory sensitivity of sea lamprey to amino acids is specifically restricted to arginine. *Chem. Senses* 17: 658.

Li, W., P. W. Sorensen and D. D. Gallaher. 1995. The olfactory system of migratory adult sea lamprey (*Petromyzon marinus*) is specifically and acutely sensitive to unique bile acids released by conspecific larvae. *J. Gen. Physiol.* 105: 569–587.

Li, X., L. Staszewski, H. Xu, K. Durick, M. Zoller and E. Adler. 2002. Human receptors for sweet and umami taste. *Proc. Natl. Acad. Sci. USA* 99: 4692–4696.

Limbird, L. E. 1986. *Cell Surface Receptors: A Short Course on Theory and Methods*. Martinus Nijhoff Publishing, Boston, MA, 196. pp.

Lipschitz, D.L. and W.C. Michel. 2002. Amino acid odorants stimulate microvillar sensory neurons. *Chem. Senses* 27: 277–286.

Madrid, R., M. Sanhueza, O. Alvarez and J. Bacigalupo. 2003. Tonic and phasic neurons in the vertebrate olfactory epithelium. *Biophys. J.* 84: 4167–4181.

Matsuoka, M., J. Yoshida-Matsuoka, N. Iwasaki, M. Norita, R. M. Costanzo and M. Ichikawa. 2001. Immunocytochemical study of $G_i2\alpha$ and $G_o\alpha$ on the epithelium surface of the rat vomeronasal organ. *Chem. Senses* 26: 161–166.

Michel, W.C. and D.S. Derbidge. 1997. Evidence of distinct amino acid and bile salt receptors in the olfactory system of the zebrafish, *Danio rerio. Brain Res.* 764: 179–187.

Michel, W.C., P. Steullet, H.S. Cate, C.J. Burns, A.B. Zhainazarov and C.D. Derby. 1999. High-resolution functional labeling of vertebrate and invetebrate olfactory receptor neurons using agmatine, a channel permeant cation. *J. Neurosci. Methods* 90: 143–156.

Michel, W.C., M.J. Sanderson, J.K. Olson and D.L. Lipschitz. 2003. Evidence of a novel transduction pathway mediating detection of polyamines by the zebrafish olfactory system. *J. Exp. Biol.* 206: 1697–1706.

Moncrief, R.W. 1967. *The Chemical Senses.* Leonard Hill Books: London, UK, 760 pp.

Moore, A. and A.P. Scott. 1992. 17α,20β-dihydroxy-4-pregnen-3-one-20-sulphate is a potent odorant in precocious male Atlantic salmon (*Salmo salar* L.) parr which have been pre-exposed to the urine of ovulated females. *Proc. R. Soc. Lond.* B 249: 205–209.

Morita, Y. and T.E. Finger. 1998. Differential projections of ciliated and microvillous olfactory receptor cells in the catfish, *Ictalurus punctatus. J. Comp. Neurol.* 398: 539–550.

Naito, T., Y. Saito, J. Yamamoto, Y. Nozaki, K. Tomura, M. Hazama, S. Nakanishi and S. Brenner. 1998. Putative pheromone receptors related to the Ca^{2+}-sensing receptor in *Fugu. Proc. Natl. Acad. Sci. USA* 95: 5178–5181.

Nelson, G., J. Chandrashekar, M.A. Hoon, L. Feng, G. Zhao, N.J.P. Ryba and C.S. Zuker. 2002. An amino-acid taste receptor. *Nature* (Lond.) 416: 199–202.

Nevitt, G. and A. Dittman. 1998. A new model for olfactory imprinting in salmon. *Integr. Biol.* 1: 215–223.

Ngai, J., A. Chess, M.M. Dowling, N. Necles, E. R. Macagno and R. Axel. 1993a. Coding of olfactory information: topography of odorant receptor expression in the catfish olfactory epithelium. *Cell* 72: 667–680.

Ngai, J., M.M. Dowling, L. Buck, R. Axel and A. Chess. 1993b. The family of genes encoding odorant receptors in the channel catfish. *Cell* 72: 657–666.

Nikonov, A.A. and J. Caprio. 2001. Electrophysiological evidence for a chemotopy of biologically relevant odors in the olfactory bulb of the channel catfish. *J. Neurophysiol.* 86: 1869–1876.

Ohno, T., K. Yoshii and K. Kurihara. 1984. Multiple receptor types for amino acids in the carp olfactory cells revealed by quantitative cross-adaptation method. *Brain Res.* 310: 13–21.

Ottoson, D. 1956. Analysis of the electrical activity of the olfactory epithelium. *Acta Physiol. Scand.* 35 (Suppl. 122): 1–83.

Parmentier, M., F. Libert, S. Schurmans, S. Schiffmann, A. Lefort, D. Eggerick, C. Ledent, C. Gerhrd, J. Perret, A. Grootegoed and G. Vassar. 1992. Expression of members of the putative olfactory receptor gene family in mammalian germ cells. *Nature* (Lond.) 355: 453–455.

Raming, K., J. Krieger, J. Strotmann, I. Boekhoff, S. Kubick, C. Baumstark and H. Breer. 1993. Cloning and expression of odorant receptors. *Nature* (Lond.) 361: 353–356.

Rodriguez, I., P. Feinstein and P. Mombaerts. 1999. Variable patterns of axonal projections of sensory neurons in the mouse vomeronasal system. *Cell* 97: 199–208.

Rolen, S. H., P. W. Sorensen, D. Mattson and J. Caprio. 2003. Polyamines as olfactory stimuli in the goldfish *Carassius auratus*. *J. Exper. Biol.* 206: 1683–1696.

Sato, K. and N. Suzuki. 2001. Whole-cell response characteristics of ciliated and microvillous olfactory receptor neurons to amino acids, pheromone candidates and urine in rainbow trout. *Chem. Senses* 26: 1145–1156.

Satou, M. 1990. Synaptic organization, local neuronal circuitry, and functional segregation of the teleost olfactory bulb. *Prog. Neurobiol.* 34: 115–142.

Satou, M. and K. Ueda. 1975. Spectral analysis of olfactory responses to amino acids in rainbow trout, *Salmo gairdneri*. *Comp. Biochem. Physiol.* 52A: 359–365.

Serizawa, S., T. Ishii, H. Nakatani, A. Tsuboi, F. Nagawa, M. Asano, K. Sudo, J. Sakagami, H. Sakano, T. Ijiri, Y. Matsuda, M. Suzuki, T. Yamamori, Y. Iwakura and H. Sakano. 2000. Mutually exclusive expression of odorant receptor transgenes. *Nat. Neurosci.* 3: 687–693.

Serizawa, S., K. Miyamichi, H. Nakatani, M. Suzuki, M. Saito, Y. Yoshihara and H. Sakano. 2003. Negative feedback regulation ensures the one receptor-one olfactory neuron rule in mouse. *Science* 302: 2088–2094.

Shephard, G.M. 1994. Discrimination of molecular signals by the olfactory receptor neuron. *Neuron* 13: 771–790.

Shoji, T., H. Ueda, T. Ohgami, T. Sakamoto, Y. Katsuragi, K. Yamauchi and K. Kurihara. 2000. Amino acids dissolved in stream water as possible home stream odorants for masu salmon. *Chem. Senses* 25: 533–540.

Sorensen, P.W., T.J. Hara and N.E. Stacey. 1987. Extreme olfactory sensitivity of mature and gonadally-regressed goldfish to a potent steroidal pheromone, 17α,20β-dihydroxy-4-pregnen-3-one. *J. Comp. Physiol.* A 160: 305–313.

Sorensen, P.W., T.J. Hara, N.E. Stacey and F.W. Goetz. 1988. F prostaglandins function as potent olfactory stimulants that comprise the postovulatory female sex pheromone in goldfish. *Biol. Reprod.* 39: 1039–1050.

Speca, D.J., D.M. Lin, P.W. Sorensen, E.Y. Isacoff, J. Ngai and A.H. Dittman. 1999. Functional identification of a goldfish odorant receptor. *Neuron* 23: 487–498.

Sutterlin, A. M. and N. Sutterlin. 1971. Electrical responses of the olfactory epithelium of Atlantic salmon (*Salmo salar*). *J. Fish. Res. Bd. Canada* 28: 565–572.

Suzuki, N. and D. Tucker. 1971. Amino acids as olfactory stimuli in freshwater catfish, *Ictalurus catus* (Linn.). *Comp. Biochem. Physiol.* 40A: 399–404.

Sveinsson. T. and T.J. Hara. 1990a. Analysis of olfactory responses to amino acids in Arctic char (*Salvelinus alpinus*) using a linear multiple-receptor model. *Comp. Biochem. Physiol.* 97A: 279–287.

Sveinsson, T. and T.J. Hara. 1990b. Multiple olfactory receptors for amino acids in Arctic char (*Salvelinus alpinus*) evidenced by cross-adaptation experiments. *Comp. Biochem. Physiol.* 97A: 289–293.

Thommesen, G. 1983. Morphology, distribution, and specificity of olfactory receptor cells in salmonid fishes. *Acta Physiol. Scand.* 117: 241–249.

Touhara, K., S. Sengoku, K. Inaki, A. Tsuboi, J. Hirono, T. Sato, H. Sakano and T. Haga. 1999. Functional identification and reconstitution of an odorant receptor in single olfactory neurons. *Proc. Natl. Acad. Sci. USA* 96: 4040–4045.

Valentinčič, T., S. Wegert and J. Caprio. 1994. Learned olfactory discrimination versus innate taste responses to amino acids in channel catfish (*Ictalurus punctatus*). *Physiol. Behav.* 55: 865–873.

Weth, F., W. Nadler and S. Korsching. 1996. Nested expression domains for odorant receptors in zebrafish olfactory epithelium. *Proc. Natl. Acad. Sci. USA* 93: 13321–13326.

Zeiske, E., J. Caprio and S.H. Gruber. 1986. Morphological and electrophysiological studies on the olfactory organ of the lemon shark, *Negaprion brevirostris* (Poey). In: *Indo-Pacific Fish Biology: Proc. 2nd Intl. Conf. Indo-Pacific Fishes, T. Uyeno, R. Arai, T. Taniuchi and K. Matsuura* (Eds). Ichthyological Society of Japan, Tokyo, pp. 381–391.

Zhang, C., S.B. Brown and T.J. Hara. 2001. Biochemical and physiological evidence that bile acids produced and released by lake char (*Salvelinus namaycush*) function as chemical signals. *J. Comp. Physiol.* B 171: 161–171.

Zhao, H., L. Ivic, J.M Otaki, M. Hashimoto, K. Mikoshiba and S. Firestein. 1998. Functional expression of a mammalian odorant receptor. *Science* 279: 237–242.

Zielinski, B. and T.J. Hara. 1988. Morphological and physiological development of olfactory receptor cells in rainbow trout (*Salmo gairdneri*) embryos. *J. Comp. Neurol.* 271: 300–311.

Zielinski, B.S. and T.J. Hara. 2001. The neurobiology of fish olfaction. In: *Sensory Biology of Jawed Fishes*, B.G. Kapoor and T.J. Hara (Eds). Science Publishers Inc., Enfield, (NH), USA, pp. 347–366.

Zielinski, B.S., J.K. Osahan, T.J. Hara, M. Hosseini and E. Wong. 1996. Nitric oxide synthase in the olfactory mucosa of the larval sea lamprey (*Petromyzon marinus*). J. Comp. Neurol. 365: 18–25.

Zippel, H.P., P.W. Sorensen and A. Hansen. 1997. High correlation between microvillous olfactory receptor cell abundance and sensitivity to pheromones in olfactory nerve-sectioned goldfish. *J. Comp. Physiol.* A 180: 39–52.

CHAPTER

3

Olfactory Discrimination in Fishes

Tine Valentinčič

ABSTRACT

The taste system of fishes detects a relatively small number of chemical stimuli at high sensitivity and enables reflex responses, such as turning and biting/snapping behaviors. The olfactory system, in contrast, detects a larger number of chemicals at relatively high sensitivity and enables learning and discrimination of the chemical stimuli. Conditioned olfactory stimuli trigger a feeding excitatory state which increases the duration and intensity of food searching behavior. Amino acids are a major class of chemical stimuli especially relevant to the stimulation of feeding behavior in teleost fishes. Fishes can discriminate any conditioned amino acid from nearly any other nonconditioned amino acid via olfaction. There are a few amino acids, such as L-isoleucine and L-valine, chemicals of similar structure, which catfish and zebrafish cannot discriminate. Catfish can also be conditioned to mixtures of amino acids. Binary and ternary mixtures of amino acids were detected initially as their more stimulatory components; additional discrimination training, consisting of 5–10 successive comparisons of conditioned with

Address for Correspondence: Tine Valentinčič Department of Biology, University of Ljubljana Večna pot 111 1000 Ljubljana, E-mail: tine.valentincic@bf.uni-lj.si nonconditioned stimulus, enabled discrimination of the conditioned mixture from its more stimulatory component alone. Conditioning and discrimination training enabled black bullhead catfish to discriminate the conditioned multimixture of 7 amino acids from its 6-component counterparts; however, catfish could not discriminate a conditioned 12-component multimixture from its 11-component counterparts. Initially, fishes perceived the main feature of an olfactory stimulus. Additional discrimination training enabled the fish to detect small differences between the main odor and its modifications caused either by minor components in the mixture or small differences between multicomponent mixtures differing by one component.

Key Words: Olfaction; Taste; Conditioning; Olfactory discrimination; Feeding behavior.

1. INTRODUCTION: SENSES AND FOOD FINDING

Most fish species depend either on vision, olfaction, taste or tactile senses to find and ingest food. Initially, the sensory input that provides the most direct information to the central nervous system (CNS) leading to the potential target(s) has priority. Different species of fishes are specialized for different ecological niches. Specialization to a specific niche determines which sense organ is best for the detection, search, and location of the food (Valentinčič, 2004). In clear water, vision provides predatory fishes with the most precise target location. Omnivorous fishes, in addition to vision, use olfaction and taste to detect and localize food. Vision enables a direct approach to the food, whereas olfaction merely signals its presence. Except at a very close range outside water currents, chemical senses do not provide directional bearings for the food site. In omnivorous fishes, chemical stimuli release food searching behavior resembling klinokinesis (Fraenkel and Gunn, 1940). Rapid swimming and turning behavior of the fish during feeding excitation greatly increases the chances of a direct encounter with the food. In fishes with barbels, such as catfish (Fig. 3.1), taste enables tropotactic location of food, which occurs generally over millimeter distances upstream and to the sides of the food item and at somewhat larger distances downstream.

Olfactory and taste stimuli are detected after chemical sensory receptors are directly exposed to water containing the stimulating substances. Unlike with vision, either the fish must swim to the stimuli or the stimuli must be transported to the fish. Substances eluted from food rapidly diffuse over millimeter distances (Jacobs, 1967) into the water

Fig. 3.1 Black bullhead catfish (*Ameirus melas*) is an omnivorous fish using taste buds on barbels and around the mouth for turning and biting/snapping reflexes. The reflex responses are controlled entirely by the taste system.

filling a small portion with relatively high concentrations. At centimeter distances, diffusion is too slow to provide for real time stimulus distribution. At the macro scale, water containing the substances spreads rapidly with turbulent currents (i.e., eddies) (Valentinčič, 2004), which are analogous to odor plumes in the air. Eddies containing chemical stimuli are usually several centimeters in diameter (Valentinčič and Caprio, 1994a). In stagnant waters, turbulent water currents drive stimulus eddies around the water with slow (e.g. <millimeter/second) velocities. In an aerated aquarium, a single eddy containing chemical stimuli passed a chemical detector within seconds (Moore et al., 1989; Valentinčič and Caprio, 1994a). From the direction of the nearest stimulus eddy, catfish could not predict the direction of the stimulus source; however, a practical solution to the chemosensory food-finding

puzzle is klinokinesis (Fraenkel and Gunn, 1940; Valentinčič, 2004). Frequently, a tropotactic turn into a nearby stimulus eddy takes a fish away from the direction of the stimulus source. A relatively long duration of the central feeding excitatory state (~2 minutes) maintains food-searching activity even after traces of stimuli eddies are temporarily lost. The fish continues swimming until it encounters the next stimulating eddy. Thus, increasing duration and intensity of food-searching activity increases the chances for fish to locate the food source.

Catfish and cyprinids have dual chemical controls, taste and olfaction, of feeding excitation. Large facial and vagal lobes, the primary gustatory nuclei of the medulla oblongata in these fishes, have abundant connections to the secondary gustatory nucleus and diencephalon (Striedter, 1990; Kanwal and Finger, 1992). Thus, taste stimuli can trigger a central feeding excitatory state and guide feeding activity even in anosmic catfish (Valentinčič and Caprio, 1994b; Valentinčič et al., 2000b). Facial and vagal lobes of sunfish (*Lepomis sp.*, Centrarchidae) are much smaller than those of catfish (Kanwal and Finger, 1992); hence most sunfish (*Lepomis sp.*, Centrarchidae) (Valentinčič, unpubl.) and rainbow trout (*Oncorhynchus mykiss*, Salmonidae) (Valentinčič et al., 1999) do not respond to taste stimuli with feeding behavior. The rainbow trout taste system is narrowly tuned as it detects a small number of amino acid stimuli (Marui et al., 1983a, b; Marui and Caprio, 1992) and alone does not release feeding excitation since either a visual or an olfactory input is required (Valentinčič and Caprio, 1997; Valentinčič et al., 1999).

A functional olfactory system exists in all fish species; however, it is ordinarily used for food finding in omnivorous and herbivorous species only. In our experiments, predatory fishes, such as walleye (*Stizostedion vitreum*, Centrarchidae) and European huchen (*Salmo hucho*, Salmonidae), which were not trained to eat nonliving foods early in life, did not use olfaction in food finding. Fry and fingerlings of predatory fishes fed exclusively on living fish and crustaceans never fed on carrion and other nonliving foods (Valentinčič, 2004). The portion of the brain that uses olfactory information was either not expressed or the olfactory information needed for feeding was ignored. As in the development of language in humans (Curtiss et al., 1974), where deprivation of early experience with speech permanently handicaps the expression of speech areas in the left hemisphere, olfactory functions in the fish brain, most probably the telencephalon, are not expressed without early experiences with nonliving food. In farms, fry of predatory fishes, such as sea bass (*Dicentrarchus labrax*) and European huchen (*S. hucho*), are transferred from feeding exclusively on living food to ingesting nonliving organic material, such as minced liver, starter feeds, and food pellets. In our laboratory, walleye (*Stizostedion vitreum*) transferred from living to nonliving food during the fingerling stage of life started to use olfaction for food finding and were later capable of discriminating amino acids (Valentinčič, 2004).

Abundant functional connections between taste receptor cells, the taste centers of the medulla and effectors enable direct turning and biting/snapping responses in catfish and cyprinids. The frequency of these reflex responses correlates directly with the number of encounters with aqueous chemical eddies containing suprathreshold amino acid stimuli (Valentinčič and Caprio, 1994a; Valentinčič, 2004). Catfish and cyprinids use visual stimuli to locate visible food and olfactory and taste stimuli to find the source of an attracting chemical trail. In response to unilateral barbel stimulation by feeding stimuli, catfish turn to and bite/snap whenever they encounter eddies comprising high concentrations of specific amino acids [e.g. L-alanine (L-Ala), L-proline (L-Pro), L-arginine (L-Arg) and L-cysteine (L-Cys)]. (Valentinčič and Caprio, 1994 a, b; Valentinčič et al., 2000b). Carp (*Cyprinus carpio*) and goldfish (*Carassius auratus*) to bite L-Ala and L-Pro reflexively (Valentinčič 2004). In addition, fishes use oral taste evaluation to either accept or reject food (Kasumyan and Nikolaeva, 1997; Pavlov and Kasumyan, 1998; Kasumyan and Prokopova, 2003). Carp also use the oropharyngeal taste system to sort foods from sediment particles (Sibbing et al., 1986; Finger, 1988; Lamb and Finger, 1995; Hansen and Reutter, 2004).

Vision enables a predator to locate mobile prey, to position at a striking distance, and to strike at the prey (Messenger, 1973). Fast swimming prey often does not detect the predator before it launches the final strike; predators do not bite unnecessarily near the prey. In adult rainbow trout, the taste controlled turning and biting/snapping movements are inhibited during prey approach as turning and biting/snapping behaviors occur only under visual and/or olfactory controls as part of the complex feeding behavior (Valentinčič and Caprio, 1997). Alevins of rainbow trout, young fish that utilize nutrients from their yolk sacs, have fully developed taste controlled biting/snapping reflexes. In spite of the fact that this behavior is not needed, the alevins bite/snap to L-Pro and L-Ala stimulation. At the onset of external feeding, on transition from alevin to the fry stage,

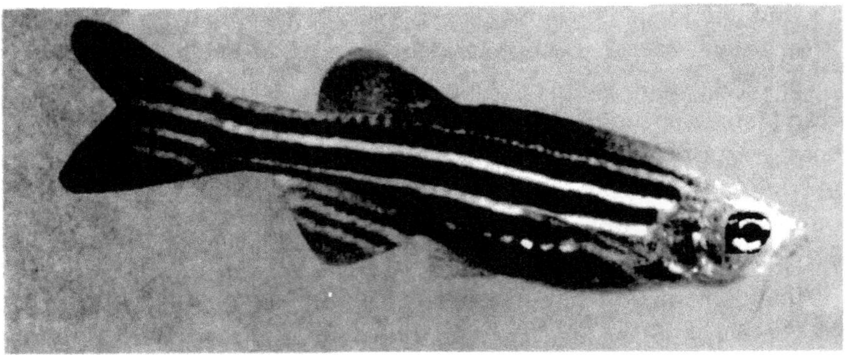

Fig. 3.2 Zebrafish (*Danio rerio*), a model fish species for molecular biological studies, is an omnivorous fish whose olfactory discrimination abilities are approximately the same as those of catfish.

rainbow trout starts to feed exogenously. Concomitantly, the taste controlled biting/snapping reflexes become inhibited for the rest of the fish's life (Valentinčič et al., 1999).

Fear inhibits feeding behavior. Duration of the state of fear differs in fish species. Channel catfish (*Ictalurus punctatus*) (Valentinčič and Caprio, 1994a) and black bullhead (*Ameiurus melas*) (Valentinčič et al., 2000b) catfishes are extremely fearful species in the laboratory, requiring more than a month of acclimation to the aquarium environment prior to resumption of regular feeding activity. At the other extreme are the imperturbable species such as carp, goldfish, and rainbow trout, which start responding to feeding stimuli within 1–3 days after transfer from a farm lake to an aquarium. Fear inhibits most behaviors including the complex feeding behavior and biting/snapping reflexes (Valentinčič and Caprio, 1994 a, b; Valentinčič et al., 1994). In contrast, hungry catfish and cyprinids, which are not frightened, regularly respond with biting/snapping reflexes and feeding behavior to appropriate taste, olfactory and visual stimuli.

Olfactory systems are broadly tuned. The large number of different molecular olfactory receptors (Ngai et al., 1993; Dugas and Ngai, 2001) that are expressed in the fish olfactory epithelium enables detection of a large number of amino acids, bile acids, and hormones (Hara, 1992). Through olfaction, fish can be conditioned to specific chemical stimuli that emanate from food (Valentinčič et al., 1994; Nikonov and Caprio, 2004). After detecting the conditioned stimulus through olfaction, the search for food becomes significantly longer and more vigorous. From studying responses to conditioned and nonconditioned olfactory stimuli, we were able to show that catfishes are capable of discriminating nearly every nonconditioned olfactory stimulus from a conditioned olfactory stimulus (Valentinčič et al., 2000b). After detection of a conditioned olfactory stimulus, the length of food search was two to five times greater than that following detection of unconditioned stimuli.

2. OLFACTION AND FEEDING

Anatomical (Scott and Leonard, 1971; Murakami et al., 1983) and electrophysiological (Campbell et al., 1969) evidence of olfactory inputs into the hypothalamus provided structural and functional elements of the olfactory control of feeding excitation. The telencephalic olfactory inputs into the hypothalamus and the direct olfactory inputs from the olfactory bulb (Laberge and Hara, 2001) into the hypothalamus possibly influence the central feeding excitatory state of fish. As in humans, odors have arousing effects (Motokizawa and Furuya, 1973). Olfactory conditioned stimuli associated with food reward release feeding behavior in fishes. In mammals, the central feeding excitatory state originates from ventromedial and lateral areas of the hypothalamus (Herrington and Ranson, 1942); however, the specific location of the feeding center within the hypothalamus of fishes is currently not known.

In a classical conditioning paradigm, amino acid solutions were delivered into aquaria ~90 seconds before the food reward. Single catfish started an increased search for food after 5–30 conditioning trials. A nonconditioned stimulus aroused the animal but the resultant food search was brief. Immediately after detecting the conditioned olfactory stimulus, catfish assumed an excited posture with all the fins erect and started swimming rapidly and excitedly. At the group level, ~20–50 trials were necessary for all the catfish (also observed for goldfish, carp, and zebrafish) to regularly respond to the conditioning stimulus with an increased food search activity (Fig. 3.3). Klinotactic swimming eventually brought the fish to the food. Olfaction, in combination with detection of a water current, enables conditioning to the upstream location of the odor source. In rivers, omnivorous fishes are potentially conditioned to find food upstream. Also, fishes such as salmon respond to specific waterborne odors which have been imprinted by instinctively swimming upstream (Hasler and Scholz, 1983).

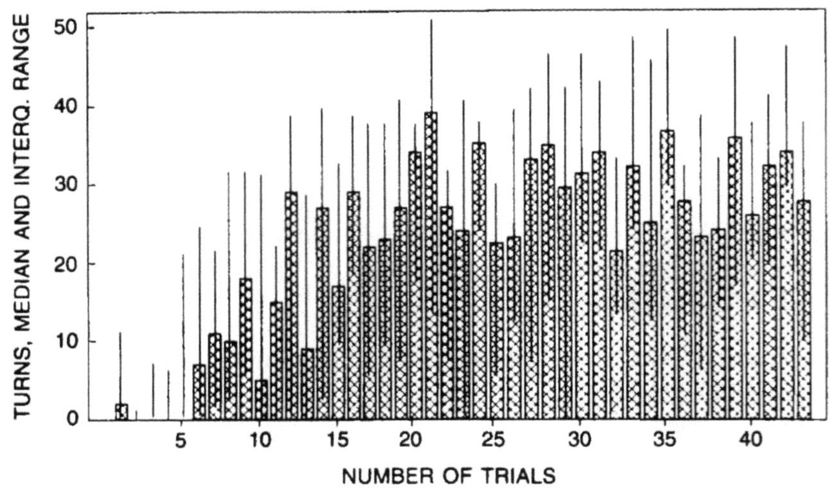

Fig. 3.3 Naive black bullhead catfish do not swim in response to single amino acid stimuli. Conditioning to either single amino acid, binary mixture with one component more stimulatory than the other, and to multimixture (this Figure, 13 amino acids) increases the food searching (swimming) activity of fishes in response to the conditioned stimulus.

3. OLFACTORY DISCRIMINATION OF AMINO ACIDS BY ICTALURID CATFISHES

We studied olfactory discrimination in two species of ictalurid catfishes, channel (*Ictalurus punctatus*) and black bullhead (*Ameiurus melas*). Electrophysiological cross-adaptation studies of amino acids in channel catfish indicated that separate receptor sites likely exist for neutral amino acids with short [glycine (Gly), L-Ala, L-serine (L-Ser)] and long [(L-leucine (L-Leu), L-isoleucine (L-Ile), L-valine (L-Val), L-norvaline (L-nVal), L-norleucine (L-nLeu), L-methionine (L-Met)] side chains, respectively, and for basic [L-Arg, L-lysine (L-Lys)] and acidic [L-aspartic acid (L-Asp) and L-glutamic acid (L-Glu)] amino acids (Caprio and Byrd, 1984). Odorant specificities of single olfactory bulb neurons for short chain neutral, long chain neutral, basic and acidic amino acids provided physiological basis for amino acid discrimination in channel catfish (Nikonov and Caprio, 2004). After conditioning to one of the amino acids, L-Ala, L-Arg, L-Lys, L-Pro, L-histidine (L-His), L-Ile, L-nLeu, or L-Cys respectively, both channel (Valentinčič and Caprio, 1994a; Valentinčič et al., 1994) and bullhead catfishes (Valentinčič et al., 2000), discriminated nearly every amino acid from the conditioned amino acid.

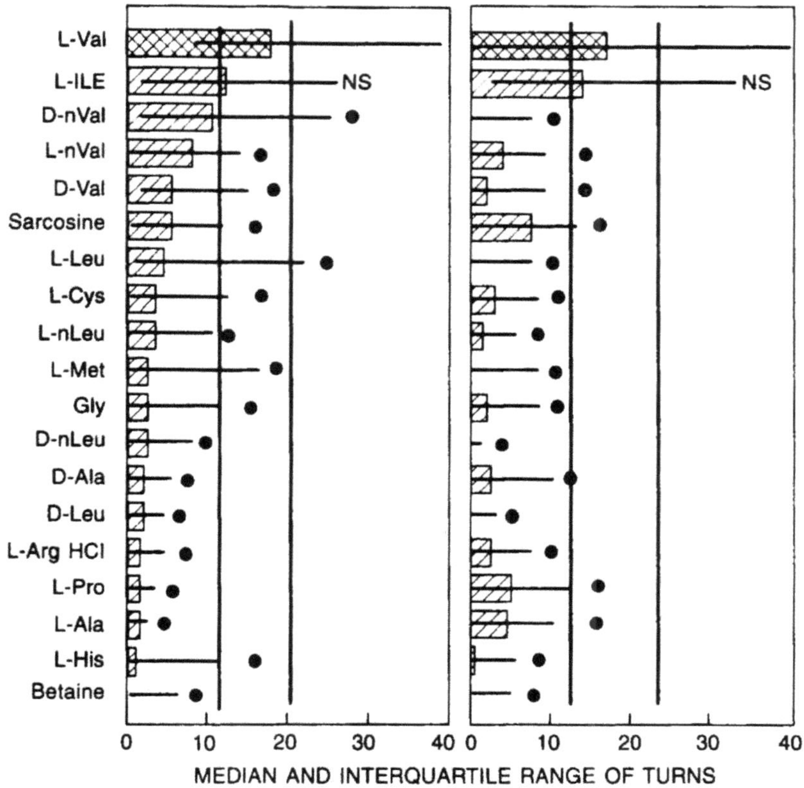

Fig. 3.4 Olfactory discrimination of amino acids from the conditioned stimulus L-Val. Bullhead catfish discriminated all amino acids except L-Ile from L-Val. Two repeated tests series are shown. Crosshatched, conditioned responses; hatched, nonconditioned responses; the two vertical lines indicate range of medians for the 18 conditioned responses; dots, significant difference ($P < 0.01$) *NS*, not significant.

Among amino acid analogues tested, channel catfish were unable to discriminate L-Pro from L-pipecolic acid. In some cases, the L-Ala conditioned channel catfish were unable to discriminate Gly from L-Ala. Bullhead catfish did not discriminate L-Ile from L-Val nor L-Val from L-Ile (Fig. 3.4), in some cases they also did not discriminate Gly and L-Ser from L-Ala.

Similar discrimination abilities were found for zebrafish (*Danio rerio*) (see Fig. 3.2) (Valentinčič et al., 2005). As expected from the nearly identical calcium imaging maps for L-Ile and L-Val of activated terminals of olfactory receptor neurons within the olfactory bulb of the zebrafish (Friedrich and Korsching, 1997), these fish were unable to behaviorally discriminate these compounds in out tests (unpubl. observations). We found small differences in olfactory discriminative abilities in black bullhead catfish from various localities. Catfishes from one locality were unable to discriminate short-chain neutral amino acids L-Ala and L-Ser, whereas catfishes from a different locality were able to do so. Even the same specimens of intact black bullhead catfish that did not discriminate L-Ala from L-Ser before their olfactory organs were extirpated bilaterally, started to discriminate these two amino acids after olfactory organ regeneration and reconditioning to L-Ala (Stenovec and Valentinčič, 2001), indicating that, in this case, discrimination depended on receptor expression rather than genetic differences.

Behavior is an all-or-none phenomenon which occurs whenever an animal detects either a key (innate) or a signal (conditioned) stimulus. Chemical stimuli release behavior only if the stimulus concentration is higher than the behavioral threshold, which is usually close to the physiological threshold of the most sensitive receptor cells. The highest concentrations of chemical stimuli are located within the center of bodies of water driven there by eddies. Two kinds of information are available, stimulus quality and concentration gradients. Olfaction enables the fish to detect the quality and differences in concentration of the olfactory stimulus. As the fish swims through the water or as the water contacts the fish, the animal detects rapidly increasing and decreasing concentrations of stimuli, i.e., transient "bumps" of stimuli. A catfish swimming through several high concentration stimuli eddies experiences several occurrences of the same smell in sequence. The duration of an encounter with a stimulus eddy is usually short (few seconds) and there is little time for receptor adaptation. Since fish normally encounter chemical gradients, the amino acid concentrations during behavioral experiments change continuously; therefore stimulus concentrations cannot be learned. Rather it is the quality of the stimulus that repeats and becomes associated with a food reward. Conditioned stimuli (signal stimuli) excite the brain and the food search begins. In the fish inner world (in the sense of Eccles, 1970), the conditioned stimulus is associated with food and it releases food expectancy. Like a hunter looking for minute signals of a wild animal, a hungry fish searches for signals representing food.

Since behavior is an all-or-none phenomenon (i.e., once the behavioral threshold is attained, the complete behavioral response occurs), olfactory stimuli presented at suprathreshold concentration release behavior in conditioned and motivated omnivorous fish irrespective of the electro-olfactogram (EOG) amplitude such stimuli would elicit. In electrophysiology stimuli releasing large amplitude of EOG are commonly termed strong stimuli and those eliciting small amplitude EOG, weak stimuli (Caprio, 1978). Both large and small amplitude EOG stimuli effectively release behavior. L-Pro is one of the substances that release very small amplitudes of EOG, yet it is a very effective olfactory stimulus. In channel and bullhead catfishes, the EOG response to L-nVal is significantly greater in magnitude than the response to L-Pro; the magnitude of the EOG response to 0.1 mM L-Pro is approximately 10% of that to 0.1 mM L-nVal (Fig. 3.5). To result in approximately equal EOG amplitudes, the concentration of L-nVal has necessarily to be a hundred thousand times lower than that of L-Pro [55]. At any concentration larger than μM, catfish can be readily conditioned to either L-nVal or L-Pro and are able to discriminate these compounds from each other and from other amino acids.

With increasing amino acid concentration, more olfactory receptor neurons (ORNs) are recruited from differentially sensitive olfactory cells. In insects such as *Drosophila* (Acebes and Ferrús, 2001; Korsching, 2002) and honeybee (Sachse and Galizia, 2003), glomerular areas of the antennal lobe activated by responding ORNs become larger with increasing stimulus concentrations. In fishes, the magnitudes of the EOG for most amino acids are also dose dependent; however, at concentrations $> \mu$M L-Pro responses are saturated. If the EOG response to L-Pro is saturated, it is highly probable that all ORNs capable of responding to this substance are fully activated. It appears that the total number of ORNs involved in the detection of specific olfactory stimuli do not determine olfactory discrimination abilities. If all ORNs specific for one substance are responding, no further increase in activated area of the olfactory bulb can be expected. Irrespective of the total areas of the olfactory bulb excited by two different amino acid stimuli, minimal qualitative differences in their chemotopic projections enable an olfactory discrimination.

The biological actions of taste and olfactory stimuli on feeding behavior differ markedly. Taste stimuli act as key stimuli and their relation to feeding is innate. Olfactory stimuli have no innate relationship with food, i.e., fishes have to associate specific olfactory stimuli with a food

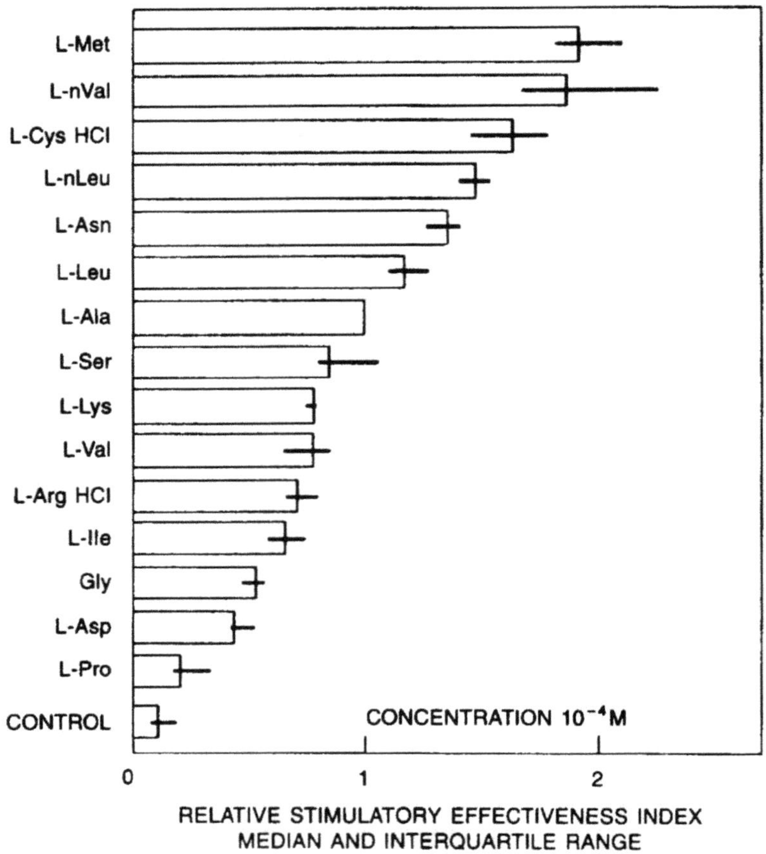

Fig. 3.5 Relative response amplitudes based on the magnitude of the peak electro-olfactogram (EOG) (relative L-Ala) to different amino acids in black bullhead catfish.

reward prior to any relationship becoming established (Valentinčič and Caprio, 1994a; Valentinčič et al., 2000b). This distinction between taste and smell has an anatomical as well as functional basis. Peripheral taste activity is transmitted to the primary gustatory nucleus located in the hindbrain, the medulla oblongata, which is a major neural center involved in reflex behaviors. In contrast, olfactory receptor axons project to the olfactory bulb that is part of the telencephalon, the portion of the CNS involved with flexible behavioral responses. From an evolutionary point of view, taste inputs are required for immediate control of food intake; therefore fast reflex responses secure successful feeding while olfactory input informs the animal about potential food sources. Learning odors of specific stimuli that signal food is critical for maximizing the search for rewarding stimuli while minimizing times of exposure to predators.

4. OLFACTORY DISCRIMINATION OF BINARY AND TERNARY MIXTURES OF AMINO ACIDS

For both aquatic and terrestrial animals, most odors are mixtures of odorants, some more complex than others. Mixtures can consist of components that vary in odorant potency. To study behavioral responses to odorant mixtures, the relative potencies of components in the mixture must be established. One method of accomplishing this task is to determine the concentrations of components that are equieffective. In humans, equal stimulatory effectiveness is determined behaviorally using intensity rating scales (Laing and Glemarec, 1992). There is a great advantage in studying mixture discrimination in fishes since equal levels of stimulatory effectiveness of amino acid stimuli can be determined physiologically (Valentinčič et al., 2000a). The underwater EOG enables monitoring olfactory activity of the entire olfactory organ (Ottoson, 1956; Silver et al., 1976). The EOG is the sum of ORN depolarizations (and hyperpolarizations) whose duration changes with stimulus concentration (Hara, 1992) and reflects the magnitude of the olfactory receptor potentials. Recent evidence from my laboratory suggests that the majority of ORNs in catfish have little or no spontaneous activity but are differentially sensitive to amino acid stimuli (Valentinčič et al., 2005). It is assumed that equal EOG amplitudes in response to two different amino acid stimuli that interact with different molecular receptors and likely different ORNs indicate similar numbers of responding ORNs. These responding ORNs in turn activate their specific glomeruli in the olfactory bulb almost equally. In insects, the more ORNs excited, the greater the odor influences the glomerular activity within the antennal lobe (Acebes and Ferrús, 2001; Korsching, 2002; Sachse and Galizia, 2003). It is assumed that the binary mixture of two equally stimulatory components exerts relatively equal influence on the specific olfactory bulb glomeruli to which they project. Binary (Valentinčič et al., 2000) and ternary mixtures (Valentinčič and Miklavc, unpubl.) of amino acids whose components have varying potencies would be detected by the fish as its more stimulatory component(s). Catfish were easily conditioned to mixtures comprising two amino acids of unequal potency, but the

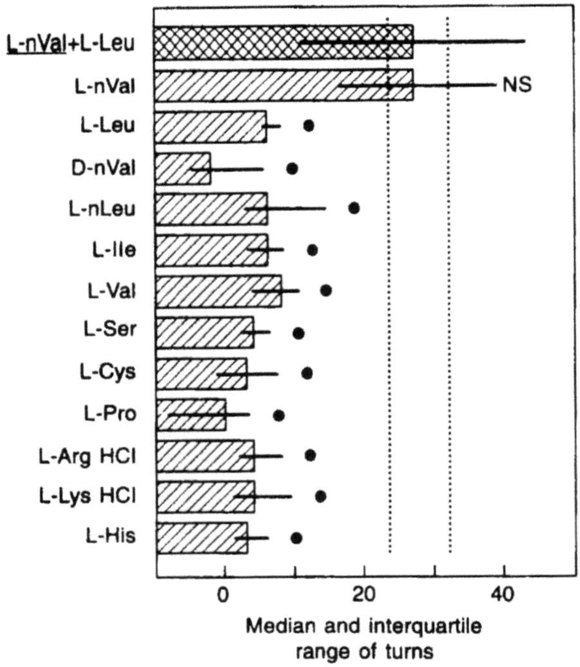

Fig. 3.6 Black bullhead catfish (*Ameiurus melas*) were conditioned to a binary mixture of L-nVal + L-Leu. L-nVal (underlined) was the more stimulatory component of the mixture; it was 30 times more concentrated than its equiresponsive concentration with L-Leu. With the sole exception of L-nVal, catfish discriminated every single amino acid from the conditioned mixture, which indicates that binary mixtures are detected initially as the more stimulatory component of the mixture. Crosshatched bars, conditioned stimulus; hatched bars, nonconditioned stimuli; vertical dotted lines, range of medians for the conditioned stimuli; dots, significant difference at $P < 0.01$; NS, nonsignificant difference.

various mixtures were initially perceived by the fish as the component that was most stimulatory to its olfactory system (Fig. 3.6).

Can catfish be conditioned to binary mixtures composed of two amino acids having equal potencies? During the continuous regeneration of olfactory receptor neurons (Farbman, 2000), the reexpression of numbers of receptor neurons for the two mixture components might be unequal. Due to unequal receptor cell expression the initially perceived more stimulatory component of the binary mixture might not always be the same. In this case the perception would flip-flop between the components; the rewarded component would change during the conditioning procedures and conditioning potentially would not occur. We observed such phenomena in our conditioning experiments with binary mixtures of equieffective amino acids (Valentinčič, unpubl.); however the evidence of catfish inability for conditioning to binary mixtures of equieffective amino acids is far from conclusive.

The strategy of establishing equipotent amino acid concentrations in mixtures that result in approximately equal EOG magnitudes for all the components enabled us to prepare different mixtures that were initially detected by any specific amino acid component of our choice (i.e., by increasing the concentration and therefore its relative potency of a component in the mixture). By using one component at 3-100 times the equieffective concentration of other mixture components, we could determine the specific component that would be initially perceived in the mixture. Conditioning enabled perception of the mixture as its more stimulatory component and additional discrimination training provided an increased selective attention (Laing and Glemarec, 1992), i.e., detection of small differences between the most stimulatory components alone and the mixture containing the more and less stimulatory components. Thus, as in human olfaction (Laing and Willcox, 1983), fish learn initially the most prominent feature of an odor, followed with additional experience, by the odor modified by the less stimulatory component.

5. BEHAVIORAL RESPONSES OF BLACK BULLHEAD CATFISH TO MULTIMIXTURES OF EQUALLY STIMULATORY AMINO ACIDS

Multicomponent mixtures (i.e., multimixtures) can be detected similarly as binary and ternary mixtures where the more stimulatory component of the mixture is perceived first and additional discrimination training enables detection of other components in the mixture (Valentinčič et al., 2000). Another possibility is that a multicomponent mixture is perceived as an entirely new entity (humans: Laing and Livermore, 1992; Livermore and Laing, 1996, 1998). We conditioned black bullhead catfish to a multimixture composed of 13 amino acids in which one of the components was presented at a 30 times greater concentration than its equistimulatory effective concentration with the other components. In all instances, catfish perceived the multimixture as different from the most stimulatory component alone. Second, we studied discrimination of multimixtures composed of equistimulatory amino acids (Valentinčič et al., 2003). The multimixtures of equistimulatory amino acids were prepared with components at their equal EOG amplitude concentrations. During approximately 50 sessions, we conditioned black bullhead catfish to multimixtures composed either of seven [L-Arg-HCl, L-Leu, L-asparagine (L-Asn), L-Ala, L-Met, L-Cys-HCl, and L-nVal] (Fig. 3.7) or twelve (L-Ala, L-Leu, L-Val, L-nVal, L-nLeu, L-CysHCl, L-Met, L-ArgHCl, L-LysHCl, L-Asp, L-Asn, and L-His) amino acids.

By removing one, two, and three amino acids from the mixture, we searched for the smallest difference in the composition of the mixture that

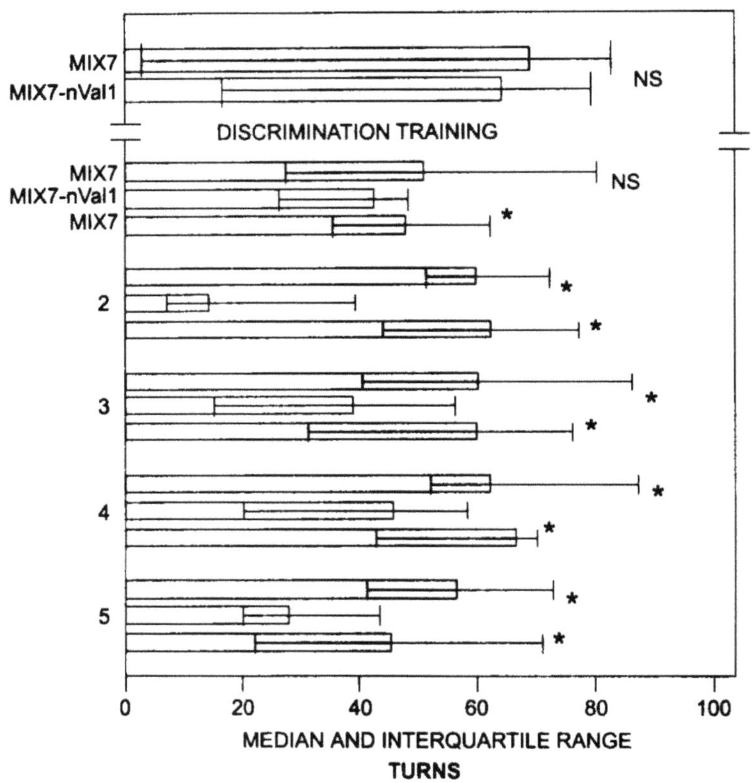

Fig. 3.7 Olfactory discrimination of multimixtures of seven and six equipotent amino acids: Initially bullhead catfish were unable to discriminate the conditioned mixture of 7 from its 6-component counterparts (mixture of 7 minus 1 amino acid). After multiple experiences with 7 minus 3 and 7 minus 2 amino acids mixtures (equivalent to discrimination training), which increases selective attention for small differences in mixture composition, bullhead catfish started to discriminate the 6-component mixture (nVal omitted) from the conditioned 7-component mixture. Dark bars are the conditioned responses to the mixture of 7 amino acids, white bars the responses to nonconditioned amino acids; asterisks indicate significance at $P < 0.05\%$. NS, nonsignificant difference.

allowed the fish to discriminate among the mixtures. Catfish discriminated behaviorally between the 7- and 7-3-component and the 12- and 12-3 component mixtures respectively; behavioral response to the nonconditioned mixture was less than one-third that to the respective conditioned mixtures. Although additional discrimination training was needed, catfish had relatively little difficulty in discriminating the 7-2 and 7-1 (Fig. 3.7) component mixtures from the 7-component conditioned mixture. The behavioral response (swimming activity) to the conditioned mixture was at least 30% greater than to the nonconditioned mixtures. Bullhead catfish however, were unable to discriminate a 12-component amino acid mixture from its 12-1-component counterparts.

Catfish detected multimixtures of amino acids as blends; the more stimulatory single amino acids were not detected within the large multimixture; qualitative differences between large multimixtures that contain seven and six equally stimulatory components enabled mixture discrimination, whereas the difference between twelve and eleven components mixtures was too small to be detected. As in binary mixtures, the most prominent odor feature in multimixtures was detected first; fine differences between mixtures differing by a single amino acid were learned later.

6. CONCLUSIONS

1. The broadly tuned olfactory organ enables fishes to detect a vast number of chemical stimuli. In herbivorous and omnivorous fishes, visual, taste, and olfactory stimuli prompt food-searching activity. If food is not visible, chemical stimuli are the main indication of its presence. Conditioned olfactory stimuli trigger a feeding excitatory state that triggers and maintains klinotactic swimming activity until the food is located.
2. After detection of the conditioned amino acid stimuli, fishes search for food longer and more intensely than after detection of nonconditioned amino acids.
3. Carnivorous fishes start using olfaction in locating food only if they learned to eat nonliving foods early in life, i.e., during fry and fingerling stages. Carnivorous fishes conditioned to eating nonlive foods can also be conditioned to discriminate amino acids.
4. Fishes conditioned to binary and ternary mixtures of amino acids detect the mixtures initially as their more stimulatory components.

Additional discrimination training consisting of 5-10 successive comparisons of conditioned with nonconditioned stimuli enabled discrimination of the binary mixture from its more stimulatory component alone.

5. Fishes can be conditioned to discriminate different multimixtures composed of 7-13 amino acids. Multimixtures are perceived as entities and the more stimulatory components in the mixture are always detected as differing from the multimixture.
6. Black bullhead catfishes discriminated multimixtures of 7 amino acids from multimixtures comprising 6 of the same amino acids present in the larger multimixture; however, they failed to discriminate a multimixture of 12 amino acids from the multimixture of 11 of the same amino acids.
7. Initially, fishes learn the most prominent odor feature of a chemical mixture; fine differences between mixtures comprising similar olfactory stimuli are learned later during discrimination training.

References

Acebes, A. and A. Ferrús. 2001. Increasing the number of synapses modifies olfactory perception in Drosophila. *J. Neurosci.* 21: 6264–6273.

Campbell, J., Bindra, D., Krebs, H. and Ferenchak, R.P. 1969. Responses of single units of hypothalamic ventromedial nucleus to environmental stimuli. *Physiol. Behav.*, 4: 183–187.

Caprio, J. 1978. Olfaction and taste in the channel catfish: an electrophysiological study of the responses to amino acids and derivatives. *J. Comp. Physiol.* [A], 123: 357–371.

Caprio, J. and R. P. Byrd. Jr. 1984. Electrophysiological evidence for acidic, basic, and neutral amino acid olfactory receptor sites in the catfish. *J. Gen. Physiol.* 84: 403–422.

Curtiss, S., W. Fromkin, S. Rigler, D. Rigler and M. Krashen. 1974. Linguistic development of a genie. *Language* 50: 528–555.

Dugas J.C. and J.N. Ngai. 2001. Analysis and characterization of an odorant receptor gene cluster in the zebrafish genome. *Genomics* 71: 53–65.

Eccles, J.C. 1970. *Facing Reality: Philosophical Adventures of Brain Scientist, Springer-Verlag*, Berlin, pp. 210

Farbman, A. I. 2000. Cell biology of olfactory epithelium. In: *The Neurobiology of Taste and Smell*, T. E. Finger, W. L. Silver and D. Restrepo (Eds). Wiley-Liss, Inc., New York, NY, pp. 131–158.

Finger, T.E. 1988. Sensorimotor mapping and oropharyngeal reflexes in goldfish, *Carassius auratus*. *Brain Behav. Evol.* 31: 17–24.

Fraenkel, G.S. and D.L. Gunn. 1940. *The Orientation of Animals*, Claredon Press, Oxford

Friedrich, R.W. and S. Korsching. 1997. Combinatorial and chemotopic odorant coding in the zebrafish olfactory bulb visualised by optical imaging. *Neuron* 18: 737–752.

Hansen, A. and K. Reutter. 2004. Chemosensory systems in fish: Structural, functional and ecological aspects. In: *The Senses of Fish: Adaptations for the Reception of Natural Stimuli*, G. Von der Emde, J. Mogdans, and B.G. Kapoor (Eds). Narosa Publishing House, New Delhi & Kluwer Academic Publishers, Dordrecht, The Netherlands, pp. 55–89.

Hara, T.J. 1992 Mechanisms of olfaction. In: *Fish Chemoreception*, T. J. Hara, (Ed.). Chapman & Hall, London, New York, pp. 150–170.

Hasler, A.D. and A.T. Scholz. 1983. *Olfactory Imprinting in Homing Salmon*. Springer-Verlag, Berlin.

Herrington, A.W. and S. W. Ranson. 1942. The spontaneous activity and food intake of rats with hypothalamic lesions. *Amer. J. Physiol.* 136: 609–617.

Jacobs, H.M. 1967. *Diffusion Processes*. Springer-Verlag, New York.

Jones, K.A. 1989. The palatability of amino acids and related compounds to rainbow trout, *Salmo gairdneri* Richardson. *J. Fish Biol.* 34: 149–160.

Kanwal, J.S. and T. Finger. 1992. Central representation and projections of gustatory systems. In: *Fish Chemoreception*, T.J. Hara (Ed.) Chapman & Hall, London, New York, pp. 79–102.

Kasumyan, O. and O.M. Prokopova. 2003. Taste preferences and the dynamics of behavioral taste response in the tench *Tinca tinca* (Cyprinidae). *J. Ichthyol.* 41: 640–653.

Kasumyan, A.O. and E.V. Nikolaeva 1997 Taste preference of *Poecilia reticulata* (Cyprinodontiformes). *J. Ichthyol.* 37: 662–669.

Korsching, S. 2002. Olfactory maps and odor images. *Curr. Opin. Neurobiol.* 12:387–392.

Laberge, F. and T.J. Hara 2001. Neurobiology of fish olfaction: a review. *Brain Res. Rev.* 36: 46–59.

Laing, D.G. and A. Glemarec. 1992. Selective attention and the perceptual analysis of odor mixtures. *Physiol. Behav.* 52: 1047–1053.

Laing, D.G. and B.A. Livermore. 1992. Perceptual analysis of complex chemical signals by humans. In: *Chemical Signals in Vertebrates*, R.L. Doty and D. Müller-Schwarze, (Eds) Plenum Press, Philadelphia, Vol. 6, pp. 587–593.

Laing, D.G. and M.E. Willcox. 1983. Perception of components in binary odour mixtures. *Chem. Senses* 7: 249–264.

Lamb, C.F. and T.E. Finger. 1995. Gustatory control of feeding behavior in goldfish. *Physiol. Behav.* 57: 483–488.

Livermore, A. and D.G. Laing. 1996. Influence of training and experience on the perception of multicomponent odor mixtures. J. Exper. Psychol. 22: 267–277.

Livermore, A. and D.G. Laing. 1998. The influence of odor type on the discrimination and identification of odorants in multicomponent odor mixtures. *Physiol. Behav.* 65: 311–320.

Marui, T. and J. Caprio. 1992. Teleost gustation. In: *Fish Chemoreception*, T.J. Hara (Ed.). Chapman & Hall, London, New York, pp. 171–192

Marui, T., R.E. Evans, B.S. Zielinski, and T.J. Hara 1983a. Gustatory responses of the rainbow trout (*Salmo gairdneri*) palate to amino acids and derivatives. *J. Comp. Physiol.* A 153: 423–433.

Marui, T., S. Harada and Y. Kasahara 1983b. Gustatory specificity for amino acids in the facial taste system of the carp, *Cyprinus carpio* L. *J. Comp. Physiol.* A 153: 299–308.

Messenger, J. B. 1973. Learning in the cuttlefish, *Sepia*. *Anim. Behav.* 21: 801–826.

Moore, P. A., G.A. Gerhardt and J. Atema 1989. High resolution spatio-temporal analysis of aquatic chemical signals using microelectrochemical electrodes. *Chem. Senses* 14: 829–840.

Motokizawa, F. and N. Furuya. 1973. Neural pathway associated with EEG arousal response by olfactory stimulation. Electroenceph. Clin. Neurophy. 35: 83–91.

Murakami, T., Y. Morita, and H. Ito, 1983. Extrinsic and intrinsic connections of the telencephalon in a teleost, *Sebastiscus marmoratus*. *J. Comp. Neurol.* 216: 115–131.

Ngai, J., M.M. Dowling, L. Buck, R. Axel, and A. Chess, 1993. The family of genes encoding odorant receptors in the channel catfish. *Cell* 72: 657–666.

Nikonov, A.A. and J. Caprio. 2004. Odorant specificity of single olfactory bulb neurons to amino acids in the channel catfish. *J. Neurophys.* 92 (In press).

Ottoson, D. 1956. Analysis of the electrical activity of the olfactory epithelium. *Acta Physiol. Scand.* 35, Suppl. 22: 1–83.

Pavlov, D.S. and A.O. Kasumyan 1998. The structure of the feeding behavior of fishes. *J. Ichthyol.* 38: 116–128.

Sachse, S. and C.G. Galizia. 2003. The coding odor-intensity in the honey bee antennal lobe: local computation optimizes odor representation. *European J. Neurosci.* 18: 2119–2132.

Scott, J.W. and C.M. Leonard 1971. The olfactory connections of the lateral hypothalamus in the rat, mouse and hamster. *J. Comp. Neurol.* 141: 331–344

Sibbing, F.A., J.W.M. Osse, and A. Terlouw. 1986. Food handling in the carp (*Cyprinus carpio*): its movement patterns, mechanisms and limitations. *J. Zool.* (Lond.) 210(A): 161–203

Silver, W.L., J. Caprio, J.F. Blackwell and D. Tucker. 1976. The underwater electro-olfactogram: A tool for the study of the sense of smell of marine fishes. *Experientia* 32: 1216–1217.

Stenovec, M. and T. Valentinčič. 2001. Catfish possessing a small portion of regenerated olfactory organ can discriminate amino acids. *Chem. Senses* 26: 1097.

Striedter, G. 1990. The diencephalon of the channel catfish, *Ictalurus punctatus*: I. Nuclear organization. *Brain Behav. Evol.* 36: 329–354.

Valentinčič T. 2004. Taste and olfactory stimuli and behavior in fishes. In: *The Senses of Fish: Adaptations for the Reception of Natural Stimuli* G. Von der Emde, J. Mogdans and B. G. Kapoor (Eds). Narosa Publishing House, New Delhi, and Kluwer Academic Publishers, Dordrecht, The Netherlands, pp. 90–108.

Valentinčič, T. and J. Caprio. 1994a. Chemical and visual control of feeding and escape behaviors in the channel catfish Ictalurus punctatus. *Physiol. Behav.* 55: 845–855.

Valentinčič, T. and J. Caprio. 1994b. Consummatory feeding behavior in intact and anosmic channel catfish Ictalurus punctatus to amino acids. *Physiol. Behav.* 55: 857–863.

Valentinčič, T. and J. Caprio. 1997. Visual and chemical release of feeding behavior in adult rainbow trout. *Chem. Senses* 22: 375–382.

Valentinčič, T., S. Wegert and J. Caprio. 1994. Learned olfactory discrimination versus innate taste responses to amino acids in channel catfish, *Ictalurus punctatus*. *Physiol. Behav.* 55: 865–873.

Valentinčič, T., C.F. Lamb and J. Caprio. 1999. Expression of a reflex snapping/biting response to amino acids prior to the first exogenous feeding in salmonid alevins. *Physiol. Behav.* 67: 567–572.

Valentinčič T., P. Miklavc, J. Dolenšek, and K. Pliberšek. 2005. Correlations between olfactory discrimination, olfactory receptor neuron responses and chemotopy of amino acids in fishes. *Chem. Senses* 30 (suppl. 1): i312–i314.

Valentinčič, T., J. Kralj, M. Stenovec, A. Koce and J. Caprio. 2000a. The behavioral detection of binary mixtures of amino acids and their individual components by catfish. *J. Exp. Biol.* 203: 3307–3317.

Valentinčič, T., J. Metelko, D. Ota, V. Pirc and A. Blejec. 2000b. Olfactory discrimination of amino acids in brown bullhead catfish. *Chem. Senses* 25: 21–29.

Valentinčič, T., S. Kralj, and V. Zgonik, 2003. Discrimination of amino acid multimixtures in catfish. *Chem. Senses* 28: A89.

Zippel, H. P., T. Largo-Schaaf and J. Caprio. 1993. Ciliated olfactory receptor neurons in goldfish (*Carassius auratus*) partially survive axotomy, rapidly regenerate and respond to amino acids. *J. Comp. Physiol.* A 173: 537–547.

CHAPTER

4

In vivo Recordings from Single Olfactory Sensory Neurons in Goldfish (*Carassius auratus*) during Application of Olfactory Stimuli

H.P. Zippel, H. Dolle, M. Foitzik, A. Hamadeh, L.G.C. Lüthje , A.M. Möller-de Beer and R. Köhnke

ABSTRACT

The goldfish olfactory system offers an advantage to scientists because the respective structures are readily accessible for experimental preparations. The olfactory epithelium lies peripherally and is covered only by a skin fold, and the pedunculated olfactory bulb can easily be prepared without lesioning telencephalic structures. A great number of naturally relevant stimuli are structurally known (see below) and can be applied in defined molarities. The present study investigated whether application of many types of biologically

Address for Correspondence: H.P. Zippel, Physiologisches Institut der Universität, Humboldtallee 23, 37073 Göttingen, Germany. E-mail: pzippel@gwdg.de relevant chemical sensory stimuli results in different response characteristics in *in vivo* recordings from the dendritic endings of single olfactory sensory neurons (surface neural recordings). Extracellular recordings were made with platinum-black electrodes. Pheromones and amino acids, amino acids, stereoisomers of amino acids, L-arg and structural analogues, and structurally similar and dissimilar non-familiar stimuli were investigated in successive series of experiments. In general, no response was recorded from 50–70% of sensory neurons during application of an olfactory stimulus. Excitatory responses were found more frequently (15–25%) than inhibitory (10–20%). Differences in responses of the same neuron to different classes of stimuli were apparent in 'generalists' and a few (or only one) in 'specialists'. Even structurally very similar stimuli such as α- and β-ionone or pairs of stereoisomers of amino acids were discriminated by approximately 60% of responding sensory neurons. From dose-response experiments it was evident that most neurons responded similarly during application of 2 log unit different concentrations of amino acids, and only in 20% of sensory neurons application of the lower concentration had a weaker effect. Whether recordings were made from ciliated or microvillous sensors is not known.

Key Words: *In vivo* recordings; Olfactory sensory neurons; Familiar, nonfamiliar stimuli; Goldfish.

1. INTRODUCTION

In contrast to land vertebrates, many types of biologically relevant chemical sensory stimuli have been both identified and structurally characterized in fish (especially the Cyprinidae) along with their effective concentrations. Another great advantage of fish as experimental models for the study of olfaction is their slightly less differentiated anatomy, including a number of unique experimental prerequisites: the olfactory epithelium is readily accessible and contains both microvillous and ciliated receptor neurons (Hansen et al., 1999); in contrast to terrestrial vertebrates, different types of receptor neurons are more equally distributed over single lamellae and the mid-line raphe; odours can be applied in a constant flow to the epithelium while the experimental animal is artificially respirated through the mouth and gills.

In the present study 'classical' *in vivo* extracellular recording techniques were used for a number of reasons: Firstly, behavioural studies (Zippel et al., 1993b; von Rekowski et al., 1995), and recordings from relay neurons in the goldfish olfactory bulb (Zippel et al., 1995, 1999, 2000, this Vol.) were also made in *in vivo* experiments in which epithelial sensory neurons were covered with mucus-containing substances such as odorant-binding proteins that may assist in delivery or removal of odorant molecules to/from the sensory neurons. Data from *in vitro* experiments are not comparable with the just-mentioned data because the normal external milieu of the cells is lost. Secondly, in *in vitro* experiments and *in vivo* electro-olfactogram (EOG) recordings, mainly rather high stimulus concentrations were applied while in the present experiments, as in the above-mentioned training and *in vivo* publications, micro- and nanomolar concentrations were used, which probably elicit fewer non-specific effects such as membrane partitioning and/or destabilization, competitive non-specific binding to olfactory sensory neurons, or non-specific effects on ion channels. Finally, in most patch-clamp studies a spontaneous occurrence of action potentials was frequently missing (see review by Schild and Restrepo, 1989), while in the present experiments interstimulus activity was the basis for determination of excitatory or inhibitory effects generated during stimulus application.

In the present study subsequent series of experiments were undertaken and the activity of single sensory neurons in the goldfish olfactory epithelium investigated during application of various pheromones, amino acids, structural analogues and stereoisomers of amino acids, and non-familiar stimuli.

1.1 Material and Methods

In vivo extracellular recordings from extremeley small single receptor neurons were made in the goldfish olfactory epithelium using platinum-black electrodes (Fig. 4.1). The platinum-plated metal-filled glass microelectrodes were made in the lab following the instructions of Gesteland (1975) and Caprio (1995). Instead of filling the electrode with Cerrelow metal, we used Woods metal (Sigma-Aldrich, Fp 70°; 50% Bi, 25% Pb, 12.5% Sn, 12.5% Cd). The electroplated tip of the electrode had a cauliflower-like shape (Fig. 4.1) with numerous 1 μm-diameter or smaller protrusions. These protrusions are obviously the essential prerequisite for successful recordings from single 1–2 μm diameter olfactory sensory neurons.

1.1.1 Animals and experimental procedure

In subsequent series of experiments goldfish (1.5–2 years old, 10–12 cm in length, 30–45 g) reared in the laboratory were used. Before operating the fish was anesthetized with methane-sulphate salt (3-aminobenzoic

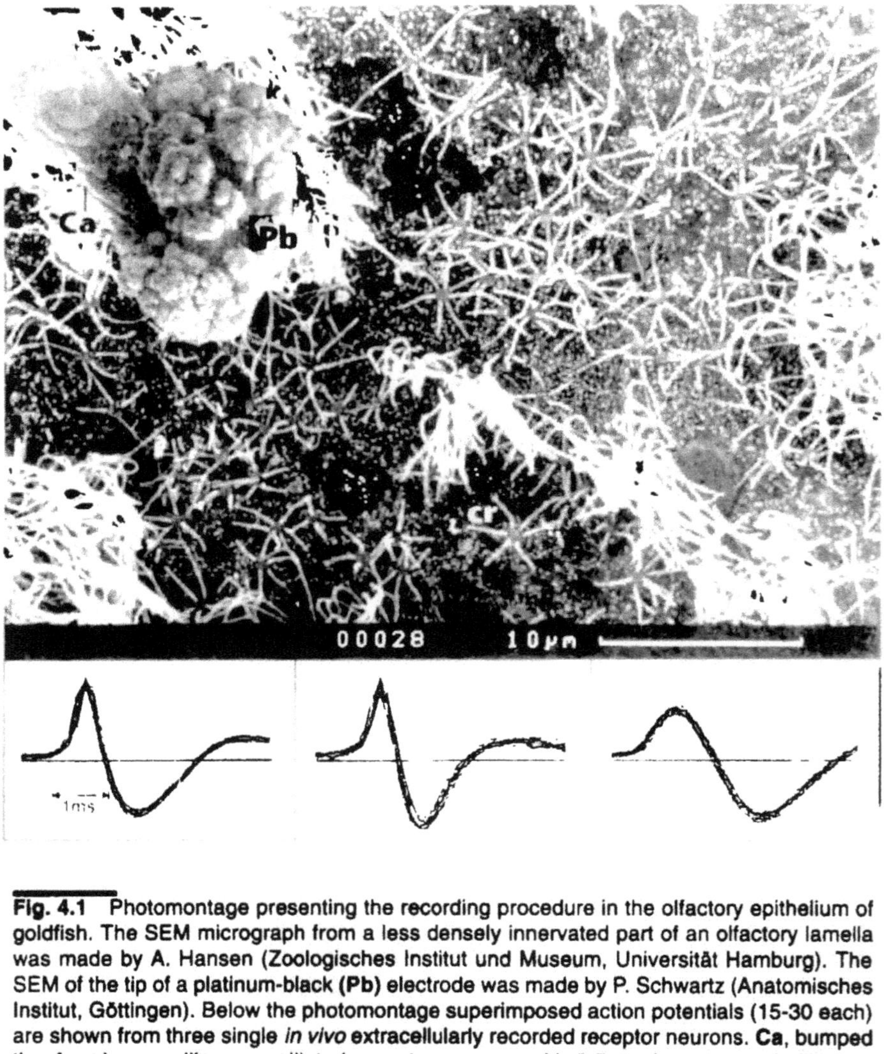

Fig. 4.1 Photomontage presenting the recording procedure in the olfactory epithelium of goldfish. The SEM micrograph from a less densely innervated part of an olfactory lamella was made by A. Hansen (Zoologisches Institut und Museum, Universität Hamburg). The SEM of the tip of a platinum-black **(Pb)** electrode was made by P. Schwartz (Anatomisches Institut, Göttingen). Below the photomontage superimposed action potentials (15-30 each) are shown from three single *in vivo* extracellularly recorded receptor neurons. **Ca**, bumped tip of a glass capillary; **cr**, ciliated receptor neurons with 3-5 µm long terminal cilia; cell body (not present in Fig.) approximately 1-2 µm in diameter. For details, see Gesteland (1975), Caprio (1995), and text (From Zippel and Lüthje, 2003).

acid ethyl ester, Sigma) 300 mg $\times$ l^{-1}, and a muscle relaxant (Horvi-Curare, Tubocurarinchloride 0.5 mg $\times$ ml^{-1}) 2 µg $\times$ g^{-1} body weight injected intramuscularly. The olfactory rosette was then exposed by simply cutting away the overlying skin fold and perfused with a laminary stream of tap-water at the rate of 0.95 ml $\times$ min^{-1}.

Olfactory stimuli were applied in sufficiently low (10^{-7}–10^{-9}M) concentrations to avoid inadequate 'blinding' effects. The 15-s stimulus exceeded the few seconds most frequently used during EOG recordings. Finally, pressure pulses between stimulus and interstimulus phases were avoided which could result in modulation of activity (Zippel and Lüthje, 2003).

Stimulus applications were interrupted by 180-s tap water applications during the interstimulus phases. These long interstimulus intervals are essential for functional recovery of epithelial sensory neurons (Zippel et al., 1995a). Recordings were made online using 'Discovery' (Brainwave software) and evaluations with Discovery subprograms which enable discrimination of individual and simultaneously recorded cells. For further details see Schild (1985), Schild and Zippel (1986), and Fischer and Zippel (1989).

1.1.2 Stimuli

Many types of biologically relevant chemical sensory stimuli have been identified in goldfish and structurally characterized along with their effective concentrations. Stimuli were purchased from Sigma (Munich, Germany) and the probable alarm pheromone from Menar Organics Limited (Gwyedd, GB). Stock solutions (1 ml) were made up at concentrations of 10^{-4}M and kept at – 18°C. Amino acids, relevant food stimuli in goldfish (Zippel et al., 1993b; von Rekowski et al., 1995), were applied at 10^{-7}M and 10^{-9}M, which elicited no recordable EOG responses (Zippel et al., 1993a, 1995, 1997a,b). Preovulatory (17,20β-dihydroxy-4-pregnen-3-one, 10^{-9}M) and ovulatory pheromones (prostaglandin $F_{2\alpha}$ and a metabolite 15-ketoprostaglandin $F_{2\alpha}$, 10^{-7}M) (Stacey and Kyle, 1983; Stacey and Sorensen, 1986; Sorensen et. al., 1988, 1990; Sorensen and Stacey, 1989; Sorensen, 1992; Stacey et al., 1994; Sorensen and Scott, 1994; Zippel et al., 1997b, 1999, 2000) were applied in concentrations that elicited maximal EOG potentials. The probable alarm pheromone hypoxanthine-3(N)-oxide (Argentini, 1976; Pfeiffer, 1982), which does not evoke a recordable EOG response at any concentration (Sorensen, pers. comm.), was applied at 10^{-7}M.

1.1.3 Receptor neurons

The olfactory epithelium of fish possesses both types of olfactory receptor neurons, microvillous and ciliated cells, which in higher vertebrates are more discretely separated into the main olfactory system (ciliated receptor cells) and the accessory olfactory system (microvillous cells). In mammals, central neural pathways exist from the accessory olfactory system to the accessory olfactory bulb and to different central nuclei, while in fish an accessory olfactory system is absent.

2. RESULTS

The different classes of stimuli were investigated in subsequent series of experiments (Table 4.1).

Table 4.1 Various classes of stimuli studied

Series 1:	L-amino acids and Pheromones: Preovulatory: 17,20ß-dihydroxyprogesterone; ovulatory: prostaglandin $F_{2\alpha}$, 15K-prostaglandin $F_{2\alpha}$; alarm pheromone: hypoxanthine-3(N)-oxide; amino acids: L-Arg, L-Gln.
Series 2:	Amino acids: Nonpolar, aliphatic R groups, L-Ala; aromatic R groups, L-Phe; polar, uncharged R groups, L-Gln, L-Ser, L-Met; negatively charged R groups, L-Asn; positively charged R groups, L-Arg.
Series 3:	L-Arg and structural analogues: Amino acid; L-Arg; organic acids: guanidoacetate, ß-guanidinoproprionate; decarboxylation: 1-amino-4-guanidobutane; esterified carboxyl group: L-arginine methyl ester; other compounds: aminoguanidine.
Series 4:	Stereoisomers of amino acids: L- and D-Ala; L- and D-Asn; L- and D-Gln; L- and D-Lys.
Series 5:	Non-familiar stimuli: Structurally rather similar stimuli: α-ionone, ß-ionone, α-irone and damascenone, α-Damascone; structurally different nonfamiliar stimuli: amylacetate, ß-phenylethanole.

2.1 Amino acids

L-amino acids were applied to olfactory sensory neurons (Series 2, Table 4.1), and dose-response relations investigated when any of the stimuli elicited a clearly recognizable response in the 10^{-7} molarity. As in the other series of experiments (Table 4.1) the number of sensory neurons not responding during application of any of the amino acids was comparatively high (50–60%) vis-à-vis olfactory bulb relay neurons, the number of excitations (Fig. 4.2) was slightly higher (~25%) than in olfactory bulb relay neurons. Inhibitions (~20%) in sensory neurons were, even during the long 15-s stimulus application, sometimes difficult to determine when the interstimulus activity was low. Greater numbers of inhibitory responses in olfactory bulb relay neurons (Zippel et al., this volume) may be the

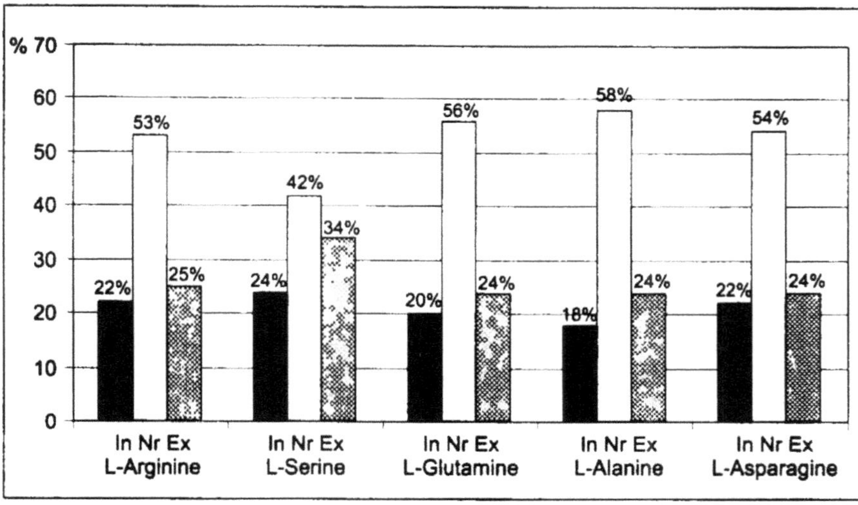

Fig. 4.2 Summary of responses recorded from single epithelial sensory neurons during L-amino acid application (10^{-7}M). **Nr**, no response: activity of sensory neuron within ± 25% compared to interstimulus activity; **In**, inhibitions: 25-100% lower; **Ex**, Excitations: 25% and higher than mean interstimulus activity. L-Arg: recordings from 291 sensory neurons; L-Ser: 248 sensory neurons; L-Gln: 202 sensory neurons; L-Ala: 185 sensory neurons; L-Asn: 187 sensory neurons.

result of lateral inhibition via granule cells. A specificity of olfactory sensory neurons for sour, neutral, or aromatic amino acids, i.e. a specificity of receptor molecules, could not be determined. Single olfactory sensory neurons could respond with excitation, inhibition, and no response during application of various amino acids. Compared to olfactory bulb neurons (Zippel et al., 1999, 2000) epithelial sensory neurons more often do not respond during stimulus application and in general respond to a smaller number of stimuli (see Fig. 4.12). These effects are the result of glomerular convergence (see Discussion below).

Highly discriminatory sensory neurons notwithstanding, lesser discriminative neurons were found in which the various amino acids elicited similar (only excitatory or only inhibitory) effects. In contrast to these discriminative (i.e., more generalizing) olfactory sensory neurons, epithelial sensory neurons were found in which the effects of applied stimuli were rather limited.

It must be emphasized however, that only a rather limited number of stimuli could be applied in the present experiments. Hence specialists need not be real specialists and could possibly respond more generally to application of a large number of stimuli.

In a certain amount (20%) of olfactory sensory neurons stimulus effects persisted for up to 30 s, presaging a rather tight bond of stimuli to membrane receptors. Other cells responded with an antagonistic pattern (10%) when stimuli were washed out. The majority (70%) of sensory neurons however, immediately returned to the interstimulus activity recorded before stimulus application.

Dose-response characteristics were only investigated in neurons in which application of the high molarities resulted in a recordable effect for at least one amino acid (see Fig. 4.3A). Figure 4.3 shows recordings from a number of sensory epithelial neurons. Cell A only responded during L-Ser application. Responses remained similar for both molarities. Cell B responded during application of 3 amino acids (L-Gln, L-Ala, and L-Phe); it was similarly excitatory during concentrations of both L-Gln and L-Ala but inhibitory to L-Phe. A lower concentration of L-Phe elicited a much weaker inhibition. Cell C responded with either excitation or inhibition during application of all the amino acids. Responses remained similar during application of the 10^{-7} and of the 10^{-9} molarities.

From dose-response experiments in 51 neurons (Fig. 4.4) it is evident that the majority of olfactory sensory neurons (40%) responded similarly to the 10^{-7} and the 10^{-9} molar concentrations. In 20% of neurons applications of lower concentration resulted in a weaker (or no recordable) response. The great variety of response characteristics found during dose-response recordings from bulbar relay neurons (Zippel et al., 2000) was not found in recordings from epithelial sensory neurons and can only be explained by intrabulbar or bulbopetal central effects (see Discussion).

2.2 Non-familiar stimuli

During application of structurally very similar (Fig. 5.6, Zippel et al., this volume) and structurally dissimilar (Table 4.1, Series 5) stimuli (phenylethanole and amylacetate) surprisingly similar data were recorded for stimulus discrimination: application of α- and β-ionone resulted in different responses in 70% of the sensory neurons, α-damascone and damascenone in 66%, and phenylethanole and amylacetate in 67%.

2.3 Stereoisomers (Table 4.1, Series 4)

In the present experiments the effects of L- and D-Ala, L- and D-Asn, L- and D-Gln, L- and D-Lys, 10^{-7}M were investigated. Both L- and D-amino acid application resulted in rather similar percent values for excitation and

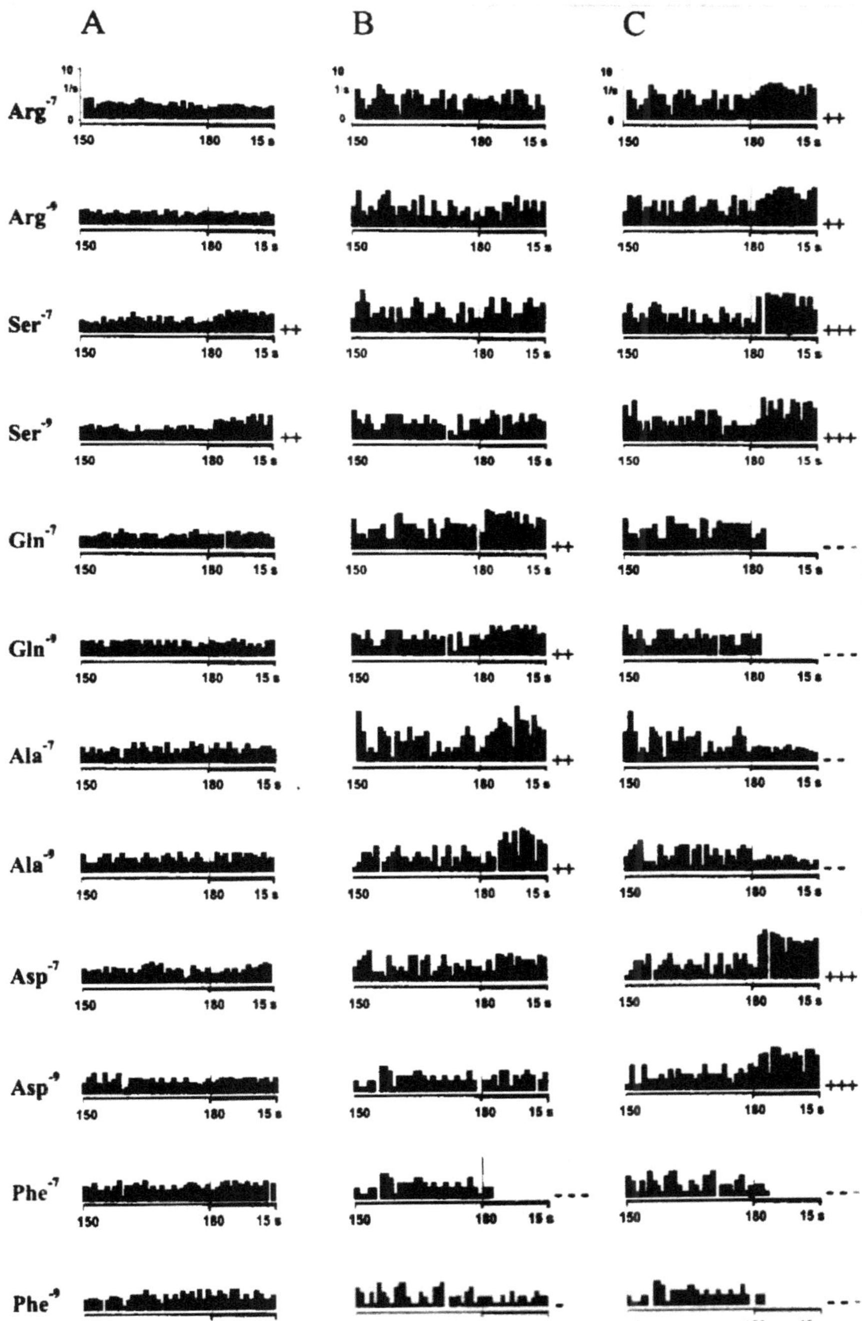

Fig. 4.3 Dose-response characteristics recorded during amino acid application in 10^{-7} and 10^{-9} molarity: Recordings present the last 30 s of the 180 s interstimulus phase and 15 s of stimulus applications; X-axis: number of potentials recorded per s. For further detail, see Figure 4.2 and Figure 4.8.

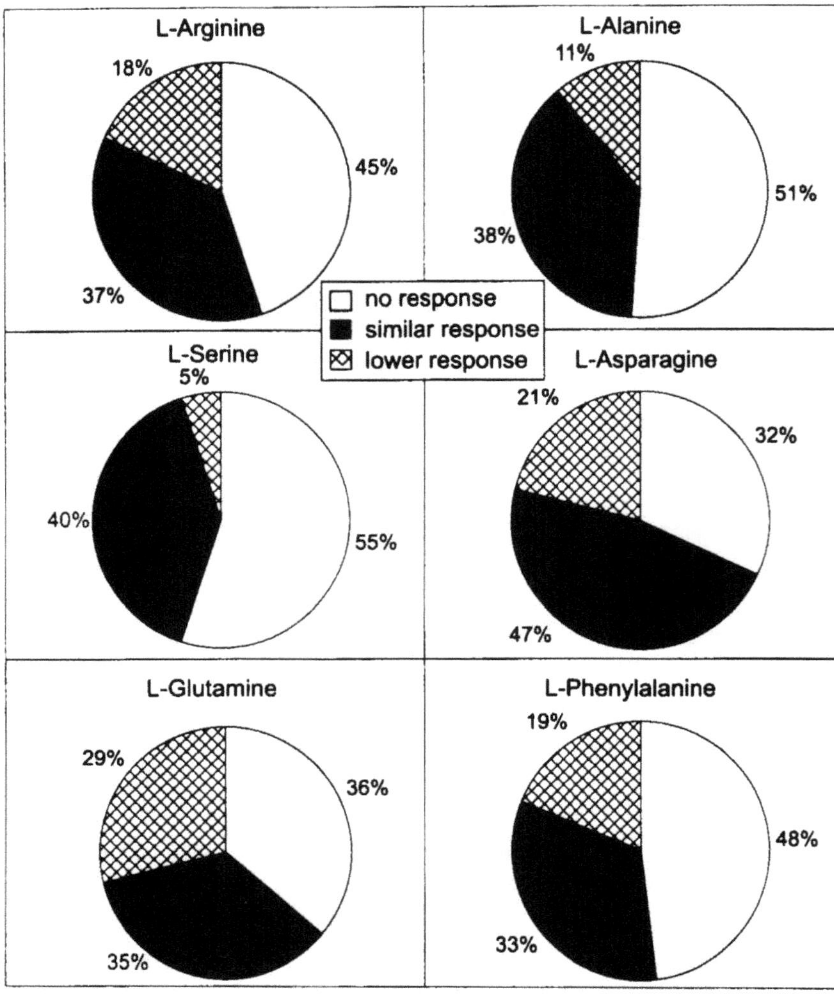

Fig. 4.4 Summary of effects recorded from 51 sensory neurons during dose-response (application of 10^{-7} and 10^{-9} molarity) experiments.

inhibition (Figs 4.5 and 4.6). The experiments therefore demonstrate a nearly similar sensitivity of epithelial sensory cells for familiar (L-amino acids) and non-familiar (D-amino acids) stimuli. In a minority (26%) of responding olfactory sensory neurons similar effects or little stimulus discrimination was recorded during application of L- and D-amino acids (Fig. 4.7). The majority of responding neurons (74%) however, discriminated L- and D-amino acids.

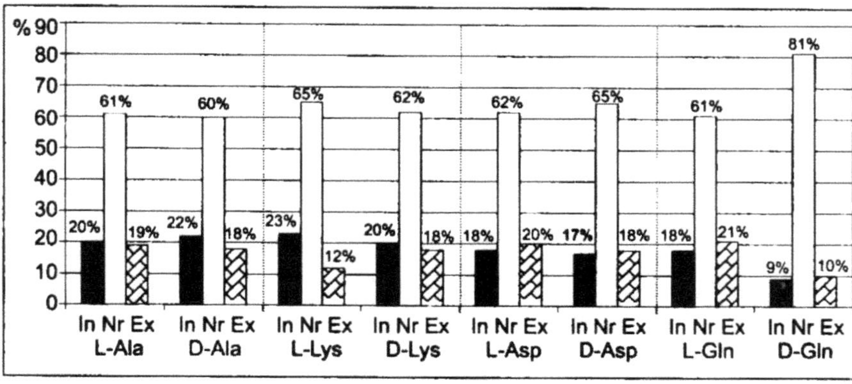

Fig. 4.5 Summary of effects recorded from single epithelial sensory neurons during stereoisomer application (10^{-7}M). Most often the epithelial sensory neurons showed no response during stimulus application. Excitations and inhibitions were more or less equally distributed. L-Ala: 100; D-Ala: 99 sensory neurons; L-Lys: 95; D-Lys: 85 sensory neurons; L-Asn: 110 D-Asn: 107 sensory neurons. For details see Fig. 4.2.

In a number of sensory neurons one amino acid, the L or the D, had no effect or similar excitatory or inhibitory effects (Figs. 4.6A and 4.7). In many other epithelial sensory cells only one (the D- or the L-amino acid) elicited an effect and showed stimulus discrimination. In a minority of cells application of L- and the D-amino acid resulted in opposing effects (Figs. 4.6C and 4.7). Opposing effects just mentioned and opposing effects recorded during application of different classes of stimuli (Figs. 4.8A-C and 4.9C) are obviously the result of activation of different molecular receptors.

2.4 Pheromones and amino acids (Table 4.1, Series 1)

Amino acids, relevant food stimuli in goldfish, were applied in a 10^{-7}M concentration. Preovulatory (17,20β-dihydroxyprogesterone, 10^{-9} M) and ovulatory pheromones (prostaglandin $F_{2\alpha}$ and the metabolite 15-ketoprostaglandin $F_{2\alpha}$, 10^{-7}M) were applied in concentrations that elicited maximal EOG potentials (Sorensen et al., 1990). The probable alarm pheromone hypoxanthine-3(N)-oxide, which does not evoke a recordable EOG response in any concentration (Sorensen, pers. comm.), was applied in a 10^{-7}M concentration.

In subsequent series of investigations of effectiveness of amino acids and pheromones we were not able to confirm a 'high correlation between microvillous olfactory receptor cell abundance and sensitivity to

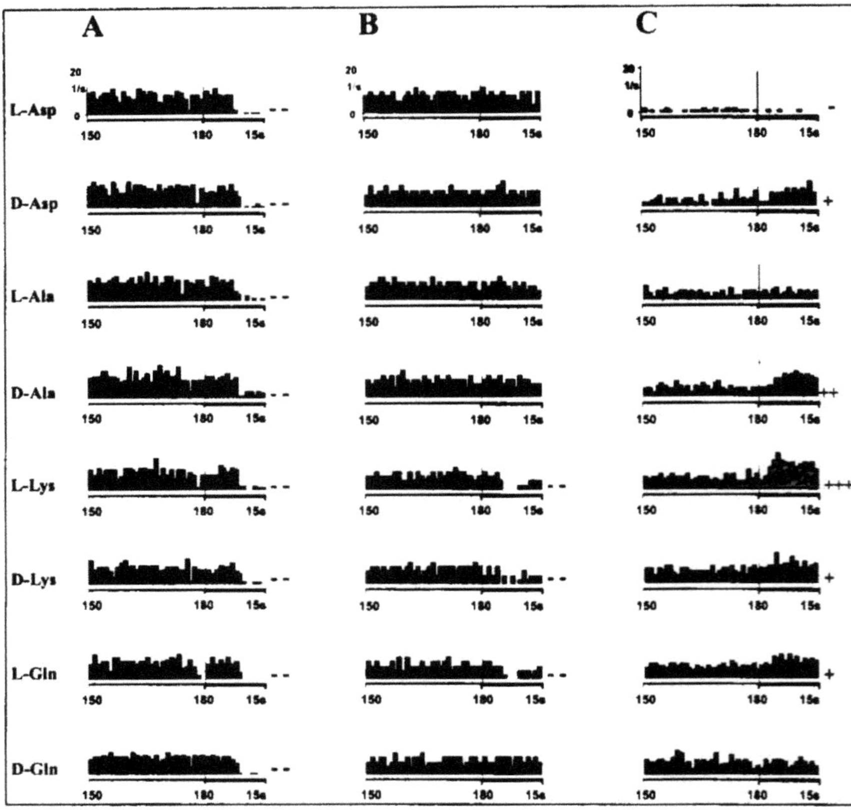

Fig. 4.6 Examples for stimulus discrimination during application of pairs of stereoisomers. Sensory neuron A responds with nearly similar delayed inhibitions during application of L- and D-amino acids. Sensory neuron B is inhibited during L- and D-Lys, and L-Gln application. Sensory neuron C responds with a great variety of response characteristics during stimulus applications. For details see Figure 4.8.

pheromones in olfactory nerve-sectioned goldfish' based on EOG, behavioural, and anatomical investigations (Zippel et al., 1997b). Data recorded 3 weeks, 4 weeks, and 4 months after olfactory nerve axotomy, and from intact control fish were essentially similar (Figs. 4.8 to 4.11). The effectiveness of amino acids and pheromones on single sensory will neurons was similar for controls and the 3-week collective in which regeneration of sensory neurons was just increasing and mainly young ciliated neurons present (Figs. 4.8 to 4.11). In contrast, a few microvillous receptor cells were observed on some lamellae but their distribution was patchy (Hansen et al., 1999).

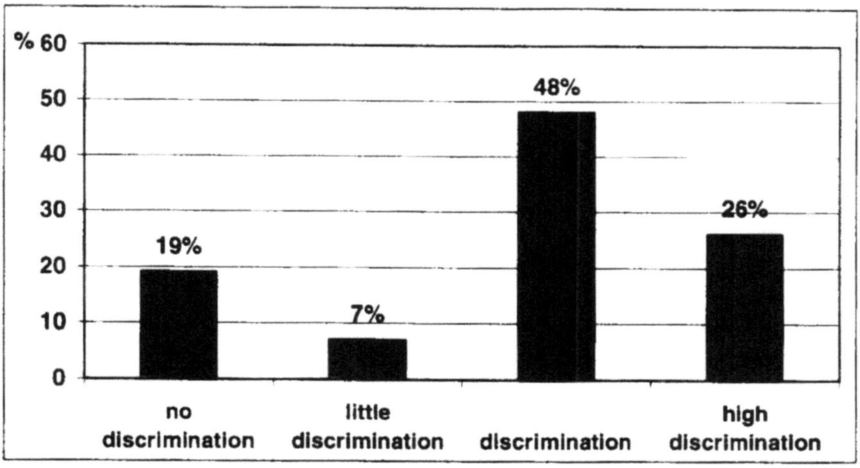

Fig. 4.7 Summary of discriminative responses recorded from 54 sensory neurons. Only cells in which all the eight stereoisomers could be applied were evaluated. No discrimination: cells either show no response during application of all the stimuli, or show quantitatively similar excitations or inhibitions (as in Fig. 4.6A); little discrimination: slightly different excitations, or slightly different inhibitions during application of the eight stimuli; discrimination: application of at least one pair of stereoisomers (such as Gln in Fig. 4.6B) must result in different responses: inhibition vs no response or excitation vs no response; high discrimination: application of at least one pair of stereoisomers must result (such as Asp in Fig. 4.6C) in reverse responses.

From Figure 4.10 it is evident that the number of excitations, inhibitions, and no responses recorded from sensory neurons with respect to amino acids and the pheromone stimuli are incredibly similar in intact controls and fish 3 or 4 weeks after olfactory axotomy.

Figure 4.11 summarizes the percentage for 'generalists' and 'specialists' in neurons in which all the stimuli could be applied. It must be emphasized that again in contrast to EOG recordings (Zippel et al., 1997b) the percent values for neurons responding during pheromone and amino acid applications were similar for intact controls and for specimens 3 and 4 weeks after bilateral axotomy. The number of specialists again is rather similar for the axotomized and intact controls, and the percent values for sensory neurons not responding to any of the 6 stimuli (20–26%) as well.

The greatest surprise in the regeneration investigations was the fact that responses recorded from single sensory neurons in intact and axotomized epithelia during regeneration were quantitatively (response rates) and qualitatively (responsiveness to amino acids and pheromones, Fig. 4.8) essentially similar, as was the distribution of generalists and specialists (Fig. 4.11).

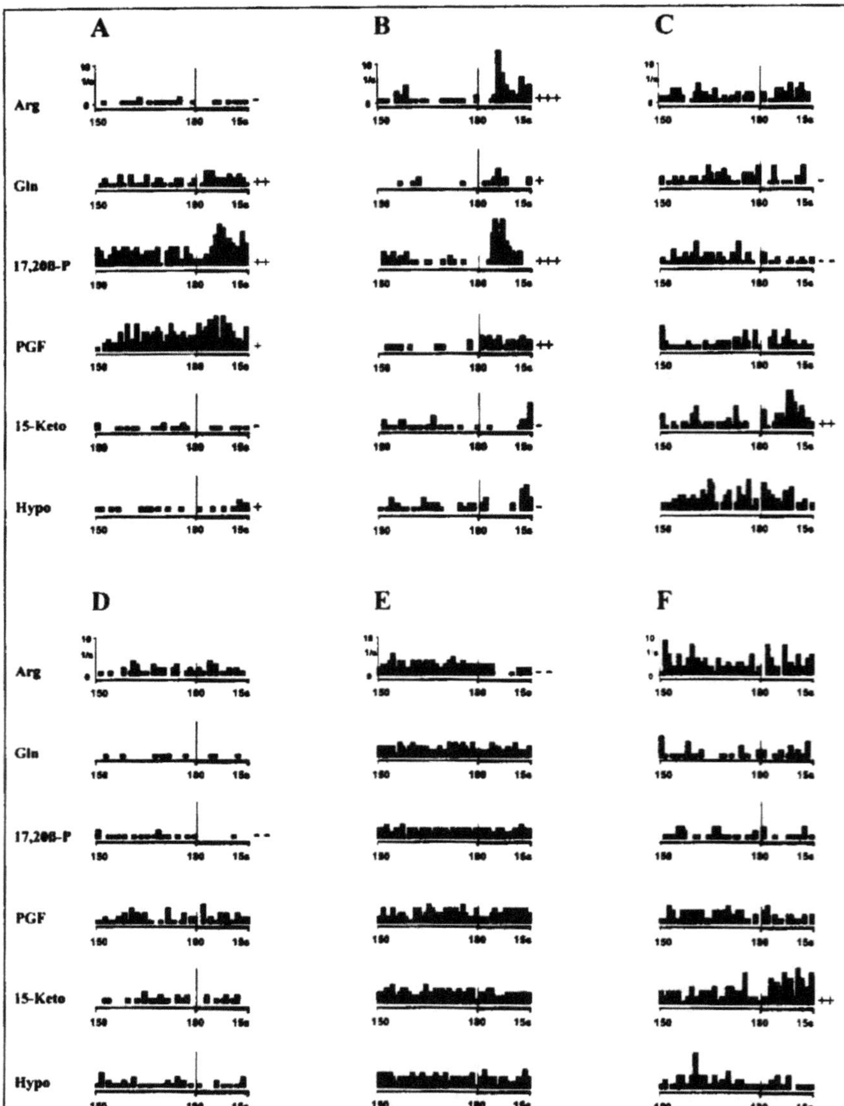

Fig. 4.8 Response characteristics recorded from six sensory neurons of intact control epithelia during pheromone and amino acid application. **A-C** 'generalists'. **A.** Slight inhibitory responses during 15K-PGF$_{2\alpha}$ application, slight excitations during PGF$_{2\alpha}$ and alarm pheromone and stronger excitations during Gln and preovulatory pheromone applications. **B.** Slight inhibition during 15K-PGF$_{2\alpha}$ and alarm pheromone application, slight excitation during Gln and strong excitations during application of the rest of stimuli. **C.** No response during Arg, PGF$_{2\alpha}$ and alarm pheromone application, inhibition during Gln (slight) and preovulatory pheromone, and strong excitation during 15K-PGF$_{2\alpha}$ application. **D-F** 'specialists'. **D.** Strong inhibition during preovulatory pheromone, **E** during Arg and **F** strong excitation during 15K-PGF$_{2\alpha}$ application. Recordings present the last 30 s of the 180 s interstimulus phases and the 15 s of stimulus application. Abbreviations listed behind abstract. + Excitation; - Inhibition; + and - activity 25–50% above/below interstimulus activity; ++ and -- 50–75% above/below interstimulus activity, +++ and --- > 75% above/below interstimulus activity. For details, see text.

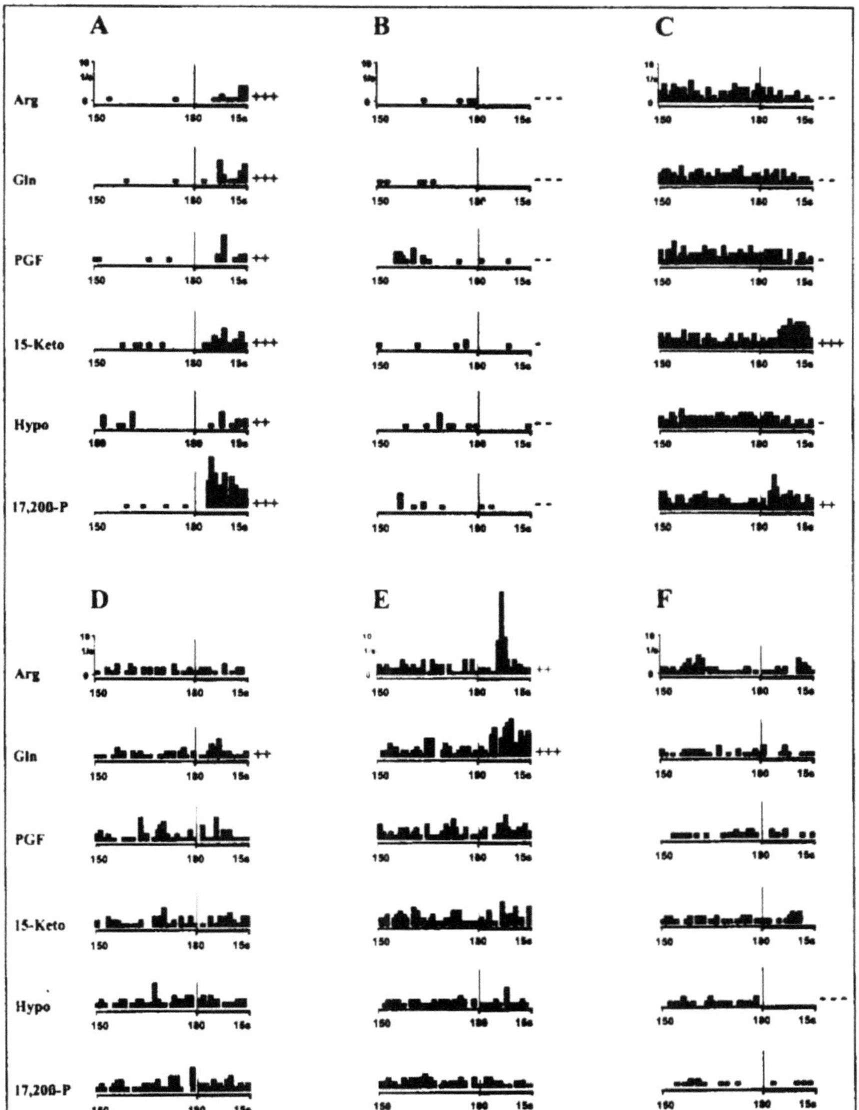

Fig. 4.9 Response charcteristics recorded from 6 sensory neurons 3 weeks after bilateral axotomy. **A-C** 'Generalists': **A.** Quantitatively different excitatory neuron. **B.** Quantitatively different inhibitory effect. **C.** Excitatory and inhibitory responses during application of different classes of stimuli. 'Specialists' **(D-F)** only respond during application of one or one class of the six stimuli. **D.** L-Gln; **E.** Both amino acids; **F.** Alarm pheromone. For details see Figure 4.8.

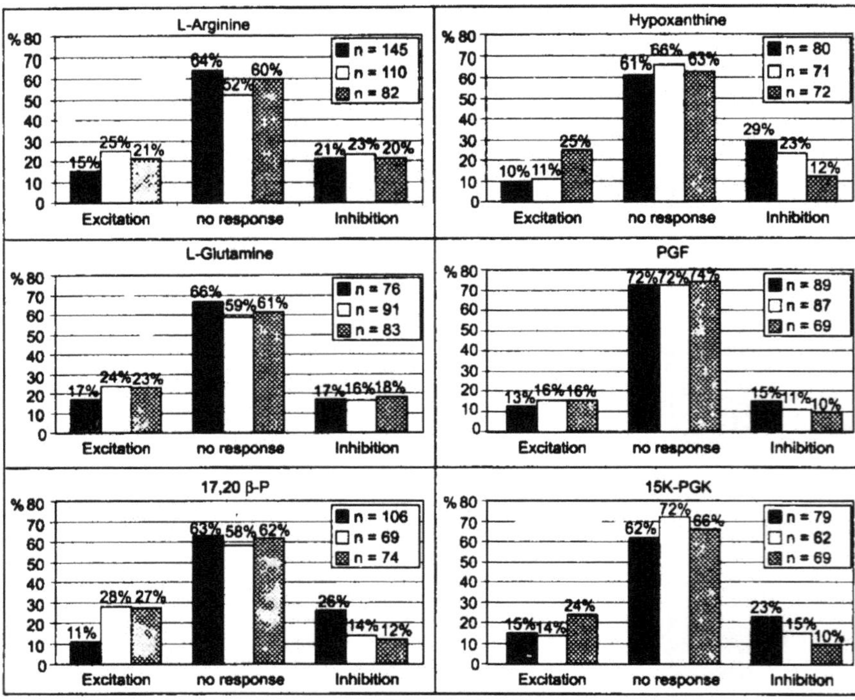

Fig. 4.10 Distribution of excitation, no response and inhibition in intact and axotomized goldfish. Black: intact controls, white: 3 weeks after, crosshatched: 4 weeks after bilateral axotomy. **N**, number of sensory neurons recorded from 13 control fish, 12 three weeks after and 9 four weeks after bilateral olfactory nerve axotomy.

Stimulus discrimination in single sensory neurons in fact is poorly understood. Recordings from 'generalists' only show a minor specificity of receptor molecules, and inhibitions and excitations elicited by various stimuli make it very likely that single sensory neurons possess more than one receptor molecule. Recent recordings from single epithelial sensory neurons during application of familiar and non-familiar, structurally extremely similar and dissimilar stimuli showing an incredibly excellent discrimination of all the mentioned stimuli, speak against negation of hypotheses based on EOG findings that application of naturally important stimuli results in greater effects on a higher number of sensory neurons than application of non-familiar stimuli.

The percent values for excitations, inhibitions, and no responses were again rather similar for the various classes of stimuli. Various stimuli can

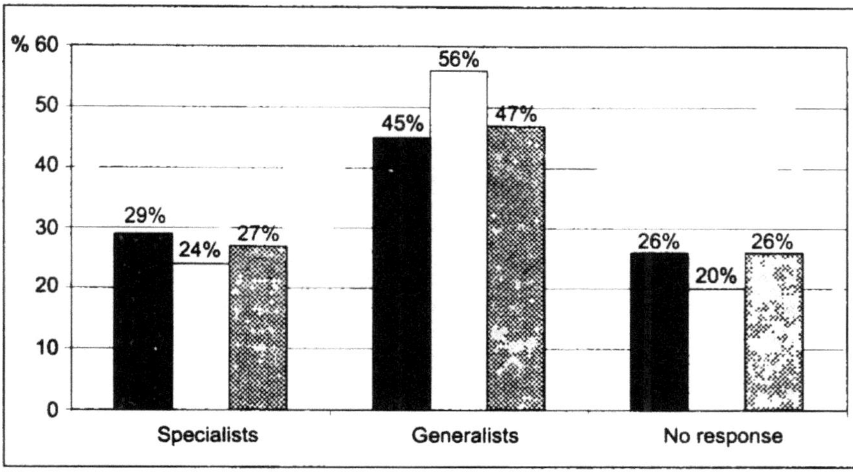

Fig. 4.11 Summary of response characteristics during application of 2 amino acids, the preovulatory and ovulatory pheromones, and alarm pheromone. Per cent values for 57 sensory neurons of intact (black), 59 sensory neurons of 3 weeks after axotomy (white), and 53 neurons 4 weeks after axotomy (crosshatched). 'Specialists' only respond during application of one stimulus (value for each single stimulus lies in the magnitude of 5–10%); 'generalists' respond during application of different classes of stimuli (see Figs. 4.8 and 4.9).

elicit excitation and inhibition in one and the same sensory neuron. Overall, excitations and inhibitions are again almost similarly distributed. Olfactory sensory neurons can respond to application of comparatively great numbers of pheromonal and non-pheromonal stimuli. In contrast to the mentioned more 'generalizing' sensory neurons, epithelial sensory neurons were found in which the effects of applied stimuli were more limited. Response characteristics recorded from single epithelial sensory neurons were extremely similar during application of familiar, biologically important, and application of non-familiar, biologically unimportant stimuli. Stimuli of very similar structures were discriminated as were stimuli of extremely dissimilar structures. Compared to epithelial sensory neurons, olfactory bulb relay neurons responded to greater numbers of olfactory stimuli. During application of L-arginine and five structural analogues (Fig. 4.12), approximately 32% of the epithelial receptor neurons and only 9% of the bulbar relay neurons showed no response to any stimulus. Roughly 30% of the sensory neurons responded during application of 1 or 2 of the six stimuli, while 20–25% of bulbar relay neurons responded during application of 2, 3 or 4 stimuli.

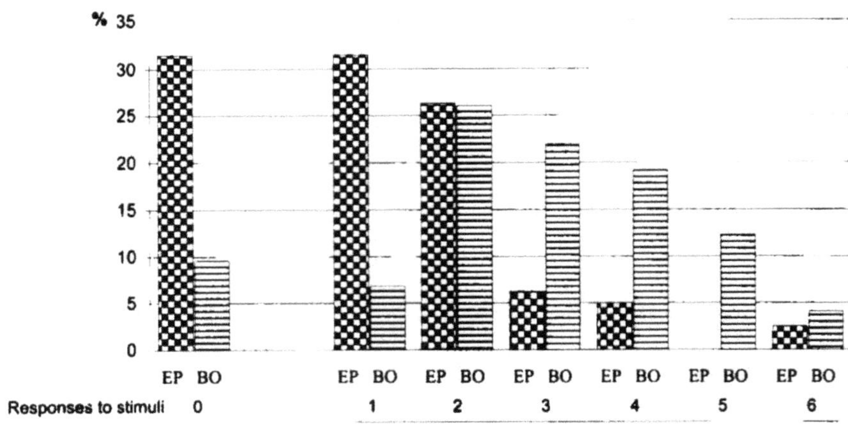

Fig. 4.12 Responses of 53 epithelial receptor neurons (EP) and 65 olfactory bulb (BO) mitral cells during application of L-arginine and five structural analogues. 32% of receptor neurons respond to none or 1 of the 6 stimuli; 27% of receptor neurons respond during application of 2 of the 6 stimuli; modulations of the interstimulus activity during application of 3-6 stimuli were found in 5% or lower percentages. Due to glomerular convergence of receptor axons only 9% of mitral cells did not respond to any of the 6 stimuli. In contrast to epithelial receptor neurons, in 20-25% of olfactory bulb mitral cells effects were recorded during application of 2-4 of the 6 stimuli and above 10% of relay neurons, still changed activity during application of 5 of the 6 stimuli. Slightly modified from Zippel and Lüthje (2003).

3. DISCUSSION

Based on the number of responses, no significant greater effectiveness of any of the amino acids could be found during application of 10^{-7}M concentrations. Low concentrations, not resulting in a recordable EOG, could however be discriminated during training (von Rekowski et al., 1995). With application of 10^{-6} molar concentrations low and rather similar amplitudes were recorded in the EOG but great variations in EOG amplitudes were only recorded when extremely high (10^{-3}–10^{-4}M) concentrations were applied (Zippel et al., 1997a, b). Since these dissimilar effects were also apparent after olfactory nerve axotomy during the time of maximal degeneration of epithelial sensory neurons (Zippel et al., 1997a), they must be classified as artificial.

In dose-response experiments with amino acids, stronger effects during application of lower concentrations or opposing effects recorded from olfactory bulb mitral cells (Zippel et al., 2000) have so far not been recorded from olfactory sensory neurons, and probably are the result of lateral inhibitory processes via bulbar granule cells or telencephalic nuclei.

The goldfish olfactory epithelium is excellently structured and distribution of the various cell types known (Hansen et al., 1999). As noted above, in the present experiments rather low concentrations of amino acids were applied which could be readily discriminated in discrimination training experiments (Zippel et al., 1993b; von Rekowski et al., 1995). The present experiments showed that olfactory receptor neurons can respond to application of a comparatively great number of stimuli. From the data presented here it is evident that single sensory neurons in the olfactory epithelium respond to stimuli in low concentrations, and can discriminate structurally incredibly similar familiar (L- and D-amino acids) and non-familiar stimuli (e.g. α- and β-ionone). We therefore cannot support the hypothesis mentioned in a number of EOG papers that application of 'important' familiar natural stimuli results in significantly greater effects compared to application of less important non-familiar stimuli. Recent results suggesting specificity of ciliated receptor neurons for amino acids and other non-pheromonal stimuli, and pheromone stimuli for microvillous receptors in EOG recordings and behavioural data (Zippel et al., 1997b) likewise cannot be supported: in contrast to the aforementioned EOG recordings response characteristics are surprisingly similar in intact control epithelia and in epithelia during maximal degeneration. Similar distributions of responses were found during amino acid and pheromone application when only a minimum number of microvillous cells were present 3 weeks after axotomy.

The heterogeneity of responses from olfactory bulb mitral cells during application of pheromones, amino acids, and non-familiar stimuli is the mirror image of effects recorded from epithelial receptor neurons; the number of sensory cells not responding during application of a single stimulus however, is much greater. Since in the present experiments and in recordings from olfactory bulb relay neurons (Zippel et al., 1999, 2000, this volume) similar stimulus concentrations were applied, the greater number of responding relay neurons is a good example of the importance of glomerular convergence. Obvious contrasting interactions between mitral cells and ruffed cells resulting in a significant 'sharpening' of centrally transmitted information however, are intrabulbar processes depending on increasing or decreasing lateral inhibitory processes via granule cells (Zippel et al., 1995, 1999, 2000). Significant differences in the time course of discrimination training and long-term memory found during presentation of familiar and non-familiar stimuli in behavioural experiments (Zippel et al., 1993b) demonstrate that behavioural differences are obviously the result of interpretative mechanisms in more central (telencephalic) nuclei.

In higher vertebrates significant effects of pheromones are behaviourally evident but the respective chemical structures presently not known. The olfactory system of higher vertebrates is divided into two subsystems: the 'general' olfactory system covering all olfactory stimuli not handled by the accessory (vomeronasal) olfactory system, mainly responsible for pheromonal stimuli. While in higher vertebrates different structures (olfactory and vomeronasal organ) are obviously responsible for pheromones and general olfactory stimuli, in fish there is no significant distribution for the respective stimuli (McLean et al., 1992). The olfactory epithelium of fish possesses both types of olfactory receptor neurons, microvillous and ciliated cells, which in higher vertebrates are more discretely separated in the main olfactory system (ciliated receptor cells) and the accessory olfactory system (microvillous cells). In higher vertebrates central pathways from the accessory olfactory system project via the accessory olfactory bulb to different central nuclei, while in fish the accessory olfactory system is absent; microvillous and ciliated receptor neurons are located in the main olfactory epithelium.

In 'classical' *in vivo* publications concerning frog olfactory receptor neurons (Duchamp et al., 1974; Getchell 1986) no significant inhibitory responses during stimulus application were recorded due to the extremely low spontaneous activity of the neurons. Contrarily; both excitatory and inhibitory effects were often recorded in the present investigations and in recordings from single epithelial neurons in catfish (Kang and Caprio, 1995). Whether these contrasting effects are mainly (or only) present in fish and/or species specific, or the result of methodological variations warrants further investigation.

In goldfish the interstimulus activity during laminary water current is surprisingly high (mean value: $2.3 \cdot s^{-1}$), and nearly the same as in olfactory bulb mitral cells (Zippel et al., 1999, 2000). Three to four min after termination of waterflow a significantly lower (mean value: $0.82 \cdot s^{-1}$) activity was recorded from single epithelial receptor neurons (Möller-de Beer and Zippel, 2001; Zippel and Lüthje, 2003). In recordings from catfish the rather high interstimulus activity was probably the result of relatively high flow rates through the epithelium (Kang and Caprio, 1995). Stimulus duration and concentrations were unfortunately based on EOG findings; inhibitory effects however were recorded during stimulus application as well. In *in vitro* recordings the activity of receptor neurons was either not present or experimentally induced by depolarization of the receptor neurons (Reisert and Mathews, 1999; Vogler and Schild, 1999). Comparison of effects with naturally important stimuli in the aforementioned publications was not possible since such stimuli were not known for the species investigated.

Sato and Suzuki (2001) recently published whole-cell response characteristics of ciliated and microvillous olfactory receptor neurons to amino acids, pheromone candidates and urine in rainbow-trout. Their results suggest that ciliated olfactory receptor neurons are 'generalists' that respond to a wide variety of odorants, including pheromones, whereas microvillous olfactory receptor neurons in rainbow trout are 'specialists' specific to amino acids. Whether in goldfish 'generalists' responding to a great variety of structurally very diverse stimuli correspond to ciliated sensory neurons, and 'specialists' responding only to preovulatory stimuli, only to ovulatory, only to the alarm pheromone, or only to amino acids are microvillous sensors as well warrants further study. From recent recordings in the regeneration olfactory epithelium in goldfish however, the distribution of specialists and generalists is similar in intact and in axotomized sensory epithelia 3 and 4 weeks after intercranial olfactory nerve axotomy. During these weeks after axotomy however, the number of microvillous sensory neurons is much smaller than the number of ciliated sensory neurons (Hansen et al., 1999).

Acknowledgements

Supported by DFG Zi 112/7-3.

References

Argentini, M. 1976. Isolierung des Schreckstoffes aus der Haut der Elritze *Phoxinus phoxinus* (L). PhD thesis, University of Zürich.

Caprio, J. 1995. Olfactory and taste recordings in fish. In: *Experimental Cell Biology of Taste and Olfaction (Current Techniques and Protocols)*, A.I. Spielman and J.G.Brand (Eds). CRC Press, Boca Raton, pp. 251–261.

Chaput, M.A., N. Buonviso and F. Berthommier. 1992. Temporal patterns in spontaneous and odour-evoked mitral cell discharges recorded in anaesthetized freely breathing animals. *Eur. J. Neurosci.* 4: 813–822.

Duchamp, A., M.F. Revial, A. Holley and P. MacLeod. 1974. Odor discrimination by frog olfactory receptors. *Chem. Senses Flavor* 1: 213–233.

Fischer, T. and H.P. Zippel. 1989. The effects of cryogenic blockade of the centrifugal, bulbopetal pathways on the dynamic and static response characteristics of goldfish olfactory bulb mitral cells. *Exper. Brain Res.* 75: 390–400.

Gesteland, R.C. 1975. Techniques for investigating single unit activity in the vertebrate olfactory epithelium. In: *Methods in Olfactory Research*, D.G. Moulton, A. Tur, J.W. Johnston, Jr. (Eds). Academic Press, London, pp. 269–322.

Getchell, T.V. 1986. Functional properties of vertebrate olfactory receptor neurons. *Physiol. Rev.* 66: 772–817.

Hansen, A., H.P. Zippel, P.W. Sorensen and J. Caprio. 1999. Ultrastructure of the olfactory epithelium in intact, axotomized, and bulbectomized goldfish, *Carassius auratus. Micros. Res. Tech.* 45: 325–338.

Kang, J. and J. Caprio. 1995. *In vivo* responses of single olfactory receptor neurons in the channel catfish, *Ictalurus punctatus. J. Neurophysiol.* 73: 172–177.

Kokemüller, A. and H.P. Zippel. 2001. Stimulus discrimination and cross-adaptation experiments in goldfish, *Carassis auratus* olfactory bulb relay neurons during application of very similar non-familiar stimuli. In: *Proc. 4th Meeting German Neurosci. Soc. and 28th Neurobiology Conf.* N. Elsner and G.W. Kreutzberg (Eds). George Thieme, Stuttgart, p. 464.

Lipschitz, D.L. and W.C. Michel. 1999. Physiological evidence for the discrimination of L-arginine from structural analogues by the zebrafish olfactory system. *J. Neurophysiol.* 82: 3160–3167.

McLean, J.H. and M.T. Shipley. 1992. Neuroanatomical Substrates of Olfaction. In: *Science of Olfaction*, M.J. Serby and K.L. Chob (Eds). Springer–Verlag, New York, pp. 126–171.

Möller-de Beer A.M., and H.P. Zippel. 2001. In-vivo-recordings from single olfactory sensory neurons in goldfish *Carassius auratus* demonstrate their bimodality. In: *Proc. 4th Meeting German Neurosci. Soc. and 28th Neurobiology Conf.* N. Elsner and G.W. Kreutzberg (Eds). Georg Thieme, Stuttgart, p. 464.

Pfeiffer, W. 1982. Chemical Signals in Communication. In: *Chemoreception in Fishes*, T.J. Hara (Ed.). Developments in Aquaculture and Fisheries Science. Elsevier, New York, pp. 307–326.

Reisert, J. and H.R. Matthews. 1999. Adaptation of the odour-induced response in frog olfactory receptor cells. *J. Physiol.* (Lond.) 519: 801–813.

Sato, K. and N. Suzuki. 2001. Whole-cell response characteristics of ciliated and microvillous olfactory receptor neurons to amino acids, pheromone candidates and urine in rainbow trout. *Chem. Senses* 26: 1145–1156.

Schild, D. 1985. A computer-controlled device for the application of odours to aquatic animals. *J. Electrophysiol.* 12: 71–79.

Schild, D. and H.P. Zippel. 1986. The influence of repeated natural stimulation upon discharge patterns of mitral cells of the goldfish olfactory bulb. *J. Comp. Physiol.* A 158: 536–571.

Schild, D. and D. Restrepo. 1998 Transduction mechanisms in vertebrate olfactory receptor cells. *Physiol. Rev.* 78: 429–466.

Sorensen, P.W. 1992. Hormones, pheromones and chemoreception. In: *Fish Chemoreception*, T.J. Hara (Ed.). Chapman & Hall, London, New York, pp. 199–221.

Sorensen, P.W. and N.E. Stacey. 1989. Differing behavioral and endocrinological effects of two female sex pheromones on male goldfish. *Horm. Behav.* 123: 317–332.

Sorensen, P.W. and A.P. Scott. 1994. The evolution of hormonal sex pheromones in teleost fish: poor correlation between the pattern of steroid release by goldfish and olfactory sensitivity suggest that these cues evolved as a result of chemical spying rather than signal specializing. *Acta Physiol. Scand.* 152: 191–205.

Sorensen, P.W., T.J. Hara, N.E. Stacey and F.W. Goetz. 1988. F prostaglandins function as potent olfactory stimulants that comprise the postovulatory female sex pheromone in goldfish. *Biol. Reprod.* 39: 1039–1050.

Sorensen, P.W., T.J. Hara, N.E. Stacey and J.G. Dulka. 1990. Extreme olfactory specificity of male goldfish to the preovulatory steroidal pheromone, 17α,20β-dihydroxy-4-pregnen-3-one. *J. Comp. Physiol.* A 166: 373–383.

Stacey, N.E. and A.L. Kyle. 1983. Effects of olfactory tract lesions on sexual and feeding behavior in the goldfish. *Physiol. Behav.* 30: 621–628.

Stacey, N.E. and P.W. Sorensen. 1986. 17α,20β-dihydroxy-4-pregnen-3-one: a steroidal primer pheromone that increases milt volume in the goldfish *Carassuis auratus. Can. J. Zool.* 64: 2412–2417.

Stacey, N.E., J.R. Cardwell, N.R. Liley, A.P. Scott and P.W. Sorensen. 1994. Hormones as sex pheromones in fish. In: *Perspectives in Comparative Endocrinology*, K.G. Davey, R.E. Peter and S. Tobe (Eds). National Research Council of Canada, Ottawa, pp. 64–72.

Vogler, Ch. and D. Schild. 1999. Inhibitory and excitatory responses of olfactory receptor neurons of *Xenopus laevis* tadpoles to stimulation with amino acids. *J. Exp. Biol.* 202: 997–1003.

von Rekowski, C., J. Schmilewski and H.P. Zippel. 1995. Behavioral experiments in goldfish confirm amino acids as olfactory meaningful natural stimuli. In: *Chemical Signals in Vertebrates* VII, R. Apfelbach, D. Müller-Schwarze, K. Reutter and E. Weiler (Eds). Pergamon Press, Oxford, pp. 491–496.

Zippel, H.P. and L.G.C. Lüthje. 2003. Recent progress in aquatic vertebrate olfaction. In: *Sensory Processing in Aquatic Environments*, S.P. Collin and N.J. Marshall (Eds). Springer–Verlag, New York, pp. 283–300.

Zippel, H.P., T. Lago-Schaaf and J. Caprio. 1993a. Ciliated olfactory receptor neurons in goldfish *Carassius auratus* partially survive nerve axotomy, rapidly regenerate and respond to amino acids. *J. Comp. Physiol.* A 173: 537–547.

Zippel, H.P., R. Voigt, M. Knaust and Y. Luan. 1993b. Spontaneous behavior training and discrimination training in goldfish using chemosensory stimuli. *J. Comp. Physiol.* A 172: 81–90.

Zippel, H.P., J. Rabba, P. Aksari, U. Paschke. 1995a. Amino acid information of olfactory bulb neurons in goldfish. In: *Chemical Signals in Vertebrates* VII, R. Apfelbach, D. Müller-Schwarze, K. Reutter and E. Weiler (Eds). Pergamon Press, Oxford, pp. 141–152.

Zippel, H.P., W. Fiedler, D. Hager, J. Kühling-Thees, K. Maier, C. von Rekowski, R. Schumann, W. Tiedemann, R.Voigt and T. Wächter. 1995b. Responses of goldfish olfactory bulb relay neurons during mucosal application of various chemical, mechanical and temperature stimuli. *Biophysics* 40: 171–192.

Zippel, H.P., A. Hansen and J. Caprio. 1997a. Renewing olfactory receptor neurons in goldfish do not require contact with the olfactory bulb to develop normal chemical responsiveness. *J. Comp. Physiol.* A 181: 425–437.

Zippel, H.P., P.W. Sorensen and A. Hansen. 1997b. High correlation between microvillous olfactory receptor cell abundance and sensitivity to pheromones in olfactory nerve-sectioned goldfish. *J. Comp. Physiol.* A 180: 39–52.

Zippel, H.P., Ch. Reschke and V. Korff. 1999. Simultaneous recordings from two physiologically different types of relay neurons, mitral cells and ruffed cells, in the olfactory bulb of goldfish. *Cell Molec. Biol.* 45: 327–337.

Zippel, H.P., M. Gloger, L.G.C. Lüthje, S. Nasser and S. Wilcke. 2000. Pheromone discrimination ability of olfactory bulb mitral and ruffed cells in goldfish *Carassius auratus*. *Chem. Senses* 25: 339–349.

CHAPTER

5

Olfactory Cross-adaptation: Not a Peripheral but a General Phenomenon

H.P. Zippel, L.G.C. Lüthje, B. Albrecht, C. Conze, N. Hessenius, U. Jakob, A. Kokemüller, K. Rindermann and H.-G. Willms

ABSTRACT

EOG cross-adaptation experiments demonstrated the fact that during permanent application of an odour, sensibility to related applied odours drastically decreased, which obviously depends on binding effects on sensory neuron receptor molecules. We were not able to show numerous full or partial cross-adaptations (CA) during application of stimuli selected from the respective electro-olfactogram (EOG, a summed potential recorded from the olfactory epithelium) papers, i.e., during application of related odours some epithelial sensory neurons and olfactory bulb relay neurons in fact showed cross-adaptation, which means that molecules of the permanently applied stimuli actually block membrane receptor molecules of the respective

Address for Correspondence: H.P. Zippel, Physiologisches Institut der Universität, Humboldtallee 23, 37073 Göttingen, Germany. E-mail: pzippel@gwdg.de epithelial sensory neurons. In a great number of sensory neurons however, such was not the case. During 15-s stimulus applications various effects were recorded compared to effects recorded after application of tap-water in the interstimulus interval: 1) no response during stimulus application, i.e., full cross-adaptation; 2) significantly reduced response, i.e., partial cross-adaptation, 3) response similar to response after water interstimulus phase, i.e., no cross-adaptation; 4) larger effects than those recorded before the cross-adaptation experiment and 5) opposing responses compared to responses recorded during water application in the 180-s interstimulus phases (excitation instead of inhibition and vice versa). Neuronal response characteristics 4 and 5 were never reported for EOG recordings. The cross-adaptation phenomenon therefore obviously depends on sensory neurons in the olfactory epithelium, relay neurons in the olfactory bulb, and hypothetically on processes in central nuclei as well.

Key Words: Cross-adaptation; Olfactory epithelium; Olfactory bulb; Single sensory and relay neurons; Goldfish.

1. INTRODUCTION

Cross-adaptation (CA) experiments have to date mainly been performed in summed recordings from the olfactory epithelium. We investigated the CA phenomenon during recordings from single epithelial sensory neurons and during recordings from single olfactory bulb relay neurons. First, the effectiveness of many biological relevant stimuli such as pheromones and amino acids, and biologically unimportant non-familiar stimuli was investigated after water application in the 180-s interstimulus phases. Thereafter one of the stimuli, or a mixture of two stimuli of the respective experimental series was applied instead of water during the interstimulus phases. The rest of the stimuli was applied again during the 15-s stimulation period and modulation of the effects recorded after water application in the interstimulus phase was determined. If in contrast to the effect recorded after water a modulation of the interstimulus activity during stimulus application was no longer apparent, this was a full cross-adaptation; reduction of the effect was a so-called partial cross-adaptation. Similar effects were recorded when the stimulus applied during the interstimulus phase had no modulatory effect on the epithelial sensory neurons/bulbar relay neurons, and the effect of one and the same stimulus was similar during the water and during the cross-adaptation experiment. In contrast to EOG recordings, stronger and even opposite effects compared to the effect recorded during the interstimulus water application were apparent in a number of CA experiments. CA experiments from bulbar mitral cells were possible for several hours and hence a greater number of stimuli (up to 12) could be applied. In subsequent series stimuli were investigated that became structurally more and more similar (Table 5.1).

For CA experiments on single epithelial sensory neurons a number of stimuli were selected from previously published EOG papers (amino acids: Caprio and Byrd, 1984; pheromones: Sorensen et al., 1990). Extracellular recordings from single sensory neurons, unlike recordings from bulbar relay neurons, are much more difficult for two reasons: the diameter of sensory neurons is extremely small (< 2 μm) and even the extremely low laminary flow of tap-water or stimulus solution (~1 ml·min^{-1}) obviously results in minor turbulences that limit the recording time from single neurons. We therefore made epithelial CA experiments with amino acids rather systematically and with pheromones more by chance (Table 5.1).

Table 5.1 Stimuli selected for 8 series of cross-adaptation experiments. Recordings from single epithelial sensory neurons (EP), from single olfactory bulb relay neurons (BO)

Series 1:	L-amino acids (BO): Nonpolar, aliphatic R groups = L-Ala, L-Val, L-Leu; Aromatic R groups = L-Phe, L-Tyr; Polar, uncharged R groups = L-Gln, L-Ser, L-Met; Negatively charged R groups = L-Lys, L-Arg
Series 2:	L-amino acids (EP): L-Arg, L-Lys and L-Met, L-Eth
Series 3:	Stereoisomeres of amino acids (BO): L- and D-Ala, L- and D-Ser, L-and D-Asn, L- and D-Gln, L- and D-Lys, L- and D-Arg
Series 4:	L-amino acids and structural analogues (BO): L-Val and Isobutyraldehyde, L-Leu and Isovaleraldehyde, L-Phe and Phenylacetaldehyde, L-Met and Methional
Series 5:	L-Arg and structural analogues (BO): L- and D-Arg, L-α-amino-β-guanidinopropionate; Organic acids: L-argininic acid, γ-guanidinobutyrate, guanidoacetate, β-guanidinoproprionate, guanidinosuccinate; Decarboxylation: 1-amino-4-guanido-butane; Esterified carboxyl group: L-arginine methyl ester; Amino and carboxyl deletion: 1-ethylguanidine sulfate; Other compounds: aminoguanidine
Series 6:	Non-familiar stimuli (BO): Structurally rather similar stimuli: α-Ionone, β-Ionone, α-Irone and Damascenone, α-Damascone: Structurally different non-familiar stimuli: Amylacetate, β-Phenylethanole
Series 7:	L-amino acids and Pheromones (BO): Preovulatory: 17,20β-Dihydroxyprogesterone, 17,20α-Dihydroxyprogesterone, 4-Pregnen-20β-ol-3-one, 17,20β,21-Trihydroxyprogesterone, Androstendione; Ovulatory: Prostaglandin $F_{2\alpha}$, and 15K-Prostaglandin $F_{2\alpha}$; Alarm pheromone: Hypoxanthine-3(N)-oxide; Amino acids: L-Arg, L-Gln
Series 8:	L-amino acids and Pheromones (EP): 17,20β-Dihydroxyprogesterone, Prostaglandin $F_{2\alpha}$, 15K-Prostaglanin $F_{2\alpha}$, Hypoxanthine-(N)-oxide, L-Arg, L-Gln

2. MATERIAL AND METHODS

Recordings from single sensory neurons in the epithelium were made with platinum-black electrodes (for details, see Zippel and Lüthje, 2003; Zippel et al., this volume). In the present experiments, rather low concentrations of amino acids were applied which could be readily discriminated in training (Zippel et al., 1993; von Rekowski et al., 1994).

In vivo extracellular recordings were made in goldfish from mitral and ruffed cells in the olfactory bulb (Zippel et al., 1999, 2000) with single tungsten microelectrodes (5-7 MΩ, A-M systems). Kosaka and Hama (1982–1983) published anatomical differences characterizing mitral cells and ruffed cells in three teleost species. In goldfish, physiological responses from the two different types of relay neurons were recorded extracellularly and simultaneously. Ruffed cells can clearly be distinguished from mitral cells morphologically and physiologically (Zippel et al., 1999, 2000). Responses to non-familiar and biologically relevant stimuli such as preovulatory, ovulatory pheromones, a probable alarm pheromone, and amino acids from food sources were investigated to determine whether pheromones and non-pheromone stimuli are recognized by various individual membrane receptors of specialized, or more generalizing sensory neurons. Responses recorded from single epithelial olfactory sensory neurons were compared with recordings obtained from olfactory bulb relay neurons using similar experimental procedures and data evaluation.

Extracellular recordings were made on-line using 'Discovery' (Brainwave software), and evaluations done using Discovery subprograms. For further details, see Fischer and Zippel (1989), Schild (1985), Schild and Zippel (1986).

For the CA experiments a great variety of stimuli were selected: non-familiar and familiar natural stimuli of greater and lesser biological importance (Table 5.1). In subsequent series of experiments structurally very similar and dissimilar stimuli were applied (Table 5.1) to investigate the question whether receptor molecules in sensory neurons and convergent information input to mitral cell dendrites in olfactory bulb glomeruli are responsible for similar responses and the cross-adaptation phenomenon. Stimuli were applied in 10^{-7} molar concentration and the preovulatory pheromone in 10^{-9} molar.

3. RESULTS

3.1 General Remarks

Since stimulus discrimination by single olfactory sensory neurons and by olfactory bulb relay neurons was surprisingly similar for non-familiar and structurally related and dissimilar stimuli (Table 5.1), data are summarized by selected examples. In general, the number of no responses was much greater during recordings from sensory neurons than during recordings from olfactory bulb relay neurons during application of stimuli in similar molarity. This is an excellent example for convergence in olfactory bulb glomeruli. This glomerular convergence is undoubtedly the explanation for the fact that during application of different stimuli mitral cells respond to a greater number of stimuli than epithelial sensory neurons (Zippel et al., this volume). The heterogeneity of responses recorded from olfactory bulb relay neurons during stimulus applications is significantly greater than the heterogeneity of responses recorded from epithelial sensory neurons, and contrasting response patterns recorded from mitral cells and ruffed cells are caused by intrabulbar processes depending on increasing or decreasing lateral inhibitory processes via bulbar granule cells (Zippel et al., this volume, 1999, 2000).

3.2 Interstimulus Activity

For CA experiments, olfactory bulb relay neurons were selected for which epithelial stimulus application resulted in clearly evident responses. Recordings from single relay neurons lasted 1.5 to 4.0 hours. After recording responses of olfactory bulbar neurons during 15-s of stimulus application (with 180-s interstimulus water phase intervals), the same stimuli were tested subsequent to one of the stimuli (i.e., the adapting stimulus) being presented to the olfactory mucosa instead of water during the 180-s interstimulus phases. Application of a stimulus instead of water during the interstimulus intervals resulted in either short lasting (1–2 min) or extensive (< 30 min) alteration of the interstimulus activity recorded during the preceding water phases.

The unexpected result that response characteristics recorded from epithelial sensory neurons and bulbar relay neurons were in general rather similar and independent from stimuli applied (with the exceptions mentioned under general remarks), makes it possible to show selected figures from the various experimental series (Table 5.1).

3.3 Sensory Neurons

Cross-adaptation data recorded during extracellular single sensory neuron experiments are given in greater detail because these findings are the basis for the main criticism of EOG summed recordings (see Discussion below). The main differences between recordings from epithelial sensory neurons (Zippel et al., this volume), and bulbar mitral cells are a slightly higher interstimulus activity and a greater number of bulbar relay neurons responding during stimulus application (Zippel et al., 1999, 2000).

3.4 Pheromone and Amino Acids

CA experiments during application of pheromones and amino acids have thus far not been made systematically. During investigations of single cell responses in the goldfish epithelium CA investigations were done when strong effects were recorded during stimulus application, and when the recordings were sufficiently stable, i.e., the amplitudes of the clustered neuron at a sufficient distance from background noise.

The sensory neuron shown in Figure 5.1 was excited during application of Gln, 15-K-$PGF_{2\alpha}$ and 17,20β, and inhibited during $PGF_{2\alpha}$ application. During the CA experiment with permanent application of the preovulatory pheromone (Fig. 5.1B) interstimulus activity drastically increased during the first run. In runs two and three interstimulus activities were only slightly lower. The activity of the sensory neuron remained far above the value recorded during water application. When in run four water was again applied instead of the preovulatory pheromone, the interstimulus activity rapidly returned to low activity after the CA experiment (Fig. 5.1B). Application of an ovulatory pheromone and the amino acid resulted in lower effectiveness, i.e. partial CA, and application of the metabolite of the ovulatory pheromone in an opposite response. CA experiments with an ovulatory pheromone (Fig. 5.1C) and an amino acid (Fig. 5.1D) also resulted in a slight increase in the activity recorded during the interstimulus phases. Reduction of the activity recorded during application of the rest of the stimuli, i.e., full or partial cross-adaptations are recorded in Figure 5.1C. During permanent amino acids application full cross-adaptations were present during preovulatory and one ovulatory pheromone stimulation, while application of $PGF_{2\alpha}$ had an effect comparable to application after water (Fig. 5.1A). That the sensitivity of the sensory neuron was still unchanged is shown in the fourth run in

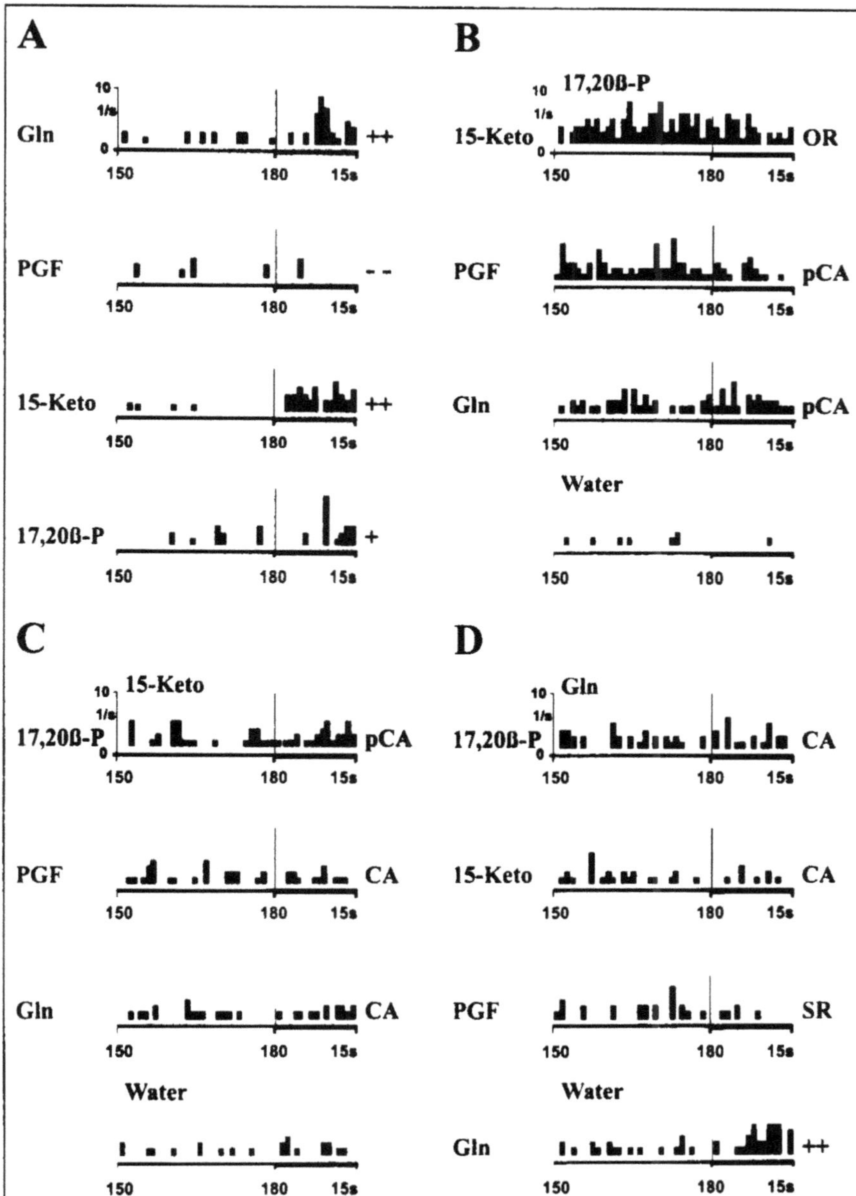

Fig. 5.1 Cross-adaptation experiment during pheromonal and amino acids applications during the 180-s interstimulus phases. **A.** Effects recorded during interstimulus water application. **B.** CA experiment with the preovulatory pheromone. **C.** CA experiment with one ovulatory pheromone. **D.** CA experiment with an amino acid. 17,20β-P, 17,20β-dihydroxy-4-pregnen-3-one; $PGF_{2\alpha}$, prostaglandin $F_{2\alpha}$; 15K-$PGF_{2\alpha}$, 15-ketoprostaglandin $F_{2\alpha}$; Hypo, hypoxanthine-3(N)-oxide; Arg, L-arg; Gln, L-gln. Recordings present the last 30 s of the 180-s interstimulus phase and 15 s of stimulus application. Excitations in comparison to preceding interstimulus activity: + = 25-50%, ++ = 50-75%, +++ > 75%; inhibitions: – = 25-50%, –– = 50-75%, ––– > 75%. CA = Cross-adaptation, pCA = partial CA, SR = similar response, AR = Amplified, stronger response, OR = Opposing response. For further details, see text.

Figure 5.1D where the effect of Gln is comparable to the effect recorded in Figure 5.1A after renewed water application during the interstimulus phase.

In Figure 5.2 the sensory neuron responded with a phasic excitation during Gln and alarm pheromone applications, a delayed strong excitation during an ovulatory (15-K-$PGF_{2\alpha}$) pheromone, and a strong phasic-tonic excitation during preovulatory pheromone application. CA experiments with the ovulatory and preovulatory pheromones only resulted in a slight reduction of the excitatory responses. During permanent application of the preovulatory pheromone as the CA stimulus (Fig. 5.2B) 15-K-$PGF_{2\alpha}$ had a delayed, albeit strong excitatory effect, while application of the rest of the stimuli resulted in slightly opposite inhibitory responses. Permanent $PGF_{2\alpha}$ application resulted in a permanent slight reduction of the interstimulus activity (Fig. 5.2C). During permanent application of the ovulatory 15-K-$PGF_{2\alpha}$ pheromone a permanent and strong excitation was present during the interstimulus phases (Fig. 5.2D). During stimulation with the other ovulatory, preovulatory pheromone and amino acid, long-lasting total inhibitions were recorded. During renewed application of tap-water in run four a strong inhibition of the interstimulus activity was evidenced. At the end of the fourth run a gradual return to the initial interstimulus activity (Fig. 5.2A) was apparent.

The sensory neuron in Figure 5.3 was strongly excited during application of the preovulatory pheromone and inhibited by the rest of the stimuli. During $PGF_{2\alpha}$ application (Fig. 5.3B) as CA stimulus a passing inhibition was recorded during the first run. In this sensory neuron stimulus efficacy remained similar compared to values recorded in Figure 5.3A.

The final example (Fig. 5.4) presents a sensory neuron that only responds during pheromone applications, excitatory to 15-K-$PGF_{2\alpha}$ and inhibitory to the rest of the stimuli. In this neuron CA experiments with the pheromones (Fig. 5.4B-D) resulted in no long-lasting changes of interstimulus activity. Total cross-adaptations were found during permanent application of the ovulatory pheromones (Fig. 5.4B,C), while in the experiment with the preovulatory pheromone (Fig. 5.4D) a cross-adaptation was present during $PGF_{2\alpha}$ application and a strong excitation elicited during 15-K-$PGF_{2\alpha}$ applications.

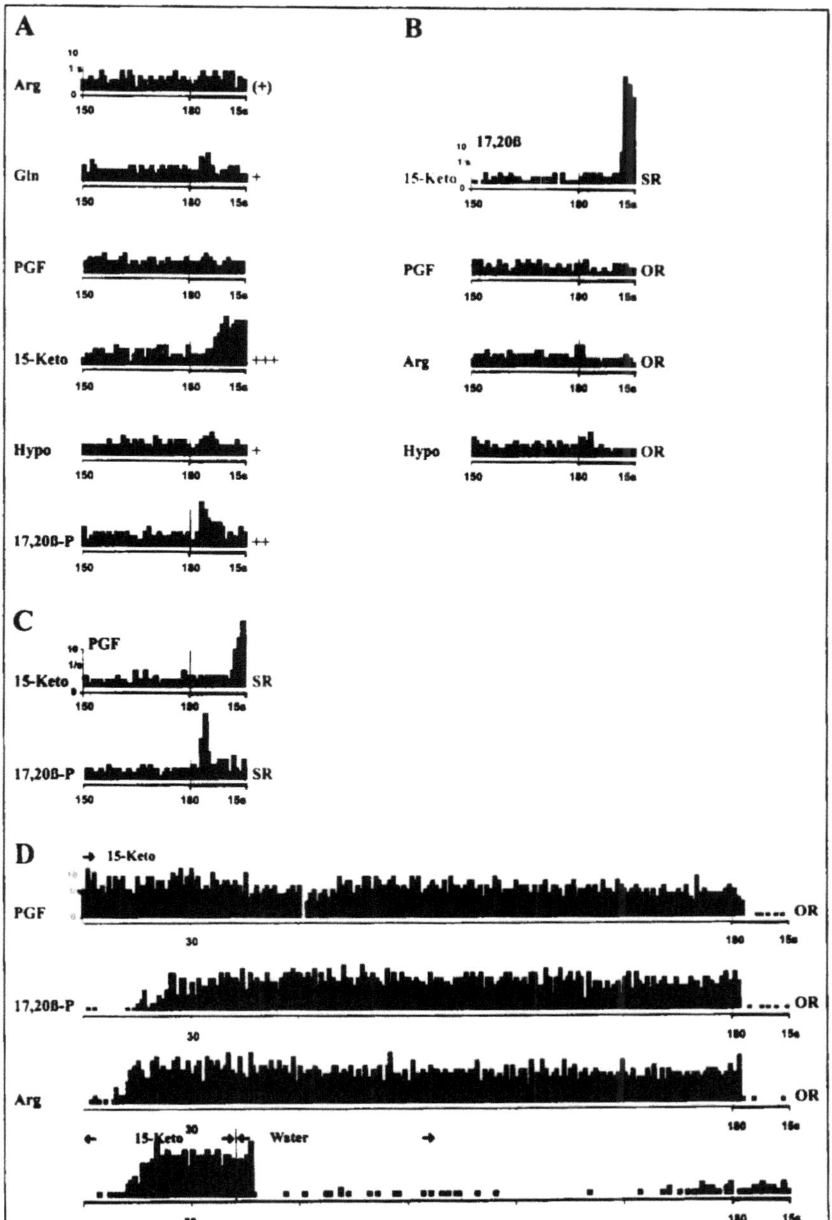

Fig. 5.2 Cross-adaptation experiment during pheromone applications. **A.** Effects after water application. **B.** CA experiment with the preovulatory pheromone. **C.** CA experiment during PGF-application. **D.** CA experiment with the second ovulatory pheromone. For explanations and details, see Figure 5.1 and text.

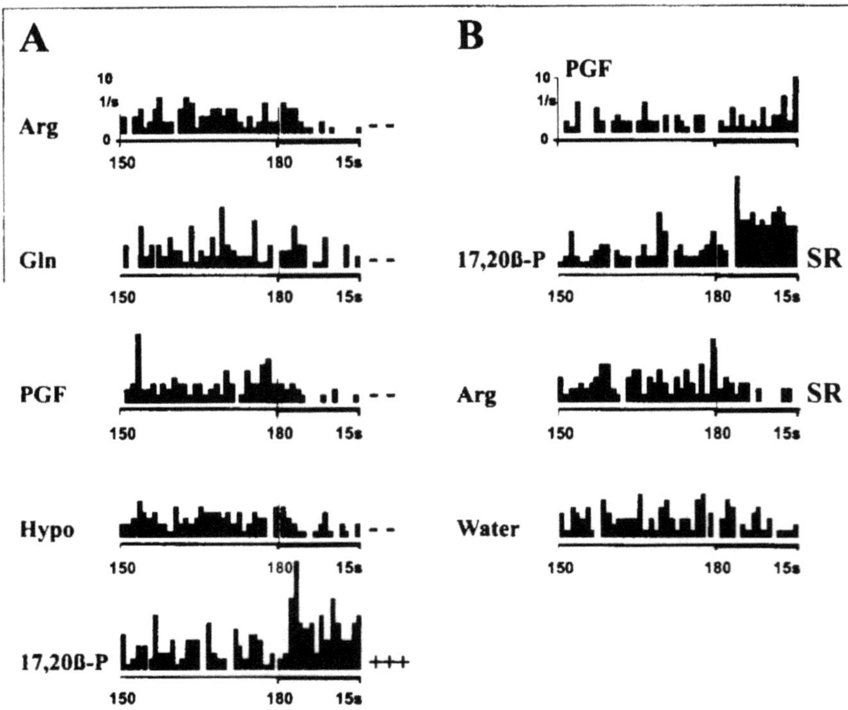

Fig. 5.3 Cross-adaptation experiment with persistent effects. **A.** Effects after water application. **B.** After PGF CA application. For details, see Figure 5.1 and text.

3.5 Amino Acids

For the single sensory neuron investigations four L-amino acids were selected which had been previously used in catfish EOG CA investigations by Caprio and Byrd (1984), two with positively charged R groups (Arg and Lys) and two with polar, uncharged R groups (Met and Eth). Effectiveness and stimulus discrimination during water application in the interstimulus phases have already been described (Zippel et al., this volume). CA experiments during application of different stimuli are somewhat difficult in the olfactory epithelium because up to 50% of epithelial sensory neurons do not respond, or respond to only one or two of the given stimuli. In the 120 CA experiments, in the majority of sensory neurons (67%) adaptation to the interstimulus activity was recorded during the CA experiment, compared to values during the preceding interstimulus water application.

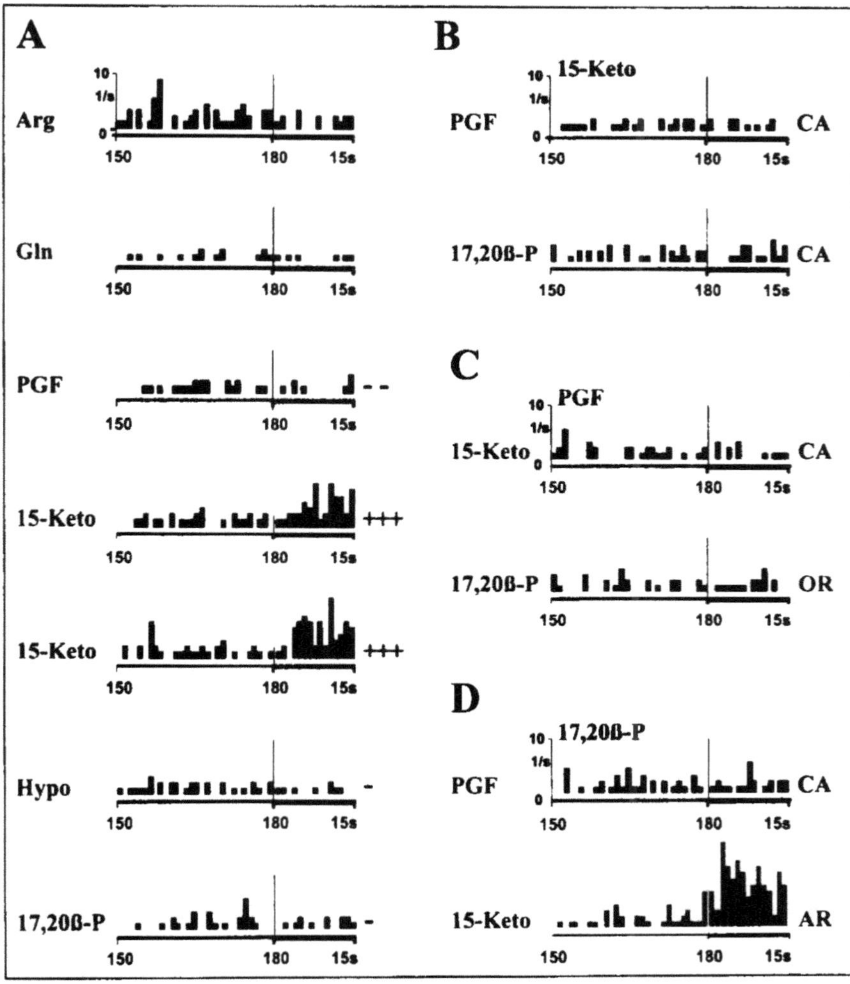

Fig. 5.4 Effects of pheromones during cross-adaptation experiments. **A.** After water application. **B** and **C.** CA during preovulatory and ovulatory pheromone stimulation during permanent application of an ovulatory pheromone. **D.** In the CA experiment with the preovulatory pheromone application of one ovulatory pheromone resulted in a cross-adaptation and the other ovulatory pheromone in a stronger effect. For details see Fig. 5.1 and text.

Response characteristics recorded from single neurons during amino acid CA experiments did not by and large differ essentially from examples (Figs. 5.1 to 5.4) shown during pheromone and amino acids application and therefore are not presented. Figure 5.5 shows in summary the per cent values of reactions recorded during CA experiments. As for the given examples for the pheromones and amino acids (Figs. 5.1 to 5.4) a great variety of response characteristics was recorded from single sensory neurons during the amino acid CA experiments. We thus can only very partly support in the single neuron experiments hypotheses interpreted on the basis of summed EOG recordings (see Discussion below). From the summary (Fig. 5.5) it is evident that the per cent values for cross-adaptation or partial cross-adaptation were not much larger for the amino acids with the positively charged R group (or amino acid with the polar uncharged R group), when the other respective amino acid was used as the cross-adaptation stimulus. Certain amounts of unchanged, similar responses, and even intensified responses and opposite effects were found, similar to the findings for olfactory bulb relay neurons (see Fig. 5.9).

3.6 Olfactory Bulb Relay Neurons

Cross-adaptation experiments were performed with a great number and a great variety of stimuli (Table 5.1). As in the single sensory neuron experiments (see Zippel et al., this volume) a great variety of responses was recorded during application of familiar and non-familiar stimuli. To our great surprise the distribution and number of effects (from cross-adaptation to opposite responses) was more or less similar and independent of the greater or lesser structural chemical similarity of stimuli (see Fig. 5.6 and Discussion). Response characteristics during stimulus application showed similar variety in bulbar relay neurons as in olfactory sensory neurons, with the exception that bulbar relay neurons more often responded to a greater number of stimuli (Zippel et al., 1999, 2000). Furthermore, the different effects recorded during CA experiments can unpredictably vary from relay neuron to relay neuron as in the olfactory epithelium (see above). For a better understanding of the principal variety of basic modulatory effects on the stimulus activity recordings, single mitral cells (Figs. 5.7 and 5.8) were selected from series 4 (Table 5.1) in which the effects recorded during water application in the interstimulus phases were rather similar, and in which the effects recorded during the CA experiments were rather similar as well. In series 4 and series 6 (Table 5.1) mixtures of two structurally rather similar stimuli (Fig. 5.6) were used to investigate the possibility whether unlike application of pure stimuli which resulted in no evident increase in cross-adaptations during application of structurally related stimuli (see Discussion) application of single components contained in the mixture would.

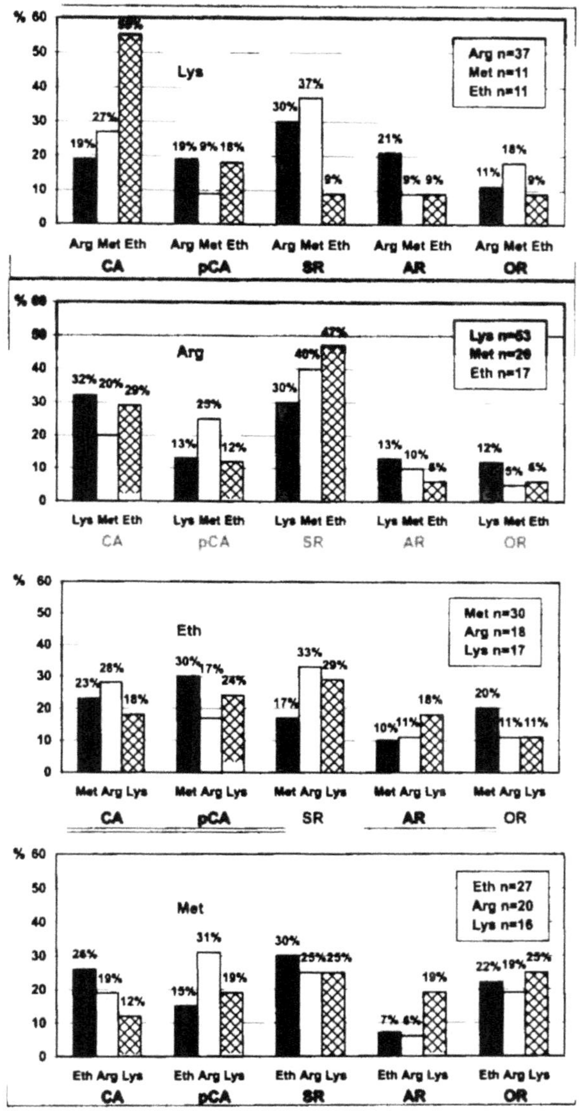

Fig. 5.5 Summary of effects recorded during cross-adaptation experiments with amino acids: CA, full cross-adaptation; pCA, partial cross-adaptation; NR, no response; SR, similar response; AR, amplified response; OR, opposite response. For details and abbreviations, see text.

Figures 5.7 and 5.8 present recordings from four different olfactory bulb mitral cells. Neuron A during the CA experiment showed cross-adaptations and partial cross-adaptations during application of all the stimuli. In neuron B effects remained similar following application and during the CA experiment. From neuron C mainly stronger effects and from neuron D strong opposite effects were recorded.

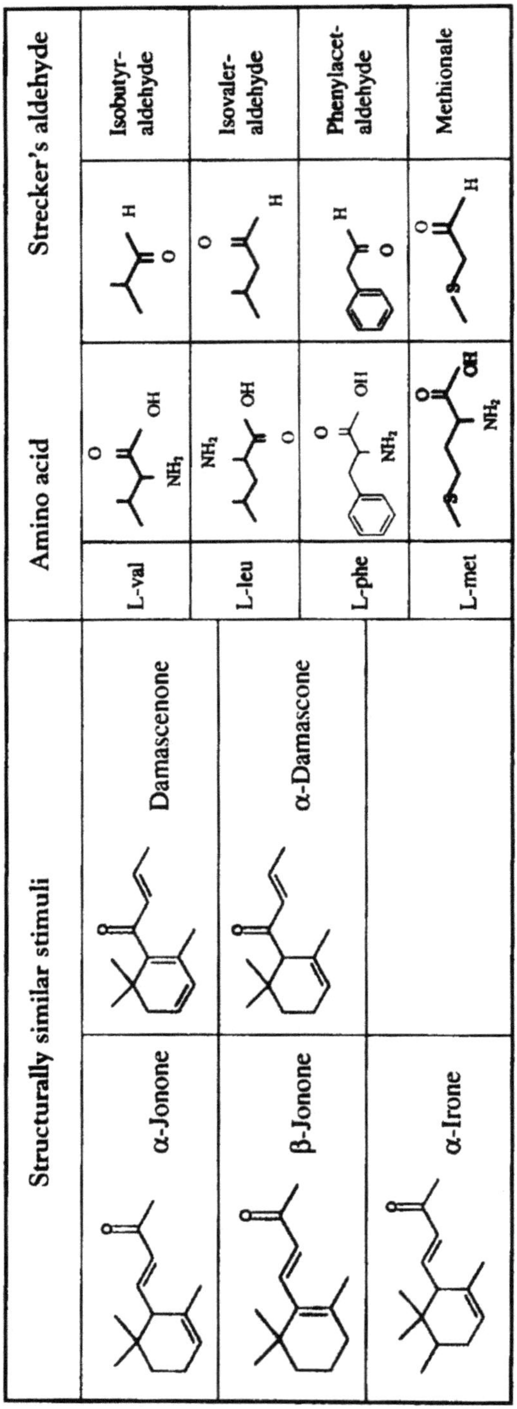

Fig. 5.6 Structures of relatively similar and very similar stimuli applied in Series 4 and 6.

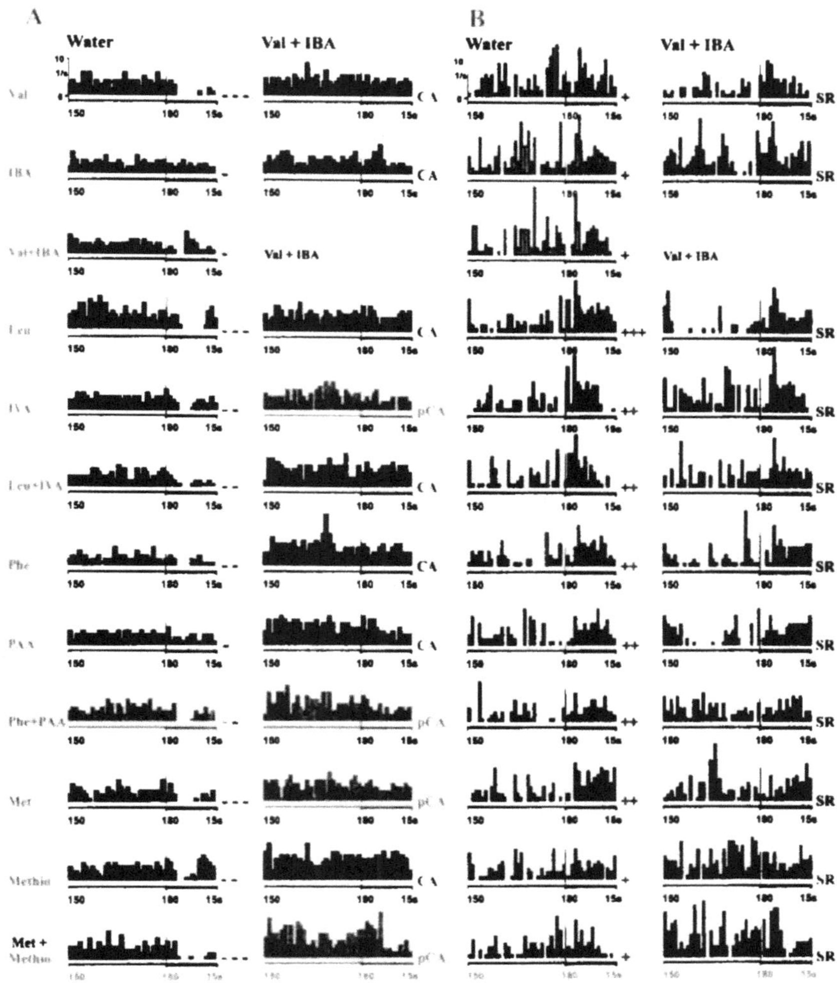

Figs. 5.7 and 5.8 *(Contd.)*

The fact that the distribution of the various effects was more or less random and varied dramatically from neuron to neuron during stimulus application after water or cross- adaptation stimulus application during the interstimulus phase, is evident from the summarization in Figure 5.9 which shows the percentage of distributions recorded in 34 CA experiments with a mixture of Val and isobutyraldehyde, and are more or less similar for the rest of the experiments in the olfactory epithelium (see

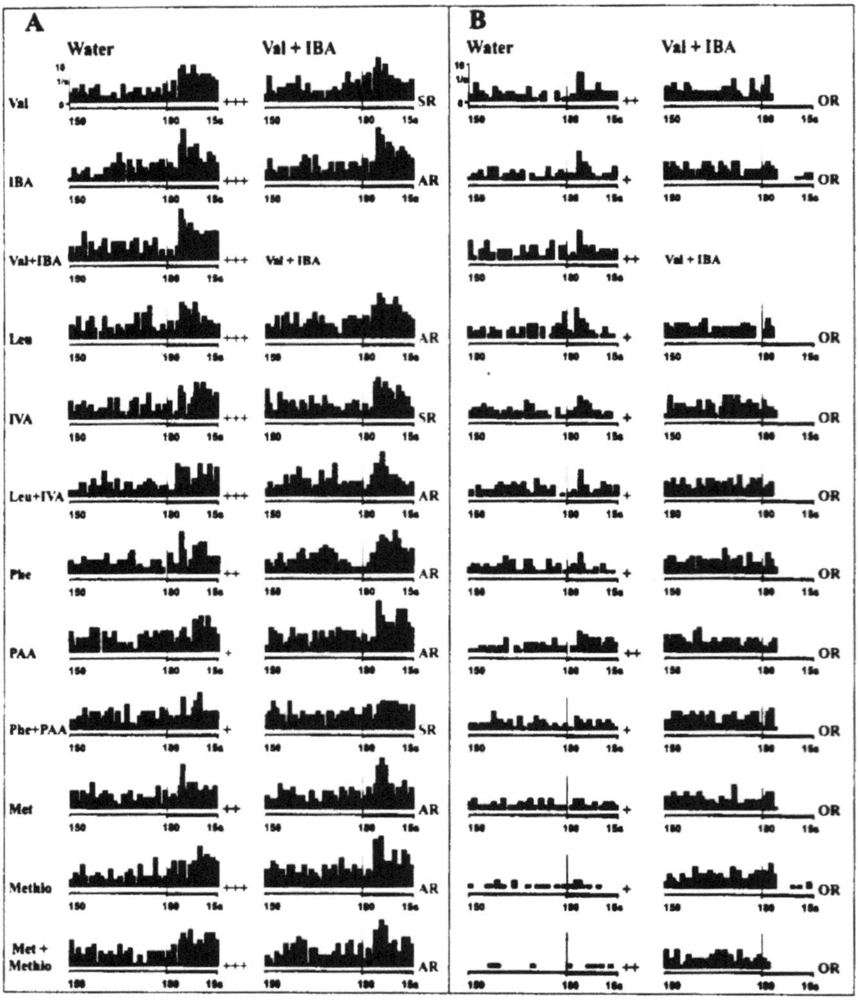

Figs. 5.7 and 5.8 Recordings from olfactory bulb mitral cells during water application and a mixture of L-Val and Isobutyraldehyde during cross-adaptation experiment. Neuron A (Fig. 5.7) shows inhibition during application of all the stimuli during 180-s water application, and total or partial cross-adaptations during the CA experiment; neuron B (Fig. 5.7) shows similar, unchanged effects during both experiments; neuron A (Fig. 5.8) mainly shows an intensification of excitations recorded after water application and neuron B (Fig. 5.8) reverse effects, inhibitions instead of excitations. For abbreviations see Fig. 5.6 and for details Fig. 5.1.

Table 5.1 and above) and the olfactory bulb (see Zippel and Lüthje, 2003 and Table 5.1).

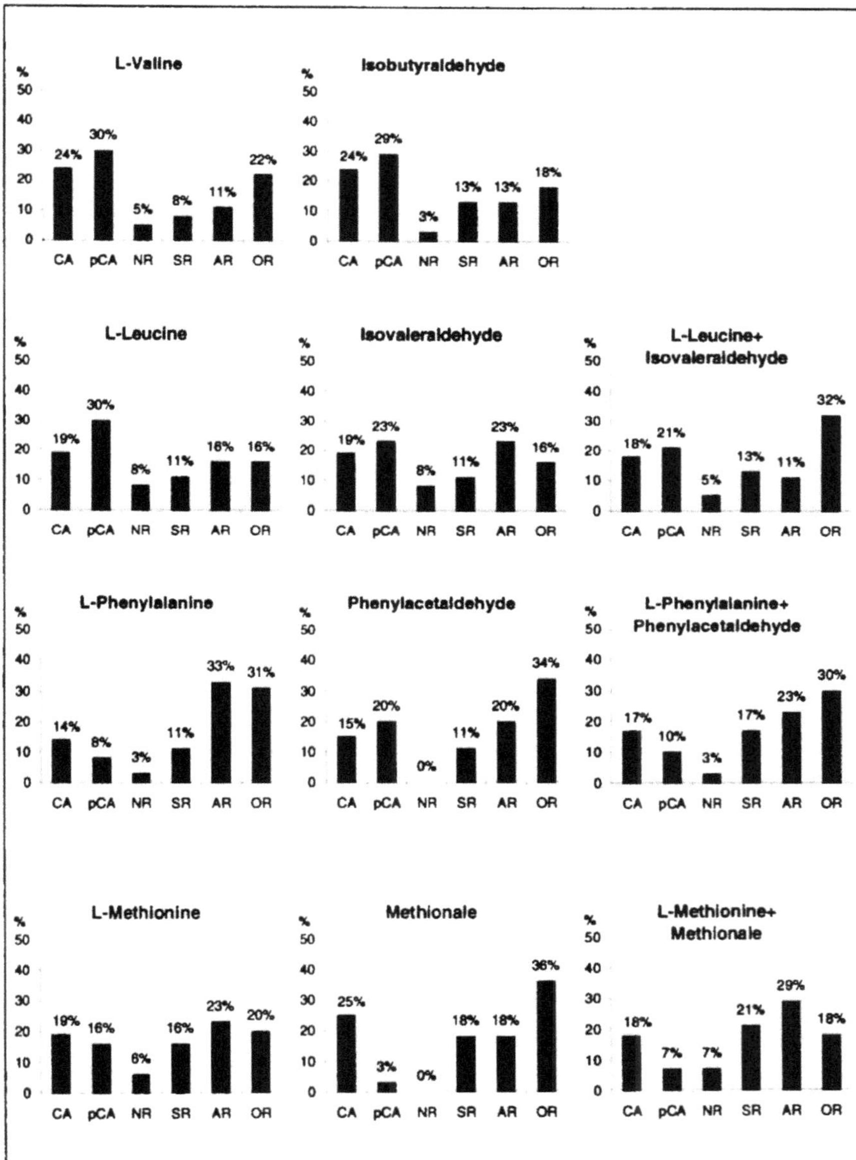

Fig. 5.9 Summary of recordings from 34 olfactory bulb mitral cells during the L-Val plus Isobutyraldehyde CA experiments. For abbreviations, see Figure 5.5 and text.

4. DISCUSSION

During CA experiments, individual mitral cells showed a great variety of responses (response types 1 to 5 mentioned above). EOG findings could be confirmed in part by the fact that prostaglandins $PGF_{2\alpha}$ and $15K\text{-}PGF_{2\alpha}$ mainly bind to different receptor molecules. The previous finding that 21-C steroids bind to one and the same class of receptor molecules (Sorensen et al., 1990) could not be confirmed. Generally speaking, each mitral cell seems to have a rather specific input from various types of epithelial sensory neurons and can during application of a great variety of stimuli, mainly show a full CA or partial CA, similar responses, some stronger reactions, or even opposite effects. Summarizing the effects recorded from sensory neurons and a representative number of mitral cells in each experimental series however, showed a more or less equal distribution of the aforesaid effects independent of the CA stimulation applied.

Physiological (Caprio and Byrd, 1984; Ohno et al., 1984; Caprio et al., 1989; Sveinsson and Hara, 1990; Kang and Caprio, 1991; Michel and Derbridge, 1997) and biochemical (Bruch and Rulli, 1988; Cagan and Zeiger, 1978) investigations demonstrate a number of different types of receptor molecules for amino acids (AA). CA experiments during application of 11 amino acids showed a similarly large range of response characteristics and a similar distribution of the above-mentioned effects, especially during application of pheromones and control stimuli. In goldfish, there are more than the four relatively independent transduction mechanisms (acidic AA, basic AA, short-chain neutral AA, and long-chain neutral AA) described for catfish on the basis of EOG-recordings (Caprio and Byrd, 1984).

In another CA series structurally similar and structurally dissimilar, non-familiar stimuli were investigated. During water application throughout the 180-s interstimulus phases, stimulation with such structurally similar stimuli as α- and β-ionone or α-damascone and damascenone, and even stereoisomeric D- and L-amino acids showed different discriminative effects recorded from at least 50% of bulbar relay neurons. Stimulus discrimination cannot simply be explained by glomerular convergence since epithelial sensory neurons such as olfactory bulb relay neurons discriminate stimuli (see Zippel et al., this volume). During one CA experiment, mixtures of familiar and non-familiar stimuli were applied in series 4 and 6 during the interstimulus phases because application of pure stimuli (see Table 5.1) did not at all result in the expected numbers of cross-adaptations. Effects and modulations of effects in series 4 and 6 however, again were similar to data recorded in the other CA experiments with pheromones and amino acids (Table 5.1). Mean values for the various modulations during CA experiments again were surprisingly similar compared to pheromone and amino acid experiments. During CA application of, e.g. a mixture of α- and β-ionone the number of cross-adaptations and partial cross-adaptations was not significantly greater during application of α-ionone and β-ionone than during application of the rest of the non-familiar stimuli (Kokemüller and Zippel, 2001) such as amylacetate and β-phenylacetate.

From recent *in vivo* CA-experiments in goldfish in which recordings were made on the basis of EOG experiments (Caprio and Byrd, 1984; Sorensen et al., 1990) from single sensory neurons in the epithelium, it is evident that a large variety of responses (response types 1 to 5 mentioned above) can also be recorded from single sensory neurons in the epithelium. The cross-adaptation effects recorded from olfactory bulb relay neurons and single epithelial sensory neurons demonstrate significant differences from EOG recordings and an incredible variety of response characteristics even during application of related and structurally very similar stimuli. The heterogeneity of responses recorded from olfactory bulb relay neurons and the unpredictability of cross-adaptations or other effects shows that the glomerular convergence of sensory axons is incredibly variable. A local projection from specific olfactory neurons to circumscribed narrow bulbar positions in zebrafish (e.g. Baier et al., 1994, Baier and Korsching, 1994; Friedrich and Korsching, 1997) and salmonid fishes (Hara and Zhang, 1998) does not necessarily mean that only a specific population of bulbar mitral cells located in the vicinity of respective glomeruli gets the input from only specific populations of epithelial sensory neurons. Firstly, recordings from different areas of the olfactory bulb show no significant responses to a specific odour (e.g. amino acids or pheromones). In contrast, the discriminative ability of single relay neurons is incredibly sensitive and can drastically vary from neuron to neuron. Secondly, a number of different 'specific' olfactory sensory neurons can converge on one glomerulus. Thirdly, the dendrites of the mitral cells divide and can terminate in several glomeruli (Oka, 1983) located every 100 μm throughout the diameter of the olfactory bulb (1,000 μm) of goldfish. For telencephalic analysis, it suffices that preferential projection of more specific information about food stimuli, the alarm pheromone and other pheromones be projected to telencephalic nuclei via different olfactory subtracts (Hamdani et al., 2000). Epithelial sensory neurons and olfactory bulb relay neurons can respond rather similar to biological relevant and non-familiar stimuli. Given the fact that biological relevant stimuli such as a number of pheromones elicit a specific behaviour, and for the fact that biological relevant stimuli such as amino acids, unlike non-familiar stimuli such as amyl acetate and coumarine, are much more rapidly learned and retained for a long term (Zippel et al., 1993; von Rekowski et al., 1995) obviously functions of central telencephalic nuclei are essentially responsible. From the cross-adaptation experiments presented in this paper it is obvious that for the CA phenomenon CA effects in epithelial sensory neurons olfactory bulb relay neurons, and obviously adaptive phenomena in central nuclei are responsible as well.

References

Baier, H. and S. Korsching. 1994. Olfactory glomeruli in the zebrafish form an invariant pattern and are identifiable across animals. *J. Neurosci.* 14: 219–230.

Baier H., S. Rotter and S. Korsching. 1994. Connectional topography in the zebrafish olfactory system: random positions but regular spacing of sensory neurons projecting to an individual glomerulus. *Proc. Natl. Acad. Sci. USA* 91: 11646–11650.

Bruch, R.C. and R.D. Rulli. 1988. Ligand binding specificity of a neutral L-amino acid olfactory receptor. *Comp. Biochem. Physiol. B: Biochem. Molec. Biol.* 91: 535–540.

Cagan, R.H. and W.N. Zeiger. 1978. Biochemical studies of olfaction: Binding specificity of radioactively labelled stimuli to an isolated olfactory preparation from rainbow trout *Salmo gairdneri. Proc. Natl. Acad. Sci. USA* 75: 4679–4683.

Caprio, J. and R.P. Byrd. Jr. 1984. Electrophysiological evidence for acidic, basic and neutral amino acid olfactory receptor sites in the catfish. *J. Gen. Physiol.* 84: 403–422.

Caprio, J., J. Dudek and J.J. Robinson. 1989. Electro-olfactogram and multiunit olfactory receptor responses to binary and trinary mixtures of amino acids in the channel catfish, *Ictalurus punctatus. J. Gen. Physiol.* 93: 245–262.

Fischer, T. and H.P. Zippel. 1989. The effects of cryogenic blockade of the centrifugal, bulbopetal pathways on the dynamic and static response characteristics of goldfish olfactory bulb mitral cells. *Exp. Brain Res.* 75: 390–400.

Friedrich, R.W. and S.I. Korsching. 1997. Combinatorial and chemotropic odorant coding in the zebrafish olfactory bulb visualized by optical imaging. *Neuron* 18: 737–752.

Hamdani, E.H., O.B. Stabell, G. Alexander and K.B. Døving. 2000. Alarm reaction in the crucian carp is mediated by the medial bundle of the medial olfactory tract. *Chem. Senses* 25: 103–109.

Hara, T.J. and C. Zhang. 1998. Topographic bulbar projections and dual neural pathways of the primary olfactory neurons in salmonid fishes. *Neuroscience* 82: 301–313.

Kang, J. and J. Caprio. 1991. Electro-olfactogram and multiunit olfactory receptor responses to complex mixtures of amino acids in the channel catfish, *Ictalurus punctatus. J. Gen. Physiol.* 98: 699–721.

Kosaka, T. and K. Hama. 1982-1983. Synaptic organization in the teleost olfactory bulb.

J. Physiol. (Paris) 78: 707–719.

Michel, W.C. and D.S. Derbridge. 1997. Evidence of distinct amino acids and bile salt receptors in the olfactory system of the zebrafish, *Danio rerio*. *Brain Res*. 764: 179–187.

Ohno, T., K. Yoshii and K. Kurihara. 1994. Multiple receptor types for amino acids in the carp olfactory cells revealed by quantitative cross-adaptation method. *Brain Res*. 310: 13–21.

Oka, Y. 1983. Golgi-electron-microscopic studies of the mitral cell in the goldfish olfactory bulb. *Neuroscience* 8: 723–742.

Schild, D. and H.P. Zippel. 1986. The influence of repeated natural stimulation upon discharge patterns of mitral cells of the goldfish olfactory bulb *J. Comp. Physiol.* A 158: 536–571.

Sorensen P.W., T.J. Hara, N.E. Stacey and F.W. Goetz. 1988. F prostaglandins function as potent olfactory stimulants that comprise the postovulatory female sex pheromone in goldfish. *Biol. Reprod.* 39: 1039–1050.

Sorensen P.W., T.J. Hara, N.E. Stacey and J.G. Dulka. 1990. Extreme olfactory specificity of male goldfish to the preovulatory steroidal pheromone, 17α,20β-dihydroxy-4-pregnen-3-one. *J. Comp. Physiol.* A 166: 373–383.

Sveinsson, T. and T.J. Hara. 1990. Multiple olfactory receptors for amino acids in arctic char *Salvelinus alpinus* evidence by cross-adaptation experiments. *Comp. Biochem. Physiol.* A 97: 289–293.

von Rekowski C., J. Schmilewski and H.P. Zippel. 1995. Behavioral experiments in goldfish confirm amino acids as olfactory meaningful natural stimuli. In: *Chemical Signals in Vertebrates* VII, R. Apfelbach, D. Müller-Schwarze, K. Reutter and E. Weiler (Eds). Pergamon Press, Oxford, pp. 491–496.

Zippel, H.P., R. Voigt, M. Knaust and Y. Luan. 1993. Spontaneous behavior training and discrimination training in goldfish using chemosensory stimuli. *J. Comp. Physiol.* A 172: 81–90.

Zippel, H.P., Ch. Reschke and V. Korff. 1999. Simultaneous recordings from two physiologically different types of relay neurons, mitral cells and ruffed cells, in the olfactory bulb of goldfish. *Cell. Molec. Biol.* 45: 327–337.

Zippel, H.P., M. Gloger, L.G.C. Lüthje, S. Nasser and S. Wilcke. 2000. Pheromone discrimination ability of olfactory bulb mitral and ruffed cells in goldfish *Carassius auratus*. *Chem. Senses* 25: 339–349.

Zippel, H.P. and L.G.C. Lüthje. 2003. Recent progress in aquatic olfaction. In: *Sensory Processing in Aquatic Environments*, S.P. Collin and N.J. Marshall (Eds). Springer-Verlag, New York, pp. 283–300.

CHAPTER

6

Review of the Chemical and Physiological Basis of Alarm Reactions in Cyprinids

Kjell B. Døving[1], El Hassan Hamdani[1], Erik Höglund[1], Alexander Kasumyan[2], Arvo O. Tuvikene[3]

ABSTRACT

The present review concerns the alarm reactions in fishes and includes the history of discovery and the attempts to isolate the alarm pheromone. A short account is given of the experiments that led to present knowledge of the origin of the alarm substance and variation in expression of the alarm reaction. The review stresses the distribution among various species, the thresholds, and the chemical nature of the alarm substances. Particular emphasis is given to new information concerning the elements of the olfactory system that mediate the alarm reaction with discussions of the peripheral sensory organ, projection of

Address for Correspondence: Kjell B. Døving, [1]Department of Biology, Division of General Physiology University of Oslo, Box 1051 Blindern, N-0316 Oslo, Norway. E-mail: kjelld@bio.uio.no

[2]Department of Ichthyology, Biological Faculty, Moscow State University, R-119 899, Moscow, Russia.

[3]Limnological Station, Institute of Zoology and Botany, Estonian Agricultural University, 61101 Rannu, Tartu County, Estonia.

the sensory neurons to the olfactory bulb, and the olfactory tract. The pathways of these neurons in the central nervous system in crucian carp are described. Some of the electrophysiological responses of the olfactory system to stimuli containing the alarm pheromone are demonstrated.

Key Words: Cypriniformes; Fright reaction; History; Alarm pheromone; Chemical nature; Olfactory system, Olfactory bulb; CNS; Pathways; Electrophysiology.

1. DISCOVERY OF THE ALARM REACTION

The alarm reaction in fishes was discovered by Karl von Frisch while performing experiments on the sense of hearing in minnows at Wolfgangsee, Austria (von Frisch, 1938). He marked a fish in order to be able to recognize it by sectioning the *N. sympaticus* close to the tail. This operation caused a darkening of the skin caudal to the sectioning. When this fish was introduced into the school of fish, the fish became frightened and swam away. Von Frisch also released another fish that was accidentally injured and when it approached the school the members rapidly dispersed. Von Frisch became keenly interested in the subject and made a series of experiments on which present-day knowledge is based. He made field experiments with minnows trained to come to a feeding tray close to the beach into which he could introduce substances via a rain gutter taken from an abandoned house nearby. His suspicious behaviour caused neighbours to alert the police! The efficient test solutions induced fright reactions within 30 to 60 s. Minnows assembled at the feeding tray fled a short distance in confusion, then crowded together and retreated. Confidence returned after a variable interval of hours or days.

2. ALARM REACTIONS IN VARIOUS CARP SPECIES

For quantitative work von Frisch observed minnows in aquaria. These aquaria were equipped with a hiding place, a food tube, running water and ten minnows per aquarium. He observed that the fish had to be conditioned to the experimental tank until they remained near the tube in expectation of food when a person approached. This conditioning time averaged about ten days. The number of fish near the feeding area was noted at intervals of 15 s for 5 min periods after each feeding. He introduced the test material 1 min after feeding the fish. The fright reaction in the aquarium was similar to the reaction observed in the field.

The latent period after introduction of an effective test substance varied from 30 s to 1 min or occasionally even up to 5 min. Von Frisch noted that an increased rate of respiratory movements preceded the fright reaction. The increased respiratory movements caused an increased flow of water through the nose.

Von Frisch ranked seven levels of fright reaction depending on the behaviour observed (von Frisch, 1941). Although arbitrary, definitions of the stages of alarm permit useful quantitative evaluation of the alarm reaction.

+++ Most intense reaction with sudden fright and rapid swimming into the hiding place, followed by an immediate emergence and rapid swimming around the tank, avoiding the feeding place for a long time.

++ Intense reaction as above but fish do not leave the hiding place. Sometimes, after several minutes, some fish return to the feeding place and timidly snatch bits of food before quickly retreating.

\+ Clearly frightened; the school retreats towards or more often into the hiding place but calms down within 5–10 min and then approaches the feeding area more or less confidently. Sometimes the fish remain longer in the hiding place but without showing extremely timid movements while there.

(+) Only slightly frightened with somewhat erratic and excited swimming; often dense crowding together and retreating to the bottom. This reaction soon calms down if nothing further happens, with occasional retreat towards the hiding place.

± Intimidated; less confident and more cautious in the area of the feeding tube; less frequent crowding together and retreating to the bottom; no retreat towards the hiding place.

$\underline{\pm}$ Slightly uneasy with some crowding together in the area of the feeding tube, but the reaction soon disappears, and the fish do not leave the feeding area.

– No reaction.

The above description drawn up for European minnow *Phoxinus phoxinus* applies equally well to many other species of schooling fish, but

some behave in a very different manner. The tench *Tinca tinca* and the crucian carp *Carassius carassius* swim excitedly with their heads against the bottom and their bodies at an angle of about 60° to the substrate. This behaviour looks inappropriate in an aquarium with a glass bottom, but is superbly apt for hiding in a natural habitat. When the fish swims at an angle against the bottom it disturbs the mud and debris and becomes hidden in the turbid water in its natural habitat. This is illustrated in the film that can be seen on the site (use QuickTime):
http://www.biologi.uio.no/genfys/groups/KD/alarm-crucian-carp.html.
Some bottom fish, gudgeon *Gobio gobio* and stone loach *Barbatula barbatula*, become motionless and thus may avoid detection by their enemies. This type of reaction is better developed in the adult than in young fish. Some fish, e.g. striped flying barb *Esomus lineatus*, flee to the water surface where they crowd together and jump out of the water. The marbled hatchetfish *Carnegiella strigata*, which normally swim close to the water surface will swim down and form a dense school in the middle of the tank upon exposure to an alarm substance.

3. ORIGIN OF ALARM SUBSTANCE

Von Frisch considered the sense of hearing a mediator of the alarm reaction. However, he showed that unhurt nervous minnows did not release alarm reactions in schools of conspecifics. Moreover, he observed that narcotized or dead fish with undamaged skin did not induce an alarm reaction. In other experiments he demonstrated that pieces of minnows or even filtered extract produced the alarm reaction ruling out the visual system as a mediator of the alarm reaction. These experiments made von Frisch suggest that the alarm reaction was released by chemical communication, and he named the effective chemical *Schreckstoff* (alarm substance, substance d'alarme) and the reaction *Schreckreaktion* (fright reaction, reaction d'effroi). In a series of tests it was shown that the alarm substance is released from injured skin. Skin from different areas of the body was compared and no differences between dorsal skin, containing melanin and yellow pigment, and ventral skin, containing guanine and red pigment, could be found. Superficial injury of the skin showed that the epidermis contains the alarm substance, but did not eliminate the corium as an additional source. The age of the minnow and its habitat had no effect on the fright reaction. Extracts made from stomach, gut, liver, spleen, and muscle did not induce an alarm reaction.

4. SPECIES SPECIFICITY

An interesting aspect of the alarm reaction is the species specificity of the reaction. Experiments demonstrating this feature were started by von Frisch and extended by Schutz (1956). They observed that the alarm reaction of one species can be induced by skin extract from heterospecifics, but is particularly strong when the skin extract is from conspecifics. However, this reaction can be intense even between species, distant both from a geographic and systematic point of view. The work of Schutz has shown that the intensity of the alarm reaction varied from 100 to 2% as a result of exposure to skin extracts from eight carp species (Table 6.1).

An alarm reaction was observed in both fry and adult bitterlings when exposed to skin extracts from bitterling *Rhodeus sericeus amarus* (Cyprinidae) and spined loach *Cobitis taenia* (Cobitidae) (Kasumyan and Ponomarev, 1986). However, in lower concentrations, adult fish exposed to skin extract from conspecifics showed a more intense alarm reaction than when exposed to skin extract from heterospecifics (spined loach) (Fig. 6.1). Interestingly, this effect of dose and species specificity was not observed in fry, in which the reaction intensity was the same for the two types of skin extract. These experiments indicate that the olfactory receptors change during development, resulting in a more sensitive and species-specific system for detecting the alarm substance in adult fish compared to fish at the fry stage.

The species specificity of an alarm reaction tells us an important message, namely that the substances must be different or comprise several substances. It also means that the olfactory receptors detecting the alarm substances must be different in these species. Most probably the alarm receptors in fry and adults are also different.

Club cells. By comparing histological sections of skin from fish species which produce an alarm substance and species that do not, Pfeiffer could draw the conclusion that the presence of a certain type of club cell in the epidermis could be associated with the presence of the alarm substance (Pfeiffer, 1960). Teleost epidermis typically shows two types of secretory cells: mucus cells and club cells. The former type of cell opens onto the epidermal surface and secretes mucus that covers the fish surface. According to Pfeiffer (1960), the club cells vary in distribution and appearance among different fish species. In bottom-living fish club cells are connected with the epidermal surface where they secrete mucus. In

Table 6.1 Species specificity of alarm reaction among Cypriniformes. Numbers give the relative response efficiency in per cent of skin extract of eight species of fish on the behaviour of seven species. Experiments on bleak could not be carried out. Data from Schutz (1956).

	Skin extract							
	Cyprinidae							*Balitoridae*
	Carp *Cyprinus carpio*	Dace *Leuciscus leuciscus*	Eurasian chub *Leuciscus cephalus*	Minnow *Phoxinus phoxinus*	Bitterling *Rhodeus amarus*	Chub *Alburnoides bipunctatus*	Bleak *Alburnus alburnus*	Stone loach *Barbatula barbatula*
Carp	100	25	50	20	20	25	20	20
Dace	4	100	50	33	10	10	10	20
Eurasian chub	10	50	100	40	5	25	20	10
Minnow	5	10	10	100	2	50	33	20
Bitterling	10	20	20	10	100	10	10	3
Chub	5	20	50	20	6	100	50	10
Bleak				data not available				
Stone loach	10	20	10	10	25	20	20	100

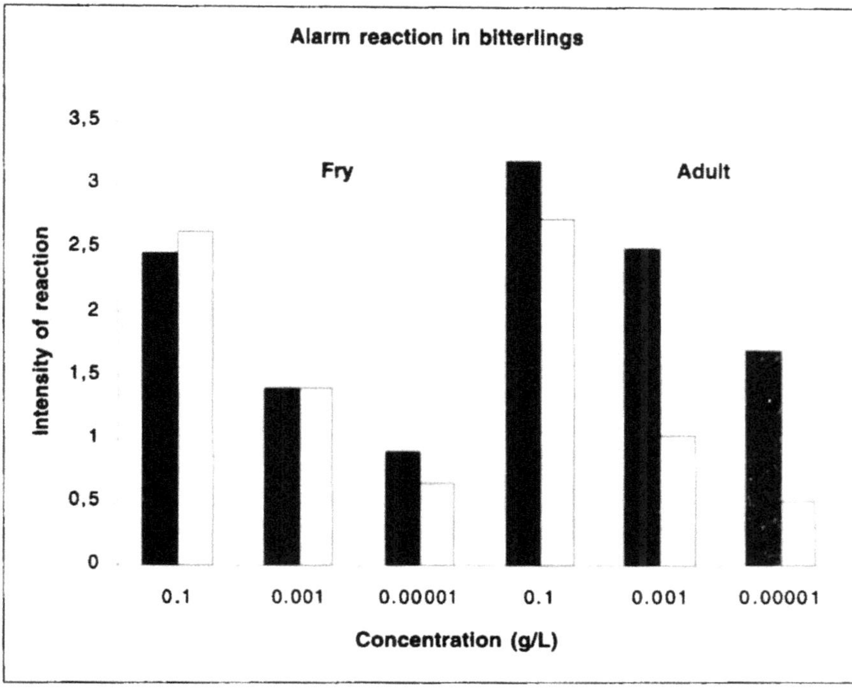

Fig. 6.1 Alarm reaction in bitterling. Bars indicate the intensity of the alarm reaction in fry and adult bitterlings to skin extracts of bitterling (black bars) and spined loach (white bars). Note the difference in efficiency of the lower concentrations of bitterling skin extract in adults compared to extract from the spined loach. Data from Kasumyan and Ponomarev (1986).

minnows and other fish having an alarm substance, the club cells are never connected with the epidermal surface. These club cells are distinct from mucus-secreting cells in general morphology and staining reaction. In mucus cells the nucleus lies peripherally but in the centre of club cells of those fish which have an alarm substance. Injury to the skin releases the contents of these cells and only in this way does the alarm substance reach the body surface. Strong evidence for connecting club cells with alarm substance was found in comparisons of extracts from samples of the skin which contained different numbers of cells. Histological studies showed that the barbel epidermis of the carp and some catfish contains no or very few small club cells. The body epidermis of these species, on the other hand, shows a high density of these cells. This histological difference was confirmed by behavioural experiments in which alarm reaction occurred to body skin but not to barbel skin.

5. THRESHOLDS AND SENSITIVITY TO ALARM SUBSTANCE

The standard procedure for preparation of skin extracts was developed by von Frisch in 1938, and applied in later experiments with the alarm substance. He took 0.2 g of fresh skin, cut it with scissors 150 times, and diluted the dermal contents in 200 ml water. The solution was shaken at 5 min intervals for 30 min and filtered. From this standard extract appropriate dilutions were made. In all tests 100 ml of liquid were poured through the feeding tube during 45 s. In his experiments 83% of 101 minnows responded to a dilution of 1:50, 74% of 117 to a dilution of 1:100, and 41% of 17 tests to a dilution of 1:500. He noted that aquarium hatched minnows responded to a dilution of 1:50,000.

Estimations of threshold concentrations of alarm substance have yielded different values. Results are summarized in Table 6.2. The threshold concentration of alarm substance for the European minnow and several other cyprinids is the water extract from homogenized fish skin at a concentration of 10^{-7} g·L^{-1}. In order to calculate the appropriate concentrations, let us assume that the alarm substance in a fish skin makes up 0.01% of the dry weight of the skin. This means that the threshold concentration of the alarm substance will be equal to 10^{-11} g·L^{-1}. Test solutions were diluted at least 10 or 100 times after injection into the aquarium. Hence the concentration of the solution reaching the olfactory organ of fishes was about 10^{-14}–10^{-15} g·L^{-1}. To express the molar threshold concentration, let us take the alarm pheromone as hypoxanthine-3-N-oxide with molecular weight near 150 Da. In this case the threshold concentration will be around 1.5×10^{-14}–10^{-15} M. If we accept that the alarm substance is of unknown chemical nature with a molecular weight about 1000, the threshold concentration would be around 10^{-15}–10^{-16} M.

6. CHEMICAL NATURE OF ALARM SUBSTANCE

The first to attempt isolation and identification of the chemical nature of the alarm substance was R. Hüttel (1941) more than 60 years ago, just after the original discovery by Karl von Frisch. Using methods of absorption on Fuller's ground and on silica gel and subsequent elution by a water-pyridine mixture and sedimentation by lead-acetate, he obtained a dry colourless powder with a smell repellent to minnow *Phoxinus*

Table 6.2 **Threshold concentrations for the alarm reactions in cyprinid fish species.** Thresholds are given in g skin wet weight per L, if nothing else is indicated. Values are concentrations for the solution injected into aquaria in behavioural tests.

Fish species	*Threshold, $g{\cdot}L^{-1}$*	*Source*
Catastomus catastomus *Catastomus macrocheilus*	10^{-2}	Pfeiffer (1963)
Conesius plumbeus *Hybognathus hankinsoni* *Mylocheilus caurinum* *Ptychocheilus oregonensis* *Rhinichthun cataractae* *Richardsonius balteatus*	10^{-3}	Pfeiffer (1963)
Tribolodon hakonensis	10^{-4}	Aoki and Kurok (1975)
Tribolodon hakonensis hakonensis *Tribolodon hakonensis taszanowskii*	5×10^{-5}	Pfeiffer (1960)
Phoxinus phoxinus	5×10^{-5}	Schutz (1956)
Phoxinus phoxinus *Carassius carassius*	2×10^{-5}	Pfeiffer (1962)
Rhodeus sericeus amarus	10^{-6}	Heintz (1954)
Phoxinus phoxinus	10^{-6} (purified pheromone)	Pfeiffer and Lamour (1976)
Phoxinus phoxinus	10^{-7}	Maljukina et al. (1974, 1977), Pashchenko and Kasumyan (1983)
Rutilus frisii *Aspius aspius* *Chalcalburnus chalcoides*	10^{-8}	Kasumyan (1982)
Ctenopharyngodon idella	10^{-8}–10^{-10}	Kasumyan and Pashchenko (1982), Pashchenko and Kasumyan (1986)
Danio malabaricus	0.4 μM 7-hydroxybiopterin	Win (2000)
Brachidanio rerio	10^{-7} cm^2 of skin	Gandolfi et al. (1968)
Phoxinus phoxinus	3×10^{-6}	Tuvikene and Freiberg (unpubl. Data)
Pimephales promelas	0.4 nM	Brown et al. (2001)
Pimephales promelas	5.8 10^4 L – the active space of 1 cm^2 skin extract	Lawrence and Smith (1989)

phoxinus. Based on nitrogen contents (25–30%) and other properties, Hüttel assumed that purine- and pterin-like substances were the main components in the powder. Using nearly the same procedure as in 1941, Hüttel and Sprengling (1943) concentrated alarm substance from 26.6 kg of skin of roach *Rutilus rutilus*, rudd *Scardinius erythrophthalmus*, and silver bream *Blicca bjoerkna*. They found by UV-spectra that the concentrate resembled a pterin and contained some other peculiarities. For further purification of the alarm substance they used methods for isolation of the pterin-like substance. From this amount of fish skin Hüttel and Sprengling isolated 1.08 g of a colourless crystal. The authors named the substance ichthyopterin with the empirical formula $C_7H_8O_3N_4$ (mol weight 196). They found that ichthyopterin resembled isoxanthopterin by analyzing the UV-absorbtion spectra and violet-blue fluorescence (450–475 nm) at an excitation wavelength of 265–390 nm.

The sample of ichthyopterin obtained by Hüttel and Sprengling was used for chromatographic analysis by Korte and Tschesche (1951) who found ichthyopterin to be a mixture of two substances with different properties. These authors assumed that one of these two substances could be 6,9-dioxi-2-amino-8-acetyl-pterin. Later this suggestion was questioned by Ziegler-Günder (1956). He demonstrated that ichthyopterin is destroyed by UV-light irradiation. This made Kauffmann (1959) continue the research and he isolated ichthyopterin from the skin of goldfish *Carassius auratus* in darkness to prevent destruction. Using various methods of analysis Kauffmann could show that the isolated ichthyopterin was identical to 6(α,β-dihydroxypropyl)-isoxanthopterine. Hüttel and Schreck (1960) arrived at the same conclusion after analysing the ichthyopterin sample from the batch obtained by Hüttel and Sprengling in 1943.

However, efforts to identify ichthyopterin have not revealed the nature of the genuine alarm substance. Behavioural bioassays using the ichthyopterin sample obtained by Hüttel and Sprengling have shown that it is not a potent alarm substance; it did not evoke fright reaction in any of the trials with minnow schools (Schutz, 1956). Using a paper chromatography approach, Schutz made a fresh new attempt to isolate the alarm substance from minnow skin but could not reach a clear conclusion. He found repellent activity among fractions having violet-blue fluorescence and others lacking fluorescence.

The majority of later attempts to establish the nature of the alarm substances followed in the wake of these studies, focusing on pterin-like substances. Reutter and Pfeiffer (1973) showed that club cells containing alarm substance from *Phoxinus phoxinus* epidermis contain fluorescent substances. Pfeiffer and Lemke (1973) isolated several fractions from *Phoxinus phoxinus* skin extract that contained fluorescent substances with molecular weight less than 500. Two of these fractions evoked fright response in giant danio *Danio malabaricus* in bioassays. The authors concluded that the alarm substance of cyprinids was a pterin-like ichthyopterin. They also found an alarm reaction of giant danio to isoxanthopterin. However, Pfeiffer (1975) described species that possess the alarm substance but lack fluorophores in their skins. He supposed that these fluorophores accompanied the genuine alarm substance in the active fractions (Pfeiffer and Lemke, 1973) as well as in the club cells (Reutter and Pfeiffer, 1973). Lastly Pfeiffer (1975) concluded that the alarm substance does not give off fluorescence and is not a pterin. The same conclusion was drawn after isolation of alarm substance from *Phoxinus phoxinus* skin extract using gel-chromatograhy on Sephadex G-15 columns (Lebedeva et al., 1975; Kasumyan and Lebedeva, 1977, 1979). Among several fractions obtained after gel-chromatograhy one was highly effective in evoking alarm response in minnows but had no fluorophores. Another fraction was inactive in behavioural trials but showed the characteristic UV absorbtion spectra, similar to that of isoxanthopterin and violet-blue fluorescence (450–475 nm) under UV light irradiation. Moreover, Kasumyan and Ponomarev (1987) found that skin extract of some cyprinid species did not contain fluorophores but contained a repellent substance.

An efficient fraction that induced alarm reaction obtained from minnow skin extract contained substances with molecular weight around 1100 Da. Among these substances one had maximum UV absorbtion at 295 nm. Using polyacrylamide electrophoresis and chromatography on a CM-Sephadex C-25 column, it was found that this substance, with a UV maximum at 295 nm, evoked no alarm reaction in minnow. The substance in the active fraction was negatively charged at pH 8.3 and bound to an anion-exchange DEAE-Sephadex. Boiling and exposure under UV light reduced the efficiency of the alarm reaction in a short time (Table 6.3). Thin-layer chromatography of the active fraction on silica gel in a system with n-butanol:acetic acid:water (4:1:1) gave one spot inducing alarm reaction. This spot had an R_f between 0.25–0.35 and was coloured after treatment by ninhydrin, indicating the presence of amino groups. The

Table 6.3 Effect of boiling and UV irradiation on the alarm pheromone

Behavioural responses to the main active fraction obtained by G-15 Sephadex chromatography. Original extract made from skin of European minnow. Alarm reaction evaluated using a 6-point scale from 0 to 5, (see Kasumyan and Tuvikene, 2003).

Exposure (min)	*Boiling*	*UV irradiation*
0	2	2
5	—	1
10	1	0.5
15	—	0
20	0	—

active spot was also coloured by the biuret reagent, indicating the presence of carbohydrate. Peptide tests revealed only slight colouration, indicating traces of peptide in the active spot of the chromatogram (Lebedeva et al., 1975; Kasumyan and Lebedeva, 1979).

The first to question the 'pterin' hypothesis that still dominated studies of the nature of the alarm substance in the 1970s was Argentini (1976) who isolated a putative alarm substance chromatographically from skin extract of *Phoxinus phoxinus* and found the alarm substance to probably be hypoxanthine-(3N)-oxide. Argentini characterized the alarm substance as a colourless, non-fluorescent substance, poorly soluble in water, showing a purine similar UV spectra and stable in water only 26 hours. Argentini synthesized hypoxanthine-(1N)-oxide and hypoxanthine-(3N)-oxide and found biological activity only for the last substance (Argentini, 1976). Hypoxanthine-(3N)-oxide was found to be as effective as skin extract in evoking behavioural responses in black tetra, *Gymnocorymbus ternetzi* (Pfeiffer et al., 1985). It was concluded that this compound is the active component or most important active component in the alarm substance. Pfeiffer (1978) also tested various pteridine, purine and pyrimidine derivatives with respect to their behavioural activity and showed that 3 of 59 compounds tested were effective in eliciting an alarm reaction; these were 2,6-diamino-4-oxodihydropteridine, isoxanthopterin and 6-acetonylisoxanthopterin.

Tucker and Suzuki (1972) studied the olfactory stimulatory effect of the skin extracts from white catfish *Ictalurus catus* and suggested that the alarm pheromone is a mixture of several compounds including some amino acids and oligopeptides.

In laboratory studies, fathead minnows *Pimephales promelas* and fine-scale dace *Chrosomus neogaeus* displayed characteristic alarm reactions when exposed to conspecific skin extract or hypoxanthine-(3N)-oxide and the functionally similar pyridine-N-oxide but not the structurally similar molecules lacking a nitrogen oxide functional group (guanine, hypoxanthine, xanthine, 4(3H)-pyrimidone and pyridine) or to a swordtail *Xiphophorus helleri* (Cyprinodontiformes) skin extract control. The field-trapping experiments supported the laboratory results (Brown et al., 2000). These data suggest that contrary to the results of Pfeiffer et al. (1985), hypoxanthine-(3N)-oxide may not be the sole active molecule in the ostariophysan alarm pheromone and the nitrogen-oxide functional group of a purine-N-oxide acts as the chief molecular trigger. As Brown and co-authors suggested (Brown et al., 2000, 2003), any compound with a nitrogen-oxide functional group may act as a potential signalling agent. In other words, the alarm pheromone may actually consist of a suite of aromatic compounds, which have in common a nitrogen oxide as a functional group. Recent results in our laboratory also indicate that hypoxanthine-(3N)-oxide is not an alarm substance (Tuvikene and Freiberg, unpubl. data). We found that hypoxanthine-(3N)-oxide with the concentration of $3 \cdot 10^{-6}$ $g \cdot L^{-1}$ dry weight did not show statistically significant alarm behaviour of European minnows. In the same settings, the minimum effective skin concentration to evoke alarm reaction in European minnow was $3 \cdot 10^{-6}$ $g \cdot L^{-1}$ (skin wet weight). Alternatively, a purine-N-oxide might be bound to a carrier molecule, such as a protein or a carbohydrate, as was suggested in previous studies (Kasumyan and Lebedeva, 1979; Kasumyan and Ponomarev, 1987).

Reed et al. (1972) found that behavioural responses of two cyprinid fishes, *Clinostomus funduloides* and *Notropis cornutus* were the same to skin extract and biogenic amine histamine. A response threshold for histamine was obtained as 0.01 ppm (~9 μM). Spectrophotofluometric analysis showed a similarity in several peaks of the emission spectra of the skin extract and histamine when tested in the same excitation wavelength. They suggested that the alarm substance could be a short-chain low molecular weight molecule, probably a polypeptide having an aliphatic rather than an aromatic ring and with a terminal amine group.

Based on the observation that heating reduced the ability of skin extract to evoke alarm behaviour, Wisenden suggested that a protein might be involved in the cyprind's alarm reaction (Wisenden, 2003).

Table 6.4 Alarm substance in Cypriniformes suggested by various authors.
Chemical name, formula, and source used for isolation of the alarm substance are given. Note that the species used in behavioural trials were not always the same as those used as sources for isolation of the alarm substance.

Suggested substance(s)	Formula	Source for alarm substance	Species used in behavioural trials	Reference
Purine or a pterin-like substance		*Phoxinus phoxinus*	*Phoxinus phoxinus*	Hüttel (1941)
Ichthyopterin, isoxanthopterin	$C_7H_8N_4O_3$	*Rutilus rutilus* *Scardinius erythrophthalmus* *Blicca bjoerkna*	–	Hüttel and Sprengling (1943)
6,9-dioxy-2-amino-8-acetyl-pterin	$C_8H_7N_5O_4$	–	–	Korte and Tschesche (1951)
Identified: 6 (α,β-dihydrooxi-propyl) isoxanthopterin	$C_9H_{11}N_5O_4$	*Phoxinus phoxinus*, *Carassius auratus*	–	Kauffmann (1959) Hüttel and Schreck (1960)
A pterin-like ichthyopterin, M.W. ≤ 500		*Phoxinus phoxinus*	*Danio malabaricus*	Pfeiffer and Lemke (1973)
Substance with amino groups banded with carbohydrate, M.W. ~ 1100		*Phoxinus phoxinus*	*Phoxinus phoxinus*	Lebedeva et al. (1975) Kasumyan and Lebedeva (1977, 1979) Kasumyan and Ponomarev (1987)
Hypoxanthine-(3N)-oxide	$C_5H_4N_4O$	*Phoxinus phoxinus*	*Phoxinus phoxinus*	Argentini (1976)
7-hydroxybiopterin	$C_9H_{11}N_5O_4$	*Danio malabaricus*	*Danio malabaricus*	Win (2000)
Oligopeptides		*Ictalurus catus*	*Ictalurus catus*	Tucker and Suzuki (1972)
Polypeptide		–	*Clinistomus funduloides* *Notropis cornutus*	Reed et al. (1972)
Protein		*Pimephales promelas*	*Pimephales promelas*	Wisenden (2003)
A suite of purine compounds sharing a common N-O functional group			*Pimephales promelas*	Brown et al. (2000)

The most recent study of the chemical nature of the alarm substance was done on the giant danio *Danio malabaricus* by T. Win as a PhD thesis for Carl von Ossietzky University, Oldenburg, Germany (Win, 2000). Her thesis can be found at http://docserver.bis.uni-oldenburg.de/publikationen/dissertation/2000/winiso00/winiso00.html.

Various methods were applied by her such as extraction and ultrafiltration, reversed-phase high-performance liquid chromatogrphy, and gel-filtration chromatography. Identification methods included UV-visible spectroscopy and TLC analysis, NMR spectroscopy, and laser-desorption/ionization mass spectroscopy. Using these methods an amount of 3 mg isolated natural pheromone was obtained and identified as 7-hydroxybiopterin $C_9H_{11}O_4N_5$. Win isolated a derivative from isoxanthopterin (7-hydroxybiopterin) which presumably differs from ichthyopterin found by Hüttel and Sprengling (1943) and found to be ineffective by Schutz (1956) as discussed above. The substance was effective at a concentration of 1×10^{-10} M or 5 ml of a 0.4 μM solution in an aquarium of 20 L.

In a recent study Brondz and co-workers (2004) recorded nervous activity from the part of the olfactory bulb sensitive to the skin extract while perfusing the olfactory epithelium with the outlet from an HPLC column. By so using the fish olfactory system as an HPLC detector, interesting peaks giving a specific increase in nervous activity were isolated. The various attempts to identify the alarm substances are listed in Table 6.4.

There is concern with all the studies on the alarm substance published thus far. They all indicate a single substance as the carrier of information. Most of these substances are stable compounds whereas behavioural studies all demonstrate that the substances in the skin extract are unstable and lose their potency if not frozen. They also fail to explain the species specificity of the alarm reaction, which tells us that the alarm substances must be species specific or comprise several substances.

7. SENSORY BASIS FOR ALARM REACTION

Soon after his discovery, von Frisch arrived at the conclusion that the sense of smell mediated the alarm reaction, since he observed that the reaction was absent in fish with cut olfactory tracts. This conclusion was supported by experiments on grass carp *Ctenopharyngodon idella*. The fish lost the ability to respond to skin extract after treatment of the olfactory epithelium with the detergent Triton X-100 that removed cilia and microvilli of the olfactory neurons. The alarm reaction reappeared when the olfactory receptor neurons had restored their sensory hairs (Kasumyan and Pashchenko, 1982; Pashchenko and Kasumyan, 1984).

In pursuing a better understanding of the sensory basis for the alarm reaction we undertook a detailed study of the organization of the olfactory system in crucian carp. A short account of these findings follows.

The olfactory system in fish consists of the sensory epithelium, an olfactory nerve composed of the axons of the sensory neurons, and the olfactory bulb in which the axons of the sensory neurons make synapses with the secondary neurons. The axons of these secondary neurons exit the bulb in different bundles. Catfishes, order Siluriformes; cods, Gadiformes; and carps, Cypriniformes have all long olfactory tracts. The tracts are divided into different nerve bundles or strands that enter the telencephalon at different regions. The layout of the olfactory system in relation to the brain as found in crucian carp is shown in Figure 6.2.

There are three types of sensory neurons: ciliated cells with long dendrites, microvillous cells with shorter dendrites and their cell bodies in the intermediate part of the epithelium, and crypt cells that have short dendrites and lie close to the surface of the sensory epithelium. The axons of these primary sensory neurons assemble or converge in different regions of the olfactory bulb and make synaptic connections with the secondary neurons. Each bundle of the olfactory tract has a specific composition of nerve fibres, suggesting that they mediate different information, as shown in cod *Gadus morhua* (Døving and Gemne, 1965; Døving, 1967). It is also evident that the majority of the fibres in the olfactory tract are the axons of secondary neurons in the olfactory system, mediating olfactory messages to the brain. However, there are also fibres that convey information out to the olfactory bulb (Døving and Gemne, 1966). The nomenclature of the olfactory tracts has roots from the classical work of Sheldon on the goldfish (Sheldon, 1912). He divided the tract into medial and lateral bundles, each having a medial and lateral partition. Thus, there is a medial and lateral portion of the medial olfactory tract (mMOT and lMOT respectively); likewise there is a medial and lateral portion of the lateral olfactory tract (mLOT and lLOT respectively). The divisions are visible to the naked eye; however, the lateral olfactory tract is divided into a number of strands that might be less striking than the division of the medial olfactory tract.

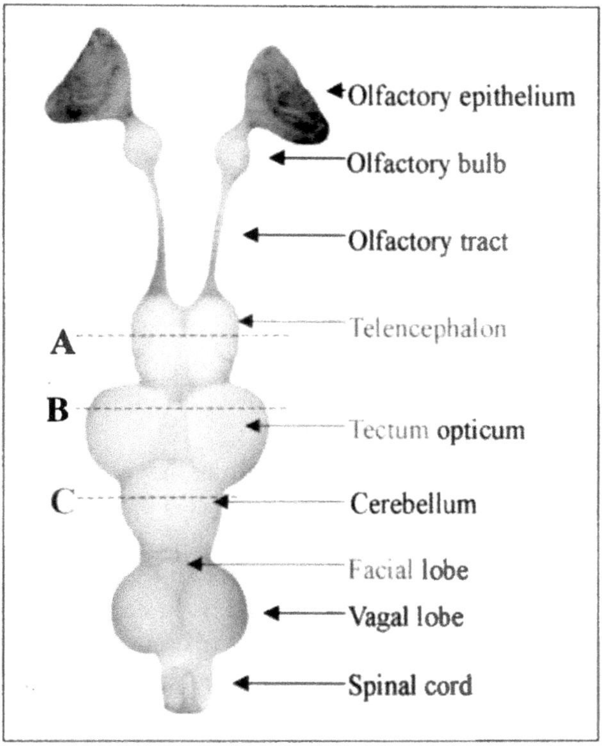

Fig. 6.2 Brain of crucian carp.

Photo of brain of a crucian carp *in situ* displaying the olfactory system. Note the long olfactory tracts. Stippled lines A, B, and C indicate planes of sections presented in Figure 6.7.

Electrical stimulation of the various portions of the olfactory tract demonstrates striking differences in functional properties (Døving and Selset, 1980). By observing the behaviour of fishes with a particular bundle or portion of an olfactory tract bundle either intact or cut it is possible to associate a portion of the tract with a specific behaviour. Such studies have been performed on crucian carp and confirm the previous studies. In short the mMOT mediates the alarm reaction (Hamdani et al., 2000); lMOT mediates reproductive behaviour (Weltzien et al., 2003); and LOT mediates feeding behaviour (Hamdani et al., 2001).

In behavioural experiments on crucian carp, with only the medial part of the medial olfactory tract intact (mMOT), the skin extract induced an alarm reaction. This reaction had the same strength as control fish with the olfactory tracts intact. In fish with only the mMOT cut the alarm reaction is absent. These results indicate that the mMOT is solely responsible for mediating the alarm reaction (Hamdani et al., 2000).

In performing these experiments, we were careful not to damage the fibres of the olfactory tracts and brain because they would be exposed to fresh water if not protected. The brain cavity was thus filled with Ringer agar and the fish kept in physiological saline. Post-mortem inspection revealed whether any damage had occurred during the experimental period. In a recent study by Ide et al. (2003) on the alarm reaction of *Brycon cephalus*, (Characiformes) the authors found no differences in behaviour irrespective of the LOT or MOT being cut. In their study the control fish had higher values of cortisol in the blood than the experimental group exposed to skin extract. This finding was contrary to that expected.

In several studies, we and other groups have shown that sensory neurons of the olfactory epithelium with specific morphology project towards a particular region of the olfactory bulb. In crucian carp ciliated sensory neurons with long dendrites project to the medial region of the olfactory bulb (Hamdani and Døving, 2002). In experiments made with the neural tracer DiI there was concomitant staining of the fibres in the mMOT, strongly suggesting that these sensory neurons with long dendrites participate in the alarm reaction. In catfish the application of DiI crystals on the ventral part of the posterior olfactory bulb resulted in staining of the ciliated sensory neurons, but concomitant staining of the olfactory tract was not investigated in these studies (Morita and Finger, 1998).

A schematic drawing indicating the organization of the olfactory system in the crucian carp is shown in Figure 6.3.

8. NERVOUS ACTIVITY OF OLFACTORY SYSTEM

Stimulation of the olfactory organ with odorants induces nervous activity in the different levels of the olfactory system that can be recorded by suitable techniques. From the sensory epithelium it is possible to record a slow potential change, an electro-olfactogram or EOG, first discovered and analyzed in detail in frog by Ottoson in 1956 (Ottoson, 1956). This potential is considered to reflect the depolarization of sensory neurons. Later studies have shown that it is also possible to record these slow potentials from the olfactory organs of fishes. In crucian carp, fresh skin extract diluted to a concentration of 10^{-12} $g \cdot L^{-1}$ and then applied will

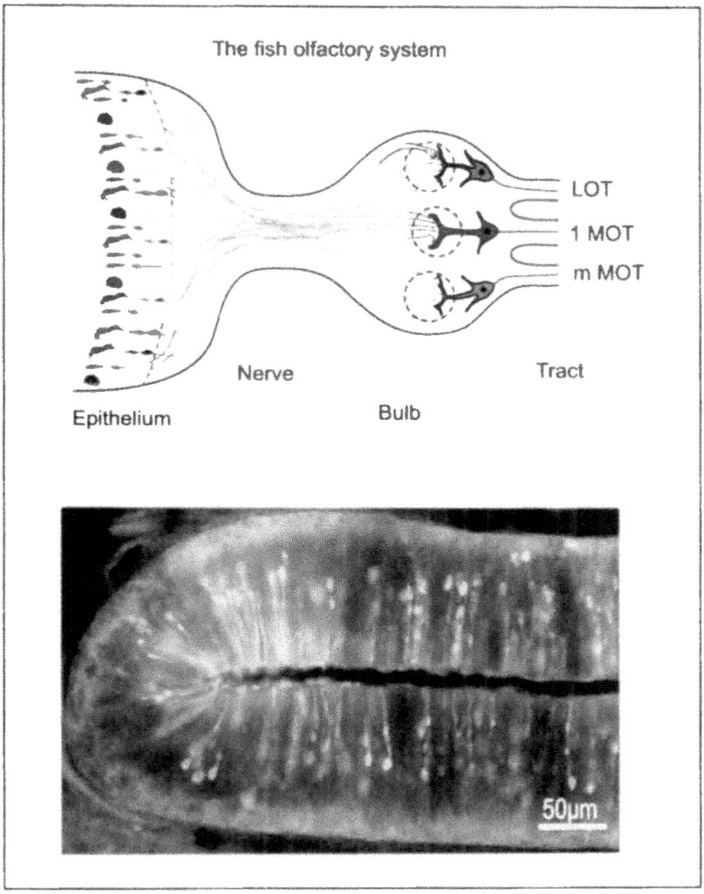

Fig. 6.3 Projection of sensory neurons to olfactory bulb.

Upper figure illustrates the tentative projection of the sensory neurons in the olfactory epithelium of crucian carp. Axons of the sensory neurons with long dendrites and cilia terminate on the relay neurons that make up the mMOT (in violet). Axons of the sensory neurons with intermediate dendrites and microvilli terminate on the relay neurons in the lateral part of the olfactory bulb and the axons of these relay neurons form the LOT (in green). Axons of the crypt cells terminate on relay neurons in the ventral region of the olfactory bulb and their axons form the lMOT (in red). *Lower photo* shows the coloration of the sensory neurons with long dendrites and cilia when DiI is placed in the posteromedial region of the olfactory bulb when only the axons in the mMOT are stained. See text for details.

still give a conspicuous response. The response to a concentration of 10^{-5} g·L^{-1} is shown in Figure 6.4B. Recordings of the EOG were made after opening the olfactory cavity by removing the skin flap over the organ. This operation causes injuries to the skin around the sensory epithelium.

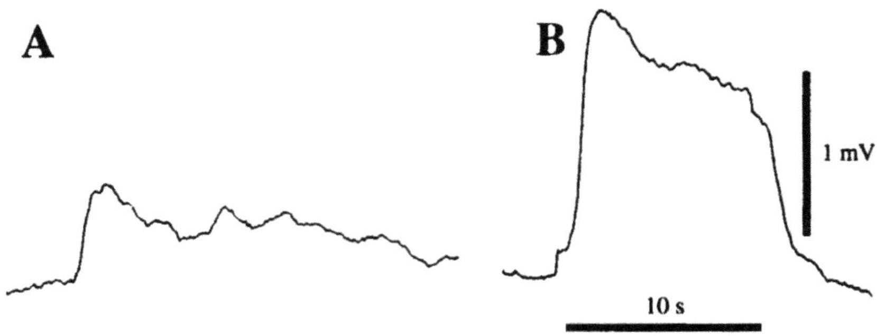

Fig. 6.4 EOG responses from crucian carp.
A. Recordings from a newly operated fish showing fluctuations probably due to spontaneous release of alarm substances from the cut skin surface. B. Recordings of the response to skin extract at concentration of 10^{-5} g·L^{-1}. Horizontal bar in B indicates the stimulation period. Notice the stable recordings as this preparation was made in a preoperated fish.

An interesting observation, related to the effect of skin extract, is that EOG experiments performed with a freshly opened organ generated an unstable baseline at the beginning of the experiment, as seen in Figure 6.4A, but 2-3 hours later the baseline stabilized. Moreover, if the fish were preoperated one week in advance, the EOG recordings were stable. The explanation for these unstable recordings seen in freshly operated animals could be the release of alarm substances from the cut skin around the olfactory epithelium. Since the wounds healed in the course of one week, recordings in a preoperated animal were stable.

Fractionation of the skin extract of crucian carp by Sephadex G-25 columns revealed interesting information when used in combination with EOG recordings and behavioural experiments (Kasumyan and Tuvikene, 2003). The most potent odorant substances responsible for the highest EOG amplitudes are fractions from region B in the chromatogram (Fig. 6.5). However, these fractions did not induce alarm reactions. On the other hand, fractions from region A induced strong alarm reactions and fractions from region C less intense alarm reactions (Fig. 6.5). These fractions did not evoke EOG responses with high amplitudes. Fractions A were from elution volumes in which standard substances with an MW between 1000 and 2000 Da appeared. Substances in region B had an MW of about 500 Da. Substances with an MW about 65 were eluted in region C. Fractions from A and C evoked EOG amplitudes about 20% and 70%

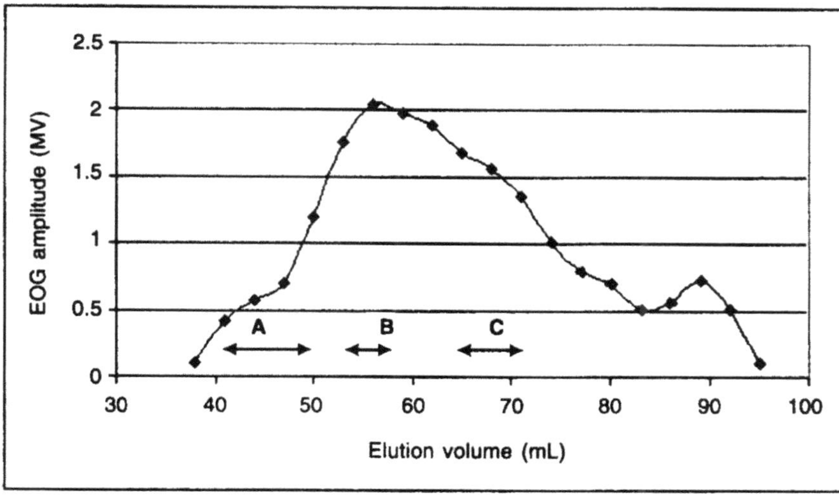

Fig. 6.5 EOG amplitudes in crucian carp to fractions of skin extract. Skin extract of crucian carp was separated in a Sephadex G-25 fine column and the fractions tested for EOG potency. Fractions at **A** evoked distinct alarm reactions, fractions at **C** evoked less obvious responses. The fraction at **B** evoked the largest EOG response but did not induce an alarm reaction. The standards used for determining the molecular weights indicated that fractions at **A, B,** and **C** contained substances of about 1500, 500 and 65 Da respectively. A portion of these data was published by Kasumyan and Tuvikene (2003).

of those evoked by fraction B. Interestingly hypoxanthine-(3N)-oxide evoked large EOG responses and judging from electrophysiological methods, more potent than free amino acids L-alanine and L-serine at the same concentrations. However, as mentioned above it is most probably not the genuine alarm substance.

Neural activity in the olfactory bulb. The anatomical studies described above indicate that the sensory neurons with long dendrites and cilia project to the posterior part of the medial region in the olfactory bulb of crucian carp. Neurons situated in this region of the bulb have been shown to be particularly sensitive to the skin extract (Hamdani and Døving, 2003).

In recordings from single neurons in the olfactory bulb two types of units were encountered. One type of unit (type I cells) was characterized by a diphasic action potential (AP) of short duration (rise time ~1 ms). Another type of unit (type II cells) displayed an AP with long duration (rise time ~1.8 ms). The AP of this latter unit was nearly always followed by a slow potential (SP), characteristic diphasic wave with a rise time of about 5 ms. The delay between AP and SP, varied between 8 and 8.5 ms. These types of units are seen in Figure 6.6. The appearance of the AP and the following SP indicated a particular type of unit. Zippel and co- workers (1999) proposed that these units represent activity of the so-called ruffed cells. Type I cells are probably secondary neurons (mitral cells). However, until correlative histological identification of the units recorded becomes available, we categorize these units as type I and type II cells. It is pertinent that only type I cells responded to application of the skin extract to the olfactory epithelium with increased activity. The firing rate of these cells increased with increase in concentration of the skin extract. Thus, these results imply that the medial part of the bulb reacts specifically to alarm substances. No other odorants used generated an increased firing rate of the neurons in this part of the bulb. These results accord with studies using other methods, suggesting that the olfactory bulb is divided into different functional zones (Friedrich and Korsching, 1997, 1998; Nikonov and Caprio, 2001). One should note that during the firing period of type I cells, type II cells were quiescent and *vice versa*, suggesting functional coupling between these relay neurons probably via granule cells.

In recordings from the posterior part of the medial region of the olfactory bulb upon stimulation of the sensory epithelium with different solutions, units of type I responded specifically to application of skin extract.

A specific chemical stimulation of the olfactory epithelium led to stimulation of specific olfactory neurons, which projected to secondary neurons in a delimited zone of the olfactory bulb. The activated secondary neurons stimulated granule cells, which in turn inhibited ruffed cells in the vicinity. Zippel and co-workers (2000a, b) suggested that it is the inhibition of the secondary neurons that decreases the inhibition of the ruffed cells via granule cells and consequently induce activation of the ruffed cells. This suggestion fits with the idea that type I cells are secondary neurons also called mitral cells and that type II cells are ruffed cells.

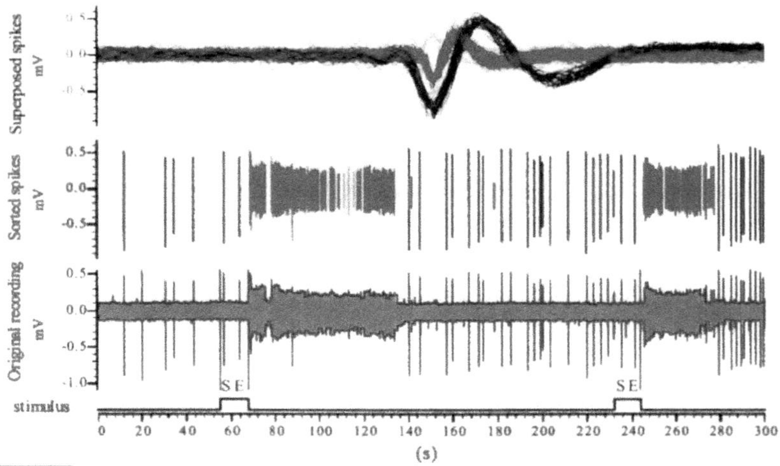

Fig. 6.6 Single unit recordings from the olfactory bulb.
Example of extracellular recording from the posterior part of the medial region of the right olfactory bulb of crucian carp upon exposure of the right olfactory organ to a solution of skin extract (SE). The lower trace shows the original recording. The middle trace shows spikes of the type I unit (red) and the type II unit (blue). Note the interaction between the two types of unit. The upper trace shows superposed spikes from the two types of units.

9. CENTRAL PROJECTIONS OF MEDIAL PART OF MEDIAL OLFACTORY TRACT

As mentioned earlier the long olfactory tracts in Cypriniformes, Siluriformes and Gadiformes can be divided in discrete bundles. We have also provided evidence that each of these nerve bundles mediates information releasing specific behaviour patterns. The mMOT mediates information releasing alarm reaction, lMOT mediates information releasing sexual behaviour and LOT mediates information leading to feeding behaviour (Døving and Selset, 1980; Hamdani et al., 2000, 2001; Weltzien et al., 2003). The fact that distinct behavioural responses can be associated with specific parts of the olfactory tract makes it a fruitful model for describing the neural substrates involved in the various behavioural patterns.

Early studies suggested that the entire telencephalon receives and processes olfactory input, naming this hemisphere to the olfactory lobes (Ariëns Kappers et al., 1936). More recent studies have demonstrated however, that secondary olfactory projections terminate in specific parts of the telencephalon, namely the ventromedial and basolateral regions (Scalia and Ebbesson, 1971; Ito, 1973; Sato et al., 1979; Oka, 1980; Bass, 1981; Murakami et al., 1983; Rooney and Laming, 1984; von Bartheld et al., 1984; Levine and Dethier, 1985; Stewart and Brunjes, 1990; Rooney et al., 1992; Sas et al., 1993; Riedel and Krug, 1997a, b).

Behavioural studies showing that mMOT mediate information releasing alarm responses (Døving and Selset, 1980; Hamdani et al., 2000), also indicate that this nerve bundle projects to discrete parts of the brain. However, few anatomical studies have been performed on the projection of mMOT only. In a study on goldfish it was suggested that all fibres in the mMOT were secondary olfactory neurons projecting from the olfactory bulb to the brain (von Bartheld et al., 1984). Some of these fibres terminated in ventral parts of the brain in the areas Vv, Vd, Vl and the ventral portion of D1. Furthermore, a projection of mMOT to the contralateral telencephalon via the anterior commissure (AC) was observed. Interestingly, these fibres terminate in the homologue regions of the contralateral telencephalon, namely Vv, Vd, Vl and the ventral portion of D1. These observations are in accordance with our preliminary results in crucian carp. Staining of the mMOT fibres on one side only shows a clear contralateral projection of mMOT (Fig. 6.7A). A similar projection pattern of mMOT has also been observed in cod (Rooney et al., 1992).

Numerous studies suggest that the olfactory fibres project beyond the telencephalon and terminate in more caudal parts of the brain (Sheldon, 1912; Holmgren, 1920; Finger, 1975; Bass, 1981; von Bartheld et al., 1984; Levine and Dethier, 1985; Rooney et al., 1992; Sas et al., 1993; Matz, 1995). In goldfish, mMOT projected through Vl of the telencephalon, passing the area preoptica and terminating near the III. ventricle in the hypothalamus (von Bartheld et al., 1984). Our preliminary results suggest a more widespread projection of mMOT in the diencephalon, and that some fibres projects to the habenula, an area that previously has been associated with nerves projecting to the brain via LOT (Finger, 1975; von Bartheld et al., 1984). We observed a diffuse projection in the optic tectum. In the section shown in Figure 6.7B we also observed stained cell bodies in the cerebellum. Furthermore, we found heavily stained cell bodies in regions situated around the IV. ventricle in the brain stem (Fig. 6.7C). Assuming that these stained cells reflect a direct, and not a secondary staining, they suggest that fibres run from these areas in the cerebellum and the brain stem out to the bulb via the mMOT. Centrifugal fibres running from the brain to the olfactory bulb were observed earlier in the lMOT (von Bartheld et al., 1984).

Electrophysiological evidences for centrifugal fibres running in all tract bundles have, however, been provided in studies on burbot, *Lota lota*. It was found that electrical shocks applied to the olfactory tract on one side evoked activity in the proximal stump of all three bundles on the contralateral side (Døving and Gemne, 1966). These findings also indicate that there is a bulb to bulb projection, findings supported by our present studies in crucian carp and anatomical studies in goldfish (von Bartheld et al., 1984). Furthermore, Sas and co-authors (Sas et al., 1993) observed centrifugal fibres with cell bodies in the brain stem i.e., *locus ceruleus* and *raphe nucleus* in the gymnotiform fish *Apteronotus leptorhyncus*. Projections from the brain stem have not been associated earlier with fibres of the mMOT.

The diffuse projection in the optic tectum from the olfactory bulb indicates that the alarm system have direct effects on the visual pathways. These projections are interesting from several points of view. In our studies on crucian carp we found that fish with the mMOT cut did not react to visual stimuli, for example turning on the room lights (Hamdani et al., 2000). Such projections might be involved in the dorsal light

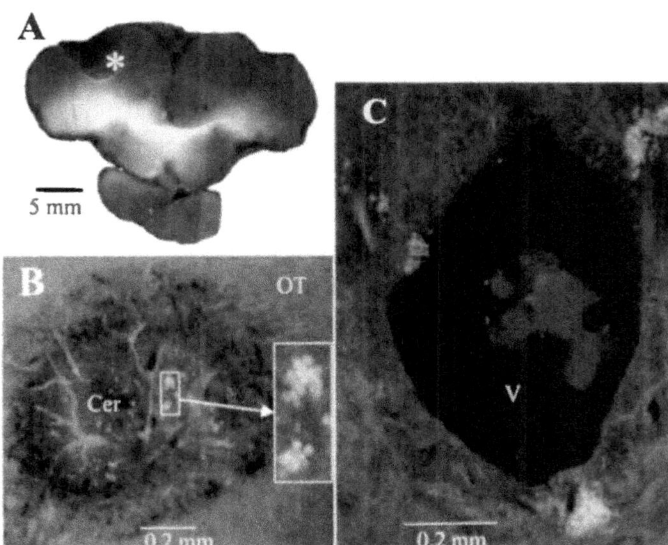

Fig. 6.7 Examples of brain regions with neurons connected to the mMOT. Fluorescence micrographs of crucian carp; transverse sections through the anterior commissure in the telencephalon **(A)**, anterior part of the optic tectum and cerebellum **(B)**, and the brain stem **(C)**. The letters refer to the planes indicated in Figure 6.2. Fish were perfused with 4% paraformaldehyde in 0.1 M phosphate buffer and DiI crystals applied on one side of the central stump of the mMOT. After three months in fixative the brains were embedded in gelatin (12%) and cut in 100 µm slices using a vibratom (section A) or cryoprotected in 30% sucrose, embedded in Tissue-Tech, then cut into 40 µm thick slices, using a cryostat (sections **B** and **C**). Sections obtained were inspected using an Olympus fluorescence microscope with a rhodamine filter. The region ipsilateral to the application site is indicated by an asterisk. **OT**, optic tectum; **Cer**, cerebellum; **V**, IV. ventricle.

response, since hypoxanthine-(3N)-oxide influences this behaviour in the black tetra, *Gymnocorymbus ternetzi* (Pfeiffer et al., 1985). It is also tempting to speculate that these projections are involved in the rapid learning processes, where any visual stimulus made in conjunction with the alarm reaction is later associated with danger and induces alarm reactions in fish (Hall and Subaski, 1995; Yunker et al., 1999; Wisenden and Harter, 2001).

10. CONCLUSIONS

More than 60 years have passed since Karl von Frisch published his discovery on alarm reaction in minnows. Subsequently a multitude of articles have described the distribution of the alarm substance among species, the alarm reaction among various groups of fishes, and how different fish species perform the alarm reaction, both in adult and juvenile specimens. There is considerable knowledge about the species specificity of alarm reaction and how the specificity depends on fish age or alarm substance concentration.

Interest in the alarm reaction in fishes continues among scientists (Smith 1992; Wisenden, 2003). Every year new articles about alarm substance and alarm reactions, including species other than carp are published. Future studies in this field could take a variety of directions. Substances that cause alarm or avoidance reactions seem more widespread than originally thought and the strategies taken by prey fish and their predators are far from resolved. The possible role of alarm substances in changing the body shape of conspecifics, as indicated in the work of Stabell and Lwin (1997) is a fascinating avenue for further research. The species specificity touched on here requires further research, both into the chemical basis of the alarm substances in various species and the performance of receptive mechanisms. Obvious questions to ask are for example: what is the molecular basis for the reception, and what are the transduction mechanisms found in the sensory neurons with long dendrites that seem to mediate the alarm reaction. Lately some understanding of which sensory neurons are responsible for the alarm substance reception and how this information is transferred to the fish brain has been garnered, but the central pathways and centres involved in the alarm reaction are still insufficiently known. It is essential to establish the nature of alarm substances in various species. In a recent study the outlet of the HPLC was led directly onto the fish olfactory organ and the neural activity of the secondary neurons in the olfactory bulb recorded as the skin extract of conspecifics was separated in the HPLC column (Brondz et al., 2004). A particular activity was associated with distinct peaks in the chromatogram. Such a strategy is a useful aid in the isolation of alarm substances.

References

Aoki, I and T. Kuroki. 1975. Alarm reaction of three Japanese cyprinid fishes, *Tribolodon hakonensis, Gnathopogon elongatus elongatus, Rhodeus occelatus occelatus. Bull. Jap. Soc. Sci. Fish.* 41: 507–513

Argentini, M. 1976. Isolerung des Schreckstoffes aus der Haut der Elritze *Phoxinus phoxinus* L. PhD. thesis. Universität Zürich, 111 pp.

Ariëns Kappers, C.U.J., G.C. Huber and E.C. Crosby. 1936. *The Comparative Anatomy of the Nervous System of Vertebrates, including Man.* Hafner Press, New York.

Bass, A.H. 1981. Telencephalic efferents in channel catfish, *Ictalurus punctatus*: projections to the olfactory bulb and optic tectum. *Brain Behav. Evol.* 19: 1–16

Brondz, I., E.H. Hamdani and K.B. Døving. 2004. Neurophysiological detector—a selective and sensitive tool in high-performance liquid chromatography. *J. Chromatography* 800: 41–47.

Brown, G.E., J.C. Adrian Jr., E. Smyth, H. Leet and S. Brennan. 2000. Ostariophysan alarm pheromones: Laboratory and field tests of the functional significance of nitrogen oxides. *J. Chem. Ecol.* 26: 139–154

Brown, G.E., J.C. Adrian Jr. and M.L. Shih. 2001. Behavioural responses of fathead minnows to hypoxanthine-3-N-oxide at varying concentrations. *J. Fish Biol.* 58: 1465–1470

Brown, G.E., J.C. Adrian Jr., N.T. Naderi, M.C. Harvey and J.M. Kelly. 2003. Nitrogen oxides elicit antipredator responses in juvenile channel catfish but not in convict cichlids or rainbow trout: conservation of the ostariophysan alarm pheromones. *J. Chem. Ecol.* 29: 1781–1796.

Døving, K.B. 1967. Comparative electrophysiological studies on the olfactory tract of some teleosts. *J. Comp. Neurol.* 131: 365–370.

Døving, K.B. and G. Gemne. 1965. Electrophysiological and histological properties of the olfactory tract of the burbot (*Lota lota* L.). *J. Neurophysiol.* 28: 139–153.

Døving, K.B. and G. Gemne. 1966. An electrophysiological study of the efferent olfactory system in the burbot. *J. Neurophysiol.* 29: 665–674.

Døving, K.B. and R. Selset. 1980. Behavior patterns in cod released by electrical stimulation of olfactory tract bundlets. *Science* 207: 559–560.

Finger, T.E. 1975. The distribution of the olfactory tracts in the bullhead catfish, *Ictalurus nebulosus. J. Comp. Neurol.* 161: 125–141.

Friedrich, R.W. and S.I. Korsching. 1997. Combinatorial and chemotopic odorant coding in the zebrafish olfactory bulb visualized by optical imaging. *Neuron* 18: 737–752.

Friedrich, R.W. and S.I. Korsching. 1998. Chemotopic, combinatorial, and noncombinatorial odorant representations in the olfactory bulb revealed using a voltage-sensitive axon tracer. *J. Neurosci.* 18: 9977–9988.

Gandolfi, G., D. Mainardi and A.C. Rossi. 1968. The fright reaction of zebra fish. *Atti. Soc. Ital. Sci. Nat. Mus. Civ. Stor. Nat. Milano* 107: 74–88.

Hall, D. and M.D. Suboski. 1995. Visual and olfactory stimuli in learned release of alarm reactions by zebra danio fish (*Brachydanio rerio*). *Neurobiol. learning Memory*. 63: 229–240.

Hamdani, E.H. and K.B. Døving. 2002. The alarm reaction in crucian carp is mediated by olfactory neurons with long dendrites. *Chem. Senses* 27: 395–398.

Hamdani, E.H. and K.B. Døving. 2003. Sensitivity and Selectivity of neurons in the medial region of the olfactory bulb to skin extract from conspecifics in crucian carp, *Carassius carassius*. *Chem. Senses* 28: 181–189.

Hamdani, E.H., O.B. Stabell, G. Alexander and K.B. Døving. 2000. Alarm reaction in the crucian carp is mediated by the medial bundle of the medial olfactory tract. *Chem. Senses* 25: 103–109.

Hamdani, E.H., A. Kasumyan and K.B. Døving. 2001. Is feeding behaviour in the crucian carp mediated by the lateral olfactory tract? *Chem. Senses* 26: 1133–1138.

Heintz, E. 1954. Actions répulsive exercées sur divers animaux par des substances contenues dans la peau ou le corps d'animaux de même espèce. *C.R. Soc. Biol. Paris* 148: 585–588.

Holmgren, N. 1920. Zur Anatomie und Histologie des Vorder- und Zwischenhirns der Knochenfische. *Acta Zool.* (Stockholm) 1: 137–315.

Hüttel, R. 1941. Die chemische Untersuchung des Schreckstoffes aus Elritzenhaut. *Naturwiss.* 29: 333–334.

Hüttel, R. and G. Sprengling. 1943. Über Ichthyopterin, einen blaufluorescierende Stoff aus Fischhaut. *Liebigs Ann. Chem.* 554: 69–82.

Hüttel, R. and D. Schreck. 1960. Notiz über Ichthyopterin. *Chem. Berichte* 93: 439–441.

Ide, L.M., E.C. Urbinate and A. Hoffmann. 2003. The role of olfaction in the behavioural and physiological responses to conspecifics skin extract in *Brycon cephalus*. *J. Fish Biol.* 63: 332–343.

Ito, H. 1973. Normal and experimental studies on synaptic patterns in the carp telencephalon, with special reference to the secondary olfactory termination. *J. Hirnforsch.* 14: 237–253.

Kasumyan, A.O. 1982. The behavior reaction to alarm pheromone and olfactory sensitivity to these signals of cyprinid fishes. In: *Pheromones and Behavior*, V.E. Sokolov (Ed.). Nauka, Moscow, pp. 53–64.

Kasumyan, A.O. and N.E. Lebedeva. 1977. On the chemical nature of the repellent of skin of the minnow. *Biologicheskye nauki* 1: 37–41.

Kasumyan, A.O. and N.E. Lebedeva. 1979. New information on the nature of alarm pheromone in cyprinids. *J. Ichthyol.* 19: 109–114.

Kasumyan, A.O. and N.I. Pashchenko. 1982. The role of olfaction in the defense reaction of the grass carp *Ctenopharyngodon idella* (Cyprinidae) to alarm pheromone. *J. Ichthyol.* 22: 122–126.

Kasumyan, A.O. and V.Y. Ponomarev. 1986. About the species specificity of alarm pheromone in fish from the order of Cypriniformes. In: *Chemical communication in animals*, V.E. Sokolov (Ed.). Nauka, Moscow, pp. 202–207.

Kasumyan, A.O. and V.Y. Ponomarev. 1987. Biochemical features of alarm pheromone in fish of the order cypriniformes. *J. Evol. Biochem. Physiol.* 23: 20–24.

Kasumyan, A.O. and A. Tuvikene. 2003. Composition and rate of release of chemical substances into water by dead fish. Paper presented at 7th Intl. Symp. Fish Physiol. Toxicology, Water Quality, Tallinn, Estonia, 2003.

Kauffmann, T. 1959. Notiz über die Konstitution des Ichthyopterins. *Justus Liebligs Ann. Chem.* 625: 133–139.

Korte, F. and R. Tschesche. 1951. Über Pteridine. V. Mitt.: Die Konstitution des Ichthyopterins. *Chem. Ber.* 84: 801–809.

Lawrence, B.J. and R.J.F. Smith. 1989. Behavioral-response of solitary fathead minnows, *Pimephales promelas,* to alarm substance. *J. Chem. Ecol.* 15: 209–219.

Lebedeva, N.Y., G.A. Malyukina and A.O. Kasumyan. 1975. The natural repellent in the skin of Cyprinids. *J. Ichthyol.* 15: 472–480.

Levine, R.L. and S. Dethier. 1985. The connections between the olfactory bulb and the brain in the goldfish. *J. Comp. Neurol.* 237: 427–444.

Maljukina, G.A., G.V. Devitsyna and E.A. Marusov. 1974. Communication based on chemoreception in fishes. *Zh. Obshch. Biol.* 35: 70–79.

Maljukina, G.A., A.O. Kasumyan, E.A. Marusov and N.I. Pashchenko. 1977. Alarm pheromone and its significance in fish behaviour. *Zh. Obshch. Biol.* 37: 123–131.

Matz, S.P. 1995. Connections of the olfactory bulb in the chinook salmon (*Oncorhynchus tshawytscha*). *Brain Behav. Evol.* 46: 108–120.

Morita, Y. and T.E. Finger. 1998. Differential projections of ciliated and microvillous olfactory receptor cells in the catfish, *Ictalurus punctatus*. *J. Comp. Neurol.* 398: 539–550.

Murakami, T., Y. Morita and H. Ito. 1983. Extrinsic and intrinsic fiber connections of the telencephalon in a teleost, *Sebastiscus marmoratus*. *J. Comp. Neurol.* 216: 115–131.

Nikonov, A.A. and J. Caprio. 2001. Electrophysiological evidence for a chemotopy of biologically relevant odors in the olfactory bulb of the channel catfish. *J. Neurophysiol.* 86: 1869–1876.

Oka, Y. 1980. The origins of the centrifugal fibers to the olfactory bulb in the goldfish, *Carassius auratus*: An experimental study using the fluorescent dye primuline as a retrograde tracer. *Brain Res.* 185: 215–225.

Ottoson, D. 1956. Analysis of the electrical activity of the olfactory epithelium. *Acta Physiol. Scand.* 35: 1–83.

Pashchenko, N.I. and A.O. Kasumyan. 1983. Some morpho-functional peculiarities of the olfactory organ in ontogenesis of *Phoxinus phoxinus* (Cypriniformes, Cyprinidae). *Zool. J.* 62: 367–377.

Pashchenko, N.I. and A.O. Kasumyan. 1984. Degenerative and restorative processes in the olfactory lining of White Amur, *Ctenopharyngodon idella* (Cyprinidae), after treatment with detergent Triton X-100. *J. Ichthyol.* 24: 112–121.

Pashchenko, N.I. and A.O. Kasumyan. 1986. Morphofunctional characteristics of the olfactory organ in Cyprinidae, 1. Morphology and functioning of the olfactory organ during the ontogenesis of the grass carp *Ctenopharyngodon idella* (Val.). *Voprosy ikhtiologii* 26: 303–316.

Pfeiffer, W. 1960. Über die Verbreitung der Schreckreaktion bei Fischen und die Herkunft des Schreckstoffes. *Z. vergl. Physiol.* 43: 578–614.

Pfeiffer, W. 1962. The fright reaction of fish. *Biol. Rev. Cambridge Phil. Soc.* 37: 495–511.

Pfeiffer, W. 1963. The fright reaction in North American fish. *Can. J. Zool.* 41: 69–77.

Pfeiffer, W. 1975. Über fluoreszierende Pterine aus der Haut von Cypriniformes (Pisces) und ihre Beziehung zum Schreckstoff. *Rev. Suisse Zool.* 82: 705–711.

Pfeiffer, W. 1978. Hydrocyclic compounds as releasers of the fright reaction in the giant danio, *Danio malabaricus* (Jerdon) (Cyprinidae, Ostariophysi, Pisces). *J. Chem. Ecol.* 4: 665–673.

Pfeiffer, W. and D. Lamour. 1976. Die Wirkung von Schreckstoff auf die Herzfrequenz von *Phoxinus phoxinus* (L.) (Cyprinidae, Ostariophysi, Pisces). *Rev. Suisse Zool.* 83: 861–873.

Pfeiffer, W. and J. Lemke. 1973. Untersuchungen zur Isolerung und Identifizierung des Schreckstoffes aus der Haut der Elrize, *Phoxinus phoxinus* (L.) (Cyprinidae, Ostariophysi, Pisces). *J. Comp. Physiol.* 82: 407–410.

Pfeiffer, W., G. Riegelbauer, G. Meier and B. Scheibler. 1985. Effect of hypoxanthine-3(N)-oxide and hypoxanthine-1(N)-oxide on central nervous excitation of the black tetra *Gymnocorymbus ternetzi* (Characidae, Ostariophysi, Pisces) indicated by dorsal light response. *J. Chem. Ecol.* 11: 507–523.

Reed, J.R., W. Wieland and T.D. Kimbrough. 1972. A study on the biochemistry of alarm substances in fish. Paper presented at Proc. 26th Ann. Conf., SE Assoc. Game Fish Commissioners, Knoxville, TN (USA) 1972.

Reutter, K. and W. Pfeiffer. 1973. Fluoreszenzmikroskopischer Nachweis des Schreckstoffes in den Schreckstoffzellen der Elritze, *Phoxinus phoxinus* (L.) (Cyprinidae, Ostariophysi, Pisces). *J. Comp. Physiol.* 83: 411–418.

Riedel, G. and L. Krug. 1997a.The forebrain of the blind cave fish *Astyanax hubbsi* (Characidae), I. General anatomy of the telencephalon. *Brain Behav. Evol.* 49: 20–38.

Riedel, G. and L. Krug. 1997b. The forebrain of the blind cave fish *Astyanax hubbsi* (Characidae), II. Projections of the olfactory bulb. *Brain Behav. Evol.* 49: 39–52.

Rooney, D.J. and P.R. Laming. 1984. Effects of olfactory bulb ablation on cardiac and ventilatory arousal responses and their habituation in the goldfish *Carassius auratus*. *Behav. Neural. Biol.* 42: 120-126.

Rooney, D., K.B. Døving, M. Ravaille-Veron, and T. Szabo. 1992. The central connections of the olfactory bulbs in cod, *Gadus morhua* L. *J. Hirnforsch.* 33: 63–75.

Sas, E., L. Maler, and M. Weld. 1993. Connections of the olfactory bulb in the gymnotiform fish, *Apteronotus leptorhynchus*. *J. Comp. Neurol.* 335: 486–507.

Satou, M., M. Ichikawa, K. Ueda, and S.F. Takagi. 1979. Topographical relation between olfactory bulb and olfactory tracts in the carp. *Brain Res.* 173: 142–146.

Scalia, F. and S.O. Ebbesson. 1971. The central projections of the olfactory bulb in a teleost (*Gymnothorax funebris*). *Brain Behav. Evol.* 4: 376–399.

Schutz, F. 1956. Vergleichende Untersuchungen über die Schreckreaktion bei Fischen und deren Verbreitung. *Z. vergl. Physiol.* 38: 84–135.

Sheldon, RE. 1912. The olfactory tracts and centers in teleosts. *J. Comp. Neurol.* 22: 177–339.

Smith, R.J.F. 1992. Alarm signals in fish. *Rev. Fish Biol. Fish.* 2: 33–63.

Stabell, O.B. and M.S. Lwin. 1997. Predator-induced phenotypic changes in crucian carp are caused by chemical signals from conspecifics. *Env. Eur. Biol. Fish.* 49: 145–149.

Stewart, J.S. and P.C. Brunjes. 1990. Olfactory bulb and sensory epithelium in goldfish: morphological alterations accompanying growth. *Brain Res. Devel. Brain Res.* 54: 187–193.

Tucker, D. and N. Suzuki. 1972. Olfactory responses to Schreckstoff of catfish. In: *Olfaction and Taste*, D. Schneider (Ed.). Wissenschaftliche Verlagsgesellschaft MBH Stuttgart, vol. 5, pp. 121–127.

von Bartheld, C.S., D.L. Meyer, E. Fiebig, and S.O. Ebbesson. 1984. Central connections of the olfactory bulb in the goldfish, *Carassius auratus*. *Cell Tissue Res.* 238: 475–487.

von Frisch, K. 1938. Zur Psychologie des Fisch-Schwarmes. *Naturwiss.* 226: 601–606.

von Frisch, K. 1941. Über einen Schreckstoff der Fischhaut und seine biologische Bedeutung. *Z. vergl. Physiol.* 29: 46–145.

Weltzien, F.-A. E. Höglund, E.H. Hamdani, and K.B. Døving. 2003. Does the lateral bundle of the medial olfactory tract mediate reproductive behavior in male crucian carp? *Chem. Senses* 28: 293–300.

Win, T. 2000. Isolation and structural identification of an alarm pheromone from the giant danio, *Danio malabaricus* (Cyprinidae, Ostariophysi, Pisces). PhD-thesis, Carl-von-Ossietzky Univ., Oldenburg.

Wisenden, B.D. 2003. Chemically mediated strategies to counter predation. In: *Sensory Processing in Aquatic Environments*, S.P. Collin and N.J. Marshall (Eds). Springer-Verlag, New York, pp. 236–251.

Wisenden, B.D. and K.R. Harter. 2001. Motion not shape facilitates association of predatory risk with novel objects by fathead minnows (*Pimephales prolemas*). *Ethology* 107: 357–364.

Yunker, W.K., D.E. Wein and B.D. Wisenden. 1999. Conditioned alarm behavior in fathead minnows (*Pimephales promelas*) resulting from association of chemical alarm pheromone with a nonbiological visual stimulus. *J. Chem. Ecol.* 25: 2677–2686.

Ziegler-Günder, I. 1956. Untersuchungen über die Purin- und Pterin-pigmente in der Haut und den Augen der Weissfische. *Z. vergl. Physiol.* 39: 163–189.

Zippel, H.P., C. Reschke and V. Korff. 1999. Simultaneous recordings from two physiologically different types of relay neurons, mitral cells and ruffed cells, in the olfactory bulb of goldfish. *Cell Molec. Biol.* (Noisy-le-grand) 45: 327–337.

Zippel, H.P., M. Gloger, S. Nasser, and S. Wilcke. 2000. Odour discrimination in the olfactory bulb of goldfish: Contrasting interactions between mitral cells and ruffed cells. *Phil. Trans. Roy. Soc. Lond. B Biol. Sci.* 355: 1229–1232.

Zippel, H.P., M. Gloger, L. Lüthje, S. Nasser and S. Wilcke. 2000. Pheromone discrimination ability of olfactory bulb mitral and ruffed cells in the goldfish (*Carassius auratus*). *Chem. Senses* 25: 339–349.

CHAPTER

7

The System of Solitary Chemosensory Cells

Anne Hansen

ABSTRACT

The system of SCCs consists of single chemosensory cells more or less randomly distributed over the body of fishes. The innervation of SCCs depends on their location on the body and includes cranial and spinal nerves. The function of SCCs in fish is known only for a few species with specialized SCC systems. Despite the morphological similarities of SCCs, taste bud cells, and cells of Schreiner organs, the problem of a possible relationship among them has yet to be resolved. However, SCCs are not confined to aquatic animals, they also exist in higher taxa and hence represent a highly conserved chemosensory modality.

Key Words: Olfaction; Taste; Ultrastructure; Electron microscope; Evolution.

Address for Correspondence: Anne Hansen, Department of Cell and Developmental Biology, University of Colorado Health Sciences Center at Fitzsimons, Aurora, CO 80045, USA. E-mail: Anne.Hansen@uchsc.edu

1. INTRODUCTION

Olfaction and taste are well-known chemosensory modalities and described in a wide variety of publications. However, another chemosensory system exists that is less well-known: the system of solitary chemosensory cells (SCC). These cells—scattered as single cells and not accumulated in a specific organ—were mentioned for the first time in 1886 by Kölliker (1886), who found them in the skin of frog larvae and called them 'Stiftchenzellen'. Morrill (1895) noticed 'Stiftchenzellen' in sea robins but it was only in 1965 that these cells were recognized as chemosensory cells. Mary Whitear postulated that Kölliker's 'Stiftchenzellen' were chemosensory cells based on her transmission electron microscopic studies (Whitear, 1965). In the following years, SCCs were described for a vast variety of fish species, including lampreys, lungfish and elasmobranchs, sturgeons, and modern teleosts (for reviews, see Kotrschal, 1991, 1996; Whitear, 1992; Kapoor and Finger, 2003; Hansen and Reutter, 2004).

2. DISTRIBUTION OF SCCs

SCCs comprise a diffuse system of sensory cells embedded in the epidermis of fish and dispersed as isolated cells across the outer body surface. They also occur in the oropharyngeal cavity, on the gill arches (Whitear, 1992), and even in the nasal epithelium of some fish (Hansen and Zippel, 1995; Hansen et al., 1999). The quantity of SCCs in a given species varies considerably. The largest number of SCCs has been found in the fin rays of the anterior dorsal fin of rocklings. This fin is modified by reduction of the skin web between the rays and increase of rays, and serves as a special sensory organ (see below). Kotrschal and Adam (1984) counted SCC densities of up to 100,000 per mm^2 in *Gaidropsarus mediterraneus* and, although the SCCs occur at much lower densities elsewhere in the skin, it has been estimated that a rockling of 20 cm total length carries between 3 to 6 million SCCs. Total numbers of SCCs found in other fish groups are much lower, yet the total number of SCCs per fish seems to be much higher than the total number of taste bud cells (Kotrschal, 1991; Peters et al., 1991). In cyprinids densities vary between 2,000 and 4,000 SCCs per mm^2. Two species of catfish revealed densities of 1,000 to 2,000 SCCs per mm^2. The lowest densities counted were found in the neon tetra, *Hyphessobrycon innesi* with 250 SCCs per mm^2 (Kotrschal, 1992). In some fish, SCCs are either absent or so poorly developed that they escape detection (e.g. stickleback, *Spinachia spinachia*, and the mudskipper,

Periophthalmus koelreuteri (Whitear, 1992). During ontogeny, SCCs occur prior to taste bud cells (Hansen et al., 2002). In the zebrafish *Danio rerio*, their numbers increase sharply after hatching to about 25 days after fertilization, and remain relatively constant thereafter (Kotrschal et al., 1997).

3. MORPHOLOGY OF SCCs

SCCs are specialized bipolar epithelial cells and therefore secondary neurons. Their morphology may vary according to the surrounding epithelium. The apical endings of SCCs protrude between other epithelial cells. They may have one stout villus of 1 – 2 μm, or two or more smaller villi (Fig. 7.1A,B,C,D). These smaller villi sometimes arise from a common base (Fig. 7.2A) (Kotrschal 1991). Apical endings may vary even in the same fish species. Kotrschal and co-authors (1997) in their studies on the ontogeny of SCCs in the zebrafish *Danio rerio*, found that in the embryo and the early larvae the apical endings of SCCs have many small villi. In mature fish however, the apical endings have almost exclusively one stout villus (Fig. 7.2B). With the methods used by that study it was not possible however, to determine whether the cells change their shape or whether the oligovillous cells die and are replaced by monovillous cells.

The cell body is usually spindle-shaped (Fig. 7.2). In thinner epithelia the SCCs may be oblique. SCCs contain many mitochondria, a pronounced Golgi system, and longitudinally arranged microtubules sometimes associated with microfilaments. Vesicles are usually abundant and occasionally also occur in the apical villus. Size and electron density of the vesicles vary in different groups of fish (50 – 70 nm in diameter in cyprinids and silurids). Vesicles of different size and electron density may occur in the same cell. The nucleus is usually embayed or lobulated (Whitear, 1992).

4. INNERVATION OF SCCs

SCCs are secondary sensory cells. Slender nerve fibers contact the SCC mostly near the base of the cell. These nerve fibers often indent the cell body so that they are almost wrapped by the SCC. Synaptic specializations are inconspicuous and resemble gustatory synapses: pre- and postsynaptic densities are fuzzy. Occasionally small vesicles are seen on the presynaptic side. The nerve fibers contacting the SCCs belong to different nerves (cranial or spinal) depending on the location of the SCC (Whitear, 1952;

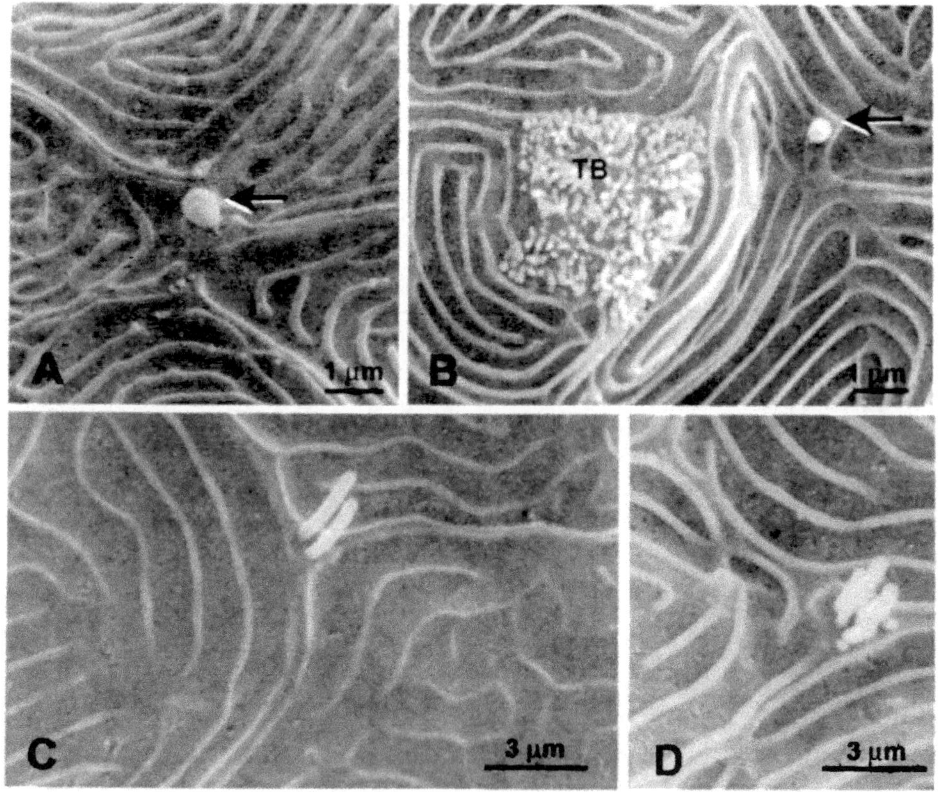

Fig. 7.1 **A – D** Scanning electron micrographs of fish epidermis. **A.** *Ictalurus punctatus*: a single stout villus (arrow) of a solitary chemosensory cell (SCC) protrudes between the epidermal cells. **B.** *Archocentrus nigrofasciatus*: a stout villus (arrow) is present next to a taste bud (**TB**). **C.** *Danio rerio*: SCC with two apical microvilli protrudes between epidermal cells of a zebrafish larva. **D.** *Danio rerio*: SCC with several apical microvilli in the epidermis of a larval zebrafish.

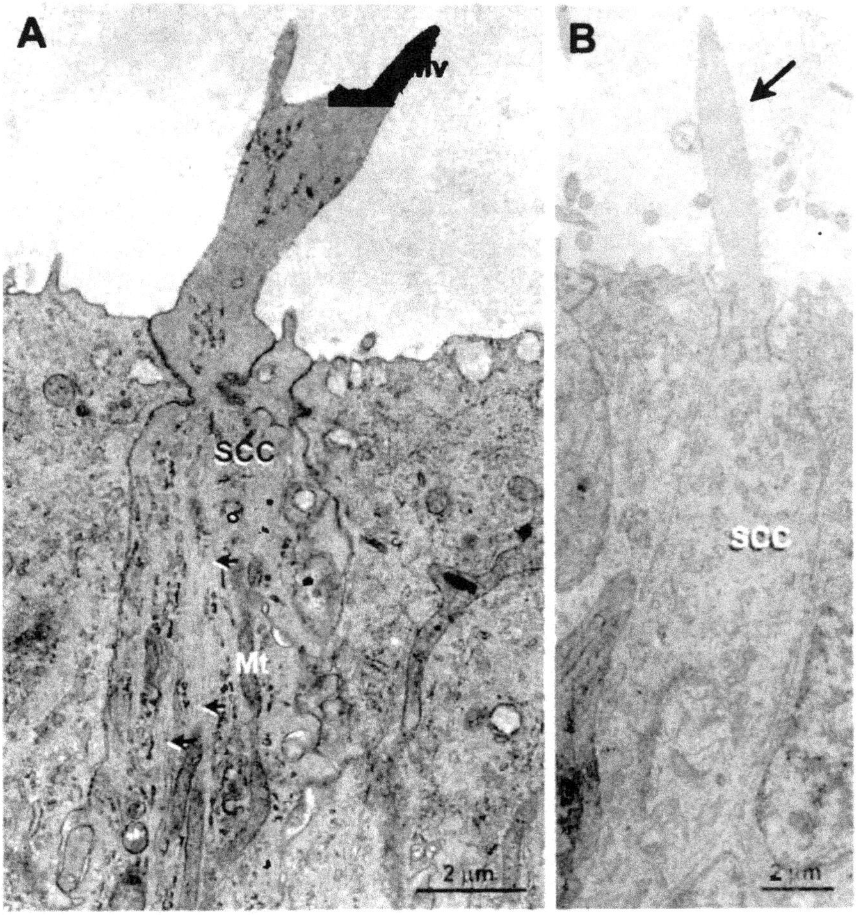

Fig. 7.2 **A – B** Transmission electron micrographs of SCCs.
A. *Carassius auratus*: SCC in goldfish epithelium. Apical ending of the **SCC** branches in several small microvilli (**Mv**). Microtubules (arrows) and mitochondria (**Mt**) are arranged longitudinally in the cell body. **B.** *Danio rerio*: SCC in the epithelium of an adult zebrafish. The slender cell body ends apically in one large villus (arrow).

Whitear and Kotrschal, 1988; Kotrschal and Finger, 1996; Kotrschal et al., 1998). SCCs on the body of teleosts are most probably innervated by branches of the spinal nerves although this has not yet been proven experimentally. In the catfish *Ictalurus melas*, the recurrent branch of the facial nerve (the VII cranial nerve) was cut. As a result the taste buds on the flank degenerated but the SCCs did not disappear (Lane and Whitear, 1977, cited in Whitear, 1992). This could either mean that the SCCs are trophically independent of the nerve or that the SCCs are innervated by spinal nerve fibers.

5. PHYSIOLOGY AND FUNCTION OF SCCs

Little is known about the physiology of SCCs. Due to their scattered distribution over the entire body it is difficult if not impossible to conduct electrophysiological experiments. So only a few studies are available on fish that have specialized SCC systems. For instance, lampreys have abundant SCCs (oligovillous cells) but no taste buds. Baatrup and Døving (1985) proved with electrophysiological methods that SCCs are chemosensory cells. These cells responded to sodium chloride, acetic acid, sialic acid, mucoid substances, and thaw water from trout. The recordings were done in areas without Merkel cells that would have been associated with tactile nerve fibers. Few other studies are available on SCCs systems without the interference of taste buds. The rockling groups (*Ciliata* and *Gaidropsarus*) have a specialized dorsal fin devoid of taste buds. The posterior part of the fin is stiff and used as a fin. The anterior part is flexible and the spines are abundantly covered with SCCs. As mentioned above, Kotrschal and coworkers (Kotrschal and Whitear, 1988; Kotrschal, 1992) estimated about 100,000 SCCs per square millimeter. The absence of taste buds made electrophysiological recordings possible. Responses were observed to diluted human saliva and mucus from other fish. As responses to typical taste stimuli such as amino acids, salts, and other acids were weak or absent, it was postulated that the system of SCCs on the dorsal fin of the rockling is used for predator avoidance, not for feeding (Peters and van Steenderen, 1987).

Innervation of the rockling dorsal anterior fin is interesting. The SCCs of this fin are innervated by the recurrent branch of the facial nerve connected to distinct dorsal areas within the facial lobe next to taste bud areas, i.e., it is centrally connected like a gustatory system but behaviorally used as a system for predator avoidance (for further details, see Finger, 1997).

Another group of fish with a specialized SCC system are the sea robins, e.g. *Prionotus* and *Trigla*. These fish belong to Scorpaeniformes, a group systematically widely separated from gadid rocklings. Sea robins have a specialized pectoral fin (Morrill, 1895; Finger and Kalil, 1985). The first three rays of the pectoral fin are free of webbing and taste buds. The fish uses the free fin rays like feet for walking and probing the ground.

Recordings showed that the fin responds to amino acids and their derivatives—similar but not identical to taste bud responses (Silver and Finger, 1984). As amino acids are the most common taste stimuli in fish, the SCCs in this case seem to form a system related to feeding.

These first three rays of the pectoral fin are innervated by the first 3 pairs of spinal nerves, nerves almost as thick as the spinal cord itself. Three accessory lobes in the spinal cord—dorsal horn enlargements—are the primary sensory nuclei, and constitute part of the somatic sensory column that leads to the ventral diencephalon. Thus the SCCs of the pectoral fin of the sea robin are centrally connected like a somatosensory system, but behaviorally used as a system that supports feeding (for further details, see Finger, 1997).

The only biochemical evidence for SCC function was reported for the channel catfish *Ictalurus punctatus*. A study on the putative arginine receptor in taste bud cells revealed that *Phaseolus vulgaris erythro-agglutinin*, a lectin used as a marker for this arginine receptor, also labeled SCCs in the barbels of catfish. Why SCCs, located close to the taste buds, express the same receptor as the taste bud cells is not known (Finger et al., 1996).

6. EVOLUTIONARY ASPECTS

Given the similarities of taste bud cells, SCCs, and the cells of Schreiner organs, ample speculations about the phylogenetic relationship of SCCs and taste bud cells have been published (for review, see Kotrschal, 1991, 1996; Finger, 1997).

SCCs show striking morphological and even cytochemical similarities to taste bud cells. Also, another specialized chemosensory system, the Schreiner organ of hagfish, consists of cells that look similar to SCCs. Schreiner organs are somatosensory systems whereas taste buds are viscerosensory systems. SCCs can either be somatosensory or viscerosensory depending on the species as described above, and it is still an unresolved question whether SCCs are precursors of taste buds or dispersed taste bud cells (Kotrschal, 1991, 1996; Finger, 1997). Braun (1998) postulated that Schreiner organs are not related to SCCs. To date it is not known whether all SCCs are homologous and/or how they are phylogenetically related to taste bud cells.

In the past SCCs were thought to be confined to aquatic animals. Recently it was shown that SCCs are also present in the developing vallate papilla (Sbarbati et al., 1998) and in the nasal cavity of juvenile and adult rodents (Finger et al., 2003). In rats and mice these cells express bitter receptors and gustducin, a G-protein involved in the transduction of bitter tastants (McLaughlin et al., 1992). Electrophysiological experiments proved that these cells respond to bitter substances, so it was postulated that they build a protective system against potentially noxious substances (Finger et al., 2003). Thus the system of SCCs is a phylogenetically old system that seems to have been conserved and is present in evolutionary younger taxa.

Acknowledgement

This work was supported in part by National Institutes of Health Grants PO1 DC00244 and RO1 DC006070 to Thomas Finger, University of Colorado Health Sciences Center at Fitzsimons, Aurora, Colorado, and P30 DC 04657 to Diego Restrepo, University of Colorado Health Sciences Center at Fitzsimons, Aurora, Colorado.

References

Baatrup, E. and K.B.Døving. 1985. Physiological studies on solitary receptors of the oral disc papillae in the adult brook lamprey, *Lampetra planeri* (Bloch). *Chem. Senses* 10: 559–566.

Braun, C.B. 1998. Schreiner organs: a new craniate chemosensory modality in hagfishes. *J. Comp. Neurol.* 392: 135–163.

Finger, T.E. 1997. Evolution of taste and solitary chemoreceptor cell systems. *Brain Behav. Evol.* 50: 234–243.

Finger, T.E. and K. Kalil. 1985. Organization of motoneuronal pools in the rostral spinal cord of the sea robin, *Prionotus carolinus*. *J. Comp. Neurol.* 239: 384–390.

Finger, T.E., B.P. Bryant, D.L. Kalinoski, J.H. Teeter, B. Böttger, W. Grosvenor, R.H. Cagan and J.G. Brand. 1996. Differential localization of putative amino acid receptors in taste buds of the channel catfish, *Ictalurus punctatus*. *J. Comp. Neurol.* 373: 129–138.

Finger, T.E., B. Böttger, A. Hansen, K.T. Anderson, H. Alimohammadi and W.L. Silver. 2003. Solitary chemoreceptor cells in the nasal cavity serve as sentinels of respiration. *PNAS* 100: 8981–8986.

Hansen, A. and K. Reutter. 2004. Chemosensory systems in fish: structural, functional and ecological aspects. In: *The Senses of Fish: Adaptations for the Reception of Natural Stimuli*, G. von der Emde, J. Mogdans and B.G. Kapoor (Eds). Narosa Publishing House, New Delhi, and Kluwer Academic Publisher, Kluwer, Dordrecht, The Netherlands, pp. 55–89.

Hansen, A. and H.P. Zippel. 1995. Solitary chemosensory cells in the olfactory organs of fish. *Chem. Senses* 20: 105.

Hansen, A., Reutter, K. and K. Zeiske. 2002. Taste bud development in the zebrafish, *Danio rerio*. *Dev. Dyn.* 223: 483–496.

Hansen, A., H.P. Zippel, P.W. Sorensen and J. Caprio. 1999. Ultrastructure of the olfactory epithelium in intact, axotomized, and bulbectomized goldfish, *Carassius auratus*. *Microsc. Res. Techn.* 45: 325–338.

Kapoor, B.G. and T.E. Finger. 2003. Taste and solitary chemoreceptor cells. In: *Catfishes*, G. Arratia, B.G. Kapoor, M. Chardon and R. Diogo, (Eds). Science Publishers, Inc. Enfield, (NH) USA & Plymouth. UK, vol. 2, pp. 753–769.

Kölliker, A. 1886. Histologische Studien an Batrachierlarven. *Z. wiss. Zool.* 43: 1–40.

Kotrschal, K. 1991. Solitary chemosensory cells - taste, common chemical sense or what? *Rev. Fish Biol. Fish.* 1: 3–22.

Kotrschal, K. 1992. Quantitative scanning electron microscopy of solitary chemoreceptor cells in cyprinids and other teleosts. *Environ. Biol. Fishes* 35: 273–282.

Kotrschal, K. 1996. Solitary chemosensory cells: Why do primary aquatic vertebrates need another taste system? *TREE* 11: 110–113.

Kotrschal, K. and H. Adam. 1984. Morphology and histology of the anterior dorsal fin of *Gaidropsarus mediterraneus* (Pisces Teleostei), a specialized sensory organ. *Zoomorphology* 104: 365–372.

Kotrschal, K. and T. E. Finger. 1996. Secondary connections of the dorsal and ventral facial lobes in a teleost fish, the rockling (*Ciliata mustela*). *J. Comp. Neurol.* 370: 415–426.

Kotrschal, K., W. D. Krautgartner and A. Hansen. 1997. Ontogeny of the solitary chemosensory cells in the zebrafish, *Danio rerio*. *Chem. Senses* 22: 111–118.

Kotrschal, K., S. Royer and J. C. Kinnamon. 1998. High–voltage electron microscopy and 3-D reconstruction of solitary chemosensory cells in the anterior dorsal fin of the gadid fish *Ciliata mustela* (Teleostei). *J. Struct. Biol.* 124: 59–69.

Kotrschal, K. and M. Whitear. 1988. Chemosensory anterior dorsal fin in rocklings (*Gaidropsarus* and *Ciliata*, Teleostei, Gadidae): somatotopic representation of the ramus recurrens facialis as revealed by transganglionic transport of HRP. *J. Comp. Neurol.* 268: 109–120.

Lane, E.B. and M. Whitear. 1977. On the occurrence of Merkel cells in the epidermis of teleost fishes. *Cell Tissue Res.* 182: 235–246.

McLaughlin, S.K., P.J. McKinnon and R.F. Margolskee. 1992. Gustducin is a taste–cell–specific G protein closely related to the transducins. *Nature* (Lond.) 357: 563–569.

Morrill, A. D. 1895. The pectoral appendages of *Prionotus* and their innervation. *J. Morphol.* 11: 177–192.

Peters, R. C. and G. W. van Steenderen. 1987. A chemoreceptive function for the anterior dorsal fin in rocklings (*Gaidropsarus* and *Ciliata*: Teleostei: Gadidae): electrophysiological evidence. *J. Mar. Biol. Assoc. UK* 67: 819–823.

Peters, R.C., K. Kotrschal and W.D. Krautgartner. 1991. Solitary chemoreceptor cells of *Ciliata mustela* (Gadidae, Teleostei) are tuned to mucoid stimuli. *Chem. Senses* 16: 31–42.

Sbarbati, A., C. Crescimanno, D. Benati and F. Osculati. 1998. Solitary chemosensory cells in the developing chemoreceptorial epithelium of the vallate papilla. *J. Neurocytol.* 27: 631–635.

Silver, W.L. and T.E. Finger. 1984. Electrophysiological examination of a non–olfactory, non–gustatory chemosense in the searobin, *Prionotus carolinus*. *J. Comp. Physiol.* 154: 167–174.

Whitear, M. 1952. The innervation of the skin of teleost fishes. Quart. *J. Microsc. Sci.* 93: 289–305.

Whitear, M. 1965. Presumed sensory cells in fish epidermis. *Nature* (Lond.) 208: 703–704.

Whitear, M. 1992. Solitary chemosensory cells. In: *Fish Chemoreception*. T.J. Hara (Ed.). Chapman & Hall, London, New York, pp. 103–125.

Whitear, M. and K. Kotrschal. 1988. The chemosensory anterior dorsal fin in rocklings (*Gaidropsarus* and *Ciliata*, Teleostei, Gadidae): activity, fine structure and innervation. *J. Zool. Lond.* 216: 339–366.

CHAPTER

8

Barbel Taste System in Catfish and Goatfish

Sadao Kiyohara and Junzo Tsukahara

ABSTRACT

Catfish and goatfish have been studied extensively because of the barbels they use to identify food items. The barbels are densely supplied with taste buds and the sensory input from them plays an essential role in the initial stages of feeding behavior such as food search and placing food in the mouth. This chapter addresses our recent findings on the barbel taste systems in the sea catfish (*Plotosus lineatus*) and goatfishes (*Parupeneus trifasciatus* and *P. pleurotaeni*).

Plotosus has 4 pairs of barbels, all similar in length. Taste buds are more densely distributed on the rostral and caudal surface of the barbels. A taste fiber bundle, a functional unit, carries information received from some longitudinal area of the barbel surface, to which the bundle is distributed to form networks. Each network is usually hexagonal, ranging 240-400 μm and 100-250 μm for maximum and minimum diameter respectively. Nerve strands leave in pairs from each network to innervate taste buds. Goatfish have a single

Address for Correspondence: Sadao Kiyohara, Department of Chemistry and BioScience, Faculty of Science, Kagoshima University, Kagoshima 890-0065, Japan. E-mail: kiyohara@sci.kagoshima-u.ac.jp pair of large barbels extending ventrally from the lower jaw. Taste buds are evenly distributed across the epithelium of the barbels and are innervated in an orthogonal pattern. One longitudinally running nerve bundle (LNB) or functional unit originates from the main trunk and divides into two circumferential nerve bundles (CNB) extending respectively medially and laterally around the barbels. At each transverse level, the CNB innervate two clusters of taste buds, each containing 14 end organs. One LNB of goatfish carries information originating from CNB fibers at a certain level of the longitudinal extent of the barbel.

A sharply defined somatotopical map is present in the facial lobe of both *Plotosus* and goatfish. The regions representing *Plotosus* barbels are sharply defined and extraordinarily enlarged as different lobules extending rostrocaudally in the facial lobe (FL). In the goatfish, the sensory inputs from the barbel terminate in a derived dorsal FL which has a highly convoluted surface forming a multitude of tubercles. These tubercles are actually recurved flexures in a convoluted continuous columnar representation of the barbel.

Key Words: Taste bud; Innervation; Topographic projection; Facial lobe; Trigeminal nerve; Somatotopy.

1. INTRODUCTION

For the gustatory system of fish and other vertebrates, peripheral gustatory inputs reach the primary taste center via three cranial nerves—facial (VII), glossopharyngeal (IX) and vagal (X). The primary taste center is organized as a pair of special visceral sensory columns located within the medulla oblongata (Herrick, 1905; Ariens-Kappers et al., 1932). Each column receives input in a rostrocaudal manner from the three cranial nerves respectively. Fish are unique vis-a-vis other vertebrates in having anatomical elaborations of the primary taste center. This primary taste center in species of fish possessing highly developed taste systems, such as catfishes or cyprinids, is subdivided into facial (FL) and vagal (VL) lobes. Behavioral experiments in catfish indicate that the FL and VL have different functions (Atema, 1971). The FL, which receives input from the facial nerve innervating taste buds located across the entire external body surface and rostral oral regions, functions in appetitive (food search and ingestive) behaviors. The VL, which receives input from glossopharyngeal (more rostral region) and vagal nerves that innervate taste buds exclusively within the oral cavity and pharynx, functions in consummatory (swallowing and rejection) behavior. This functional difference between the FL and VL is also supported by anatomical findings showing different reflex connections of both lobes within the brainstem (Morita and Finger, 1985a).

Particular species of fish with highly developed FLs or VLs have specialized taste end organs in which a large number of taste buds are concentrated. For example, catfish possess pairs of barbels around the jaws, the length and number of which are species specific. Goatfish have a single pair of large barbels extending downward from their lower jaw. The sensory input from the barbels of catfish and goatfish terminates in a large somatotopically organized FL that plays an essential role in initial stages of feeding behavior such as food search and placing food in the mouth. On the other hand, cyprinids such as carp have a large muscular palatal organ in the roof of the pharyngeal cavity. Input from the palatal organ terminates in a large somatotopically organized VL. The cyprinids utilize senses other than taste to locate food and usually take food in the mouth with nonedible material such as particles of substrate. The palatal organ serves in later stages of feeding behavior such as intraoral food selection and carrying food toward the esophagus (Sibbing, 1982).

This chapter describes our recent findings on the barbel taste system in the sea catfish (*Plotosus lineatus*) and goatfishes (*Parupeneus trifasciatus* and *P. pleurotaeni*), including the structure of barbel taste buds, their distribution, innervation, and central representation in the FL as well as the general morphology of the barbel and primary taste center (Kiyohara et al., 1986, 1996, 2002; Kiyohara, 1988; Marui et al., 1988; Sakata et al., 2001).

2. ANATOMY OF BARBELS

Sea catfish have four pairs of barbels: nasal, maxillary, lateral mandibular, and medial mandibular (Fig. 8.1A). They are basically the same in length and structure. Unlike goatfish, sea catfish do not actively trail food substances with their barbels. Instead they usually keep them passively extended forward. When they touch potential food with their barbels or lips, they quickly ingest it (Satô, 1937a). The surface of the barbels is composed of a stratified, squamous epidermis covering the loose connective tissue of the dermis (Fig. 8.1C). The epidermis of each barbel is differentiated into three regions—rostral, intermediate or lateral, and caudal, as shown in Fig. 8.1C. The rostral surface of each barbel faces inward or toward the lip side of the forward extended barbels. The rostral epidermis is thicker than the intermediate and caudal. The dermis, with

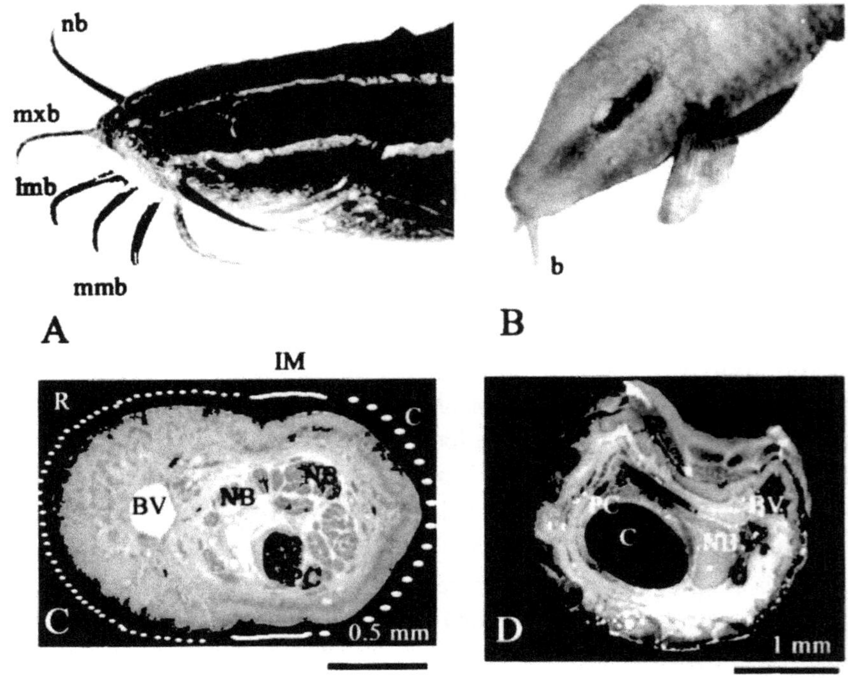

Fig. 8.1 **A.** The catfish, *Plotosus lineatus*, has four pairs of barbels, called nasal (nb), maxillary (mxb), lateral mandibular (lmb), and medial mandibular (mmb). **B.** The goatfish *Parupeneus trifasciatus* possesses a pair of elongate, stiff barbels as major sensory organs (Courtesy of Mr.Y. Hirai). **C.** Toluidine blue stained cross section through a barbel of *Plotosus*. Epidermal surface of the barbel divided into three regions: rostral (R), intermediate (IM), and caudal (C). **D.** Transversely cut barbel of goatfish. The structure of barbels is essentially similar between *Plotosus* and goatfish, but each component is greater in size in goatfish than in *Plotosus*. BV, blood vessel; NB, nerve bundle; C, cartilage; PC, perichondrium. Reproduced from Sakata et al. (2001) and Kiyohara et al. (2002).

its associated blood vessels and nerve fiber bundles, surrounds the nerve trunk and is encapsulated in the perichondrium (Fig. 8.1C). The large nerve trunk is located in the caudal region of the barbel, with some bundles also located in the rostral region. No intrinsic musculature exists in the barbels.

The goatfish (*Parupeneus trifasciatus* and *P. pleurotaeni*) have a large pair of barbels (like a goatee) extending downward from their lower jaw (Fig. 8.1B). The barbels are fairly rigid and are moved rapidly to both probe and stir the substrate (Satô, 1938). Once a potential prey item is encountered, the fish strikes at that particular location on the bottom and ingests the food item. The anatomical constituents of barbels in the goatfish are essentially similar to those in catfish, as mentioned above, but each constituent is better developed in goatfish than in *Plotosus* (Fig. 8.1D). No intrinsic musculature exists in the barbels, similar to *Plotosus*, but sets of muscles located in the lower jaw connect through their tendons with the most proximal part of the barbel cartilage and control barbel movement.

3. STRUCTURE OF BARBEL TASTE BUDS

Both the catfish and goatfish have many taste buds, not only on the barbels but also in various regions of the skin and mouth cavity. In *Plotosus*, external buds are distributed across the entire body surface from the lips to the caudal fin, as in other catfishes (Atema, 1971). In goatfish, they are limited to the head region and are relatively scarce except on the barbels (Satô, 1938). Taste buds vary in size, especially height, even in the same species according to their location as, shown in the minnow *Pseudorasbora parva* (Kiyohara et al., 1980). In both *Plotosus* and goatfish, barbels possess the largest buds, suggesting their importance as taste organs. The opening of the taste buds to the external environment, the so-called taste pore and receptor area, is unusually large in goatfish. The taste pores in goatfish reach upward of 25 μm, compared to less than 10 μm for the taste pores of catfish (Fig. 8.2A-D). Taste buds on the barbel of goatfish are larger than any taste buds observed so far in other species of fish (Fig. 8.2C). The maximum height, maximum width, and diameter of the pore, measured on the enlarged picture of sections through the longitudinal axis of the taste bud, are 85.98 ± 4.68 μm, 55.57 ± 4.83 μm, 24.81 ± 1.65 μm respectively (n = 11). In *Plotosus*, maximum height is similar, but the other parameters are less than 50% those in goatfish. Thus the buds in *Plotosus* appear elongated and distally tapered (Fig. 8.2A).

The apical surface of barbel taste buds is slightly depressed from the general epithelial surface in both groups of fish (Fig. 8.2A-D). Two types of cell projections are found at the apex of the taste buds, rod-shaped processes and microvilli, as observed in other species of fish (Grover-Johnson and Farbman, 1976; Reutter, 1978; Kitoh et al., 1987; Royer and Kinnamon, 1996; Reutter and Hansen, 2005). But this distinction is rather difficult in goatfish since the rod-shaped projections are small. Each taste bud contains at least three types of cells – tubular (t) or light cells,

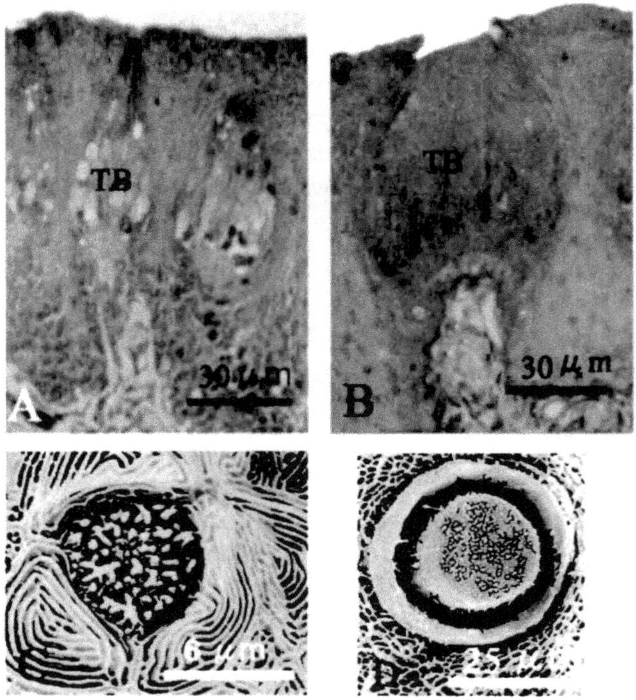

Fig. 8.2 A and B. Cross sections of *Plotosus* (A) and goatfish (B) barbel showing taste buds (TB). C and D. Scanning electron micrographs of the barbel surface of *Plotosus* (C) and goatfish (D) showing the apical processes of the taste bud cells extending outward through a single taste pore. Note the taste pore is much larger in goatfish than in *Plotosus*. Reproduced from Sakata et al. (2001) and Kiyohara et al. (2002).

filamentous (f) or dark cells, and basal cells, according to the characteristics of their cytoplasm or their position in the taste bud (Fig. 8.3). T-cells and f-cells are longitudinally elongated. T-cells have a rod-shaped process (large receptor villus) at the apex and are usually surrounded by f-cells with microvillar processes. One noticeable difference between catfish and goatfish is the number of elongated cells contained in each bud. The number is approximately 80–100 in *Plotosus* and 350–450 in goatfish. Thus enlargement of goatfish taste buds is due to an increase in number of elongated cells rather than size of each cell.

4. DISTRIBUTION OF TASTE BUDS IN BARBELS

In *Plotosus* and goatfish, taste buds are distributed across the entire epithelium of the barbel along its entire length and density of buds

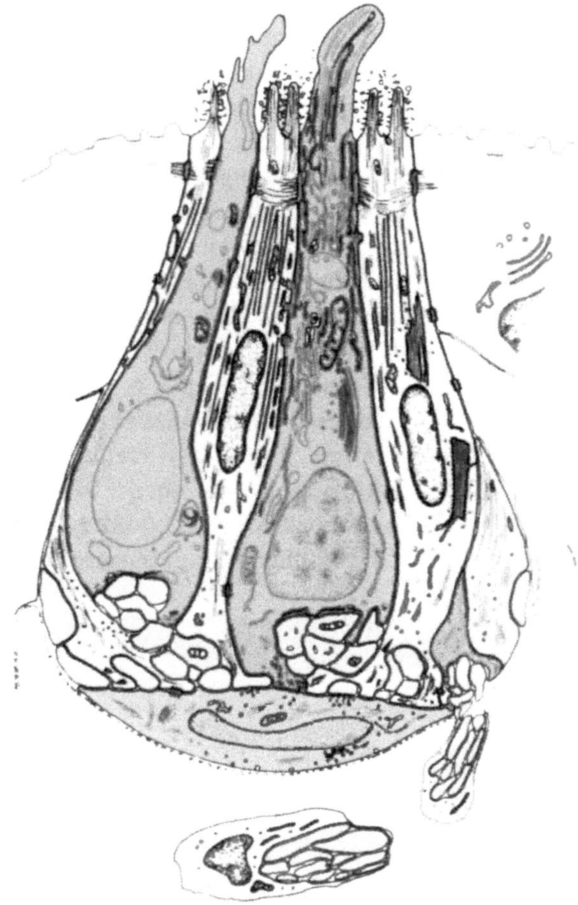

Fig. 8.3 Schematic representation of fish taste buds showing t-cell (light cell, red), f-cell (dark cell, blue), basal cell (green), and nerve fibers (yellow) of the bud's nerve fiber plexus. Reproduced from Kitoh et al. (1987).

increases toward the tip of the barbel. The density is generally higher in goatfish than in *Plotosus*. For comparison, taste bud density was measured in a catfish and a goatfish of 16 cm in total length. In *Plotosus*, taste bud density increased gradually from 100 mm^{-2} at the basal portion to over 200 mm^{-2} at the tip of the barbel. In the goatfish, density increased gradually from 150 mm^{-2} at the basal portion to over 250 mm^{-2} at the tip of the barbel.

The distribution of barbel buds showed a sharp contrast between *Plotosus* and goatfish. In *Plotosus*, the density varied considerably depending on the surface in the proximodistal extent. At each level of the barbel, taste bud density was highest on the rostral surface, moderately high on the caudal, but notably low on the intermediate epidermis (Fig. 8.4). Thus each barbel has buds with the highest density on the lip-side, which is thought to contact food more often than the intermediate and caudal sides. This pattern of distribution of buds is also found on barbels of other species of catfish, such as channel catfish and *Arius felis* (Kiyohara and Caprio, 1996), and of the gadid fish, *Ciliata mustela* (Kotrschal et al., 1993). In contrast, the goatfish has densely packed, large taste buds with an interbud spacing of approximately 50 µm (Fig. 8.5). At the basal region of the barbel, the taste bud-bearing epithelium forms patches intermingled with areas of non-sensory epithelium (Fig. 8.5A). Frequently, taste buds are distributed in groups of 12–16 sensory end organs. In the middle and distal portions of the barbel, taste buds are distributed evenly across the entire epithelium (Fig. 8.5B, C). This distribution pattern allows goatfish

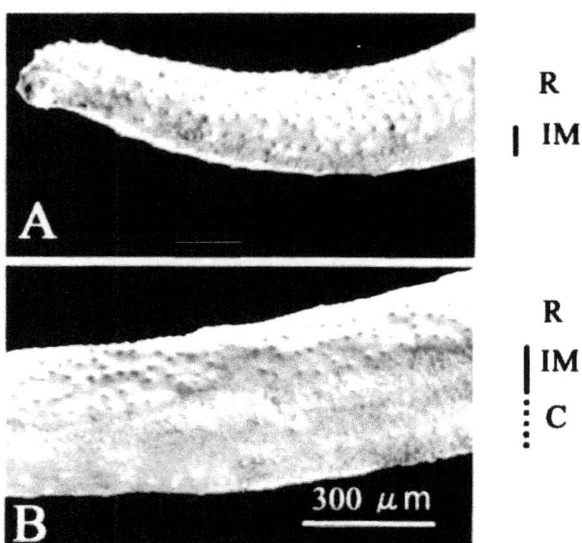

Fig. 8.4 Scanning electron micrographs of the surface of the nasal barbel of *Plotosus* showing distribution of taste pores of taste buds with different densities in the rostral (R), intermediate (IM) and caudal (C) epidermis. Taste buds are most concentrated in the rostral epidermis. **A.** Tip of barbel, **B.** Rostral and intermediate surfaces of middle part of barbel. Reproduced from Sakata et al. (2001).

to detect food by inserting the distal part of the barbel into the substrate and stirring it vigorously. A similar distribution pattern is also found in barbels of the silvereye, *Polymixia japonica* (Satô, 1937b).

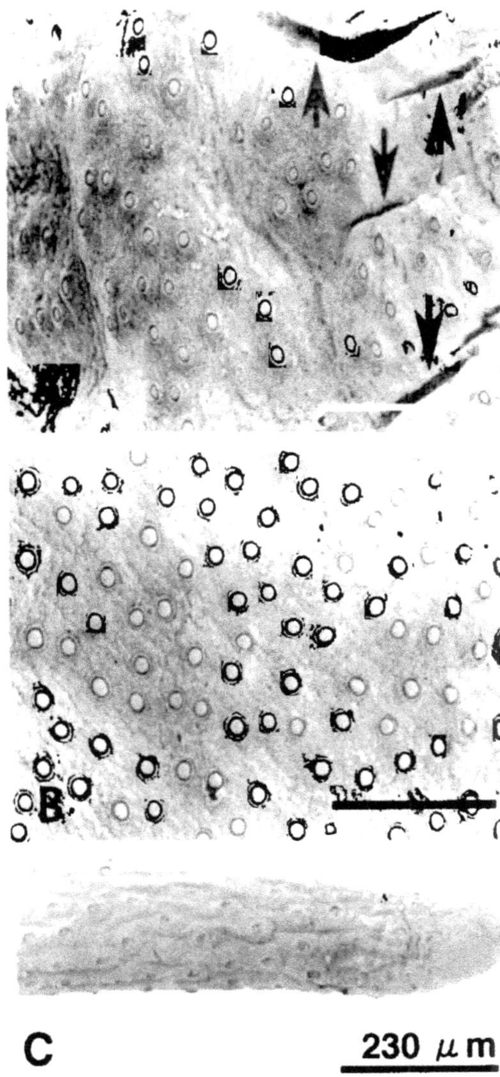

Fig. 8.5 Scanning electron micrographs of surface of barbel in goatfish showing distribution of taste pores of taste buds in basal (**A**), middle (**B**) and tip (**C**) regions. At each region, the taste pores are uniformly distributed and their density increases distally. Arrows in **A** indicate artifacts. Reproduced in part from Kiyohara et al. (2002).

5. DISTRIBUTION OF NERVE FIBERS TO INNERVATE TASTE BUDS IN THE BARBELS

Depending on the peripheral location of taste buds, they are innervated by either the facial, glossopharyngeal, or vagal nerves. Innervation of taste buds and their morphology has been studied extensively in various species (Grover-Johnson and Farbman, 1976; Joyce and Chapman, 1978; Reutter, 1978; Royer and Kinnamon, 1996). Nerve fibers enter the basal portion of a barbel taste bud as intragemmal fibers and form synaptic connections with at least two morphological classes of taste bud cells: t-cells (light) and basal cells (Reutter, 1978, 1992). Fibers also frequently pass between taste buds and their surrounding epithelium as perigemmal fibers, and in the epithelium between taste buds as extragemmal fibers (Finger and Böttger, 1990). To reveal the entire distribution patterns of nerve fibers in the barbels, fibers were labeled in *Plotosus* and goatfish with the carbocyanine dye 1,1'diocadecyl-3, 3,3', 3'-tetramethylindocarbocyanine perchlorate (DiI) and sections of barbels or surface epithelium were viewed by epifluorescence using compound, dissecting, and laser scanning confocal microscopes (Sakata et al., 2001; Kiyohara et al., 2002). The results are detailed below.

5.1 *Plotosus*

Each barbel of *Plotosus* is supplied by a nerve trunk, which receives fibers from corresponding peripheral rami containing both trigeminal and facial fibers. The nerve trunk enters the caudal region of each barbel at its base and sends many bundles to the rostral side as it runs toward the tip. When DiI is placed on the stump of the caudal trunk in a cut piece of barbel, fluorescent-labeled fibers in bundles can be followed along the length of the barbel under a dissecting fluorescent microscope. The nerve bundles reach the dermis, ramifying repeatedly to form networks under the epidermis (Figs. 8.6 and 8.7). The networks are most abundant under the rostral surface, moderately abundant under the caudal surface, and sparse under the intermediate surface (Fig. 8.6A, B). The networks are usually hexagonal in shape (Fig. 8.7A). Their size is smaller in the rostral than in the caudal surface of the barbel and becomes smaller toward the tip. The maximum and minimum widths of the hexagonal arrays are 240–400 μm and 100–250 μm respectively. Figure 8.7A is a surface view of these networks and the nerve bundles beneath them. This picture was obtained from a three-dimensional reconstruction of labeled fibers within

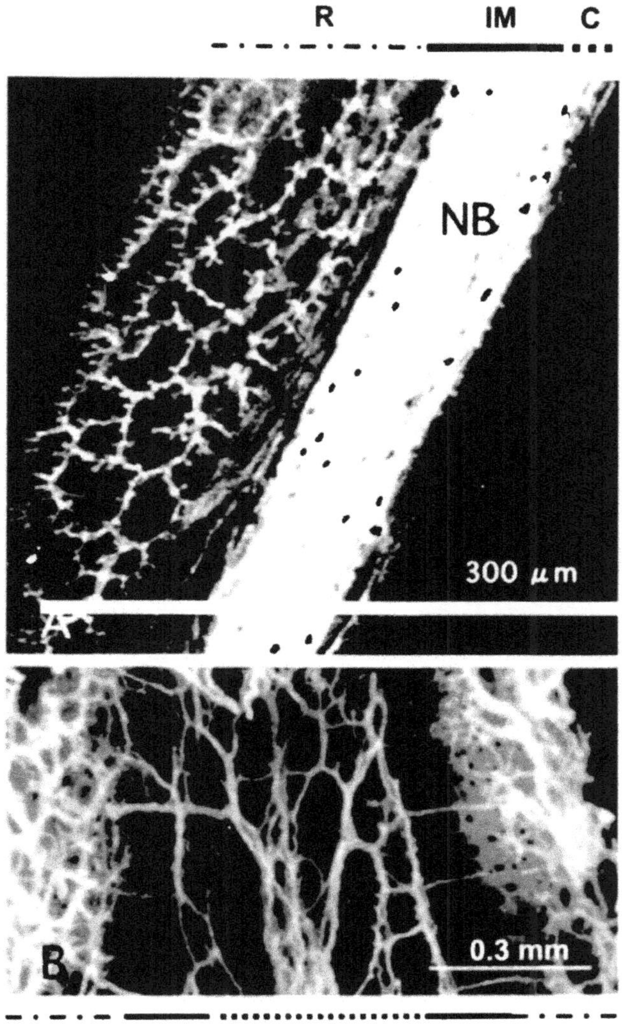

Fig. 8.6 Distribution of nerve fibers under the epidermis of the maxillary barbel of *Plotosus* as identified by DiI application to the barbel nerves. **A.** Confocal microscopy showing rostral and intermediate surface of the middle part of the barbel. Note many hexagonal nerve networks on the rostral side. On the right side, corresponding to the intermediate surface, a heavily labeled trunk, which lies under the epithelium, is seen and this labeling prevents surface observation of this side. **B.** Distribution of nerve fibers in the rostral, intermediate, and caudal surface. Note the nerve networks and fiber bundles are heavy in the rostral, medium in the caudal, and relatively sparse in the intermediate region. This material was prepared first by incising the rostral part longitudinally along the length of a cut piece of barbel and removing the central part from the piece. Then the remaining part of the cut piece was spread in a flat plane with the surface above and photographed by a standard epifluorescence microscope. NB, nerve bundle; R, rostral side; IM, intermediate side; C, caudal side. Reproduced from Sakata et al. (2001).

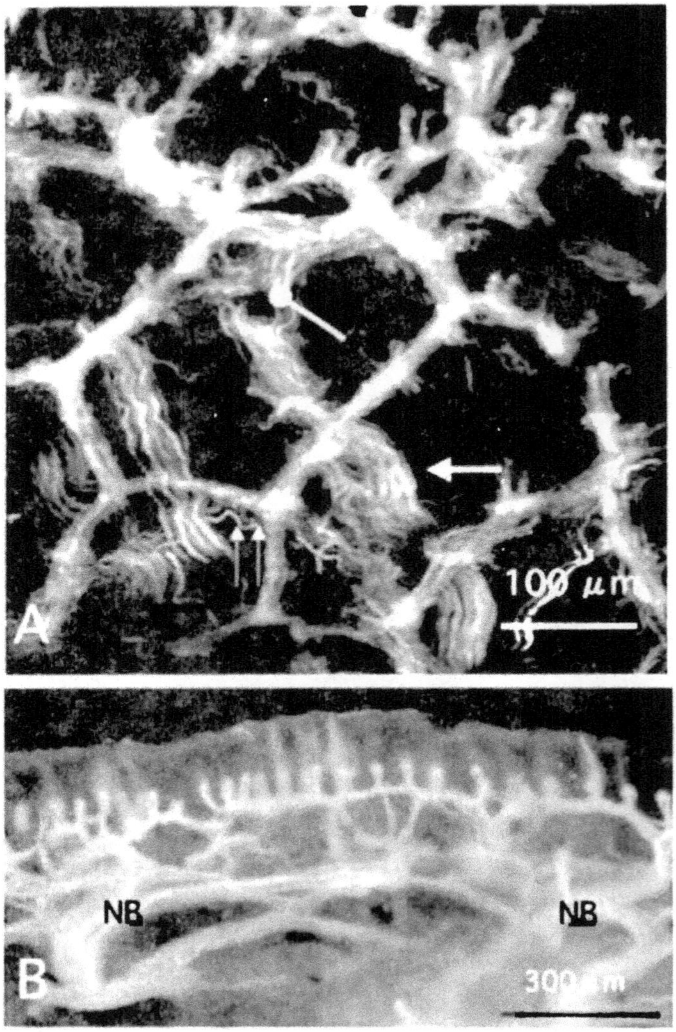

Fig. 8.7 Enlarged view of the barbel surface showing the distribution patterns of nerve bundles (NB) in *Plotosus* as identified by DiI application to the barbel nerves. **A.** Rostral surface of the maxillary barbel. Note clear hexagonal nerve networks between the dermis and epidermis. A large nerve bundle (triangular arrow) appears in the deeper region and ramifies (knobbed arrow) sending fibers to the above network. Coarse fibers are also found to terminate in the dermis (double trianglular arrows). Reconstructed photograph from 300 confocal pictures taken with 1μm intervals. **B.** Lateral view of the surface of a longitudinally cut section of the barbel. Upper, rostral surface, with many taste buds weakly labeled. Each taste bud receives a small nerve strand, which originates perpendicularly from the nerve networks. In the deeper region, large NBs are located and send fibers to the aforesaid networks. Reproduced from Sakata et al. (2001).

approximately 380 μm from the surface of a barbel. A large bundle (triangular arrow) is located in the dermis and runs longitudinally in the barbel. This bundle ramifies (knobbed arrow) into small bundles which send fibers to the overlying networks. A few thick fibers (double arrows) are also found to run between the large bundles and the networks and appear to end as free nerve endings in the dermis. The relationships of these networks and their underlying nerve bundles are illustrated in a longitudinal section of a barbel in Figure 8.7B.

Nerve strands travel perpendicularly from the networks toward the epidermis, where they enter taste buds (Fig. 8.7B). Most of these strands originate in pairs from the network (Fig. 8.8B). This paired appearance of strands coincides with an SEM view of the surface of a barbel (Fig. 8.8A). In Figure 8.8A, taste pores labeled with red spots are arranged in pairs. The number of strands originating from a network varies between 20 and 50. At the transition from the corium papilla to a taste bud each strand is further divided into two substrands (Fig. 8.9). Each substrand crosses the basement membrane separately beneath a taste bud through either of the two most basolateral sides. Some of these intragemmal fibers from the substrands ascend 3-5 μm to form terminal varicosities (Fig. 8.9). Some fibers also separate from the nerve plexus of a taste bud and are distributed between the taste bud and the surrounding epithelium as perigemmal fibers. They are usually thick in diameter and sometimes reach nearly to the level of the apical taste pore. Single fibers frequently emerge from networks and are distributed as extragemmal fibers in the epidermis. Some of these fibers reach the epithelial surface, but more often extend approximately two-thirds of the way toward the surface of the epithelium. Frequently, varicosities occur along the length of these extragemmal fibers. The intermediate epidermis appears to have fewer free nerve endings than the rostral or caudal side. Taste buds are labeled as fluorescent haze and cells in the taste buds are occasionally labeled (presumably transcellularly) as described for the channel catfish (Finger and Böttger, 1990) and gadid fish, *Ciliata mustela* (Kotrschal et al., 1993). The degree of these labelings appears to depend essentially on the period or the distance of the application of DiI.

5.2 Goatfish

After DiI-staining, bundles of axons can be similarly followed along the length of the goatfish barbel and seen to ramify beneath the skin to

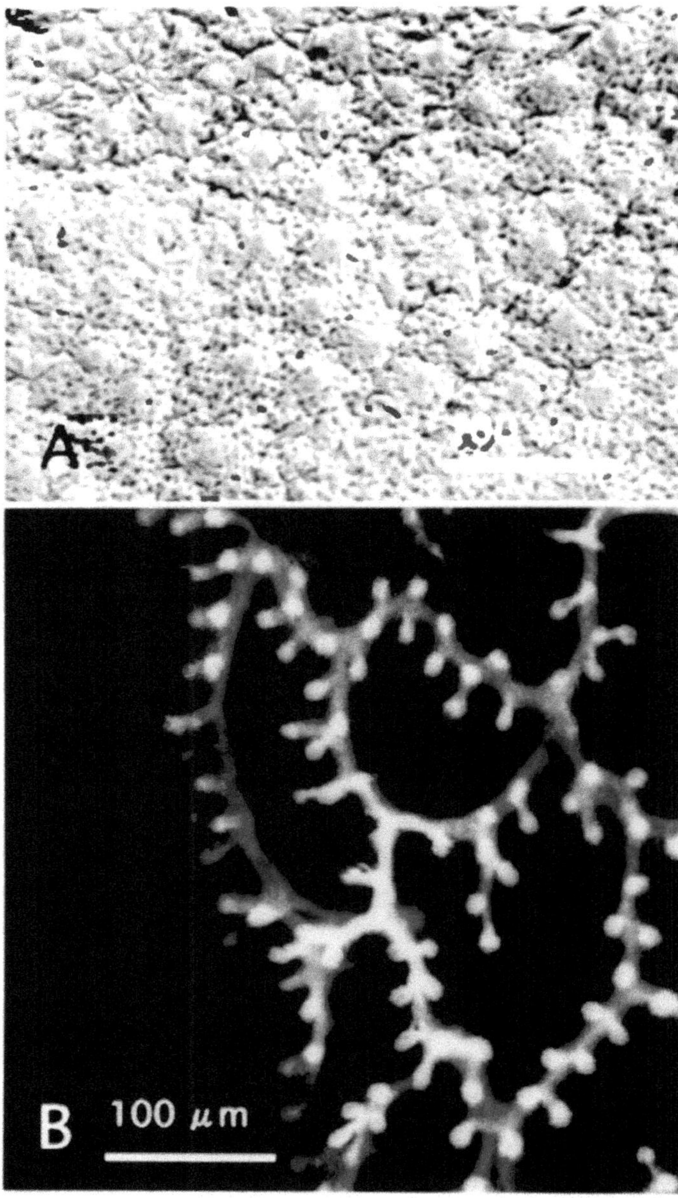

Fig. 8.8 Distribution of taste buds and their innervation in the rostral region of the nasal barbel of *Plotosus*. **A.** Scanning electron microscopy of the rostral surface of the barbel. Taste pores painted red. Note that taste buds appear to be distributed in pairs. **B.** Nerve networks and nerve strands under the rostral surface as identified by DiI application to the barbel nerves. This photograph was obtained by superimposing two confocal pictures taken at a 17μm interval. Labeled nerve fibers are shown in red or yellow. Yellow image is superficial to red image. Note that most of the nerve strands appear in pairs from the underlying networks. Reproduced from Sakata et al. (2001).

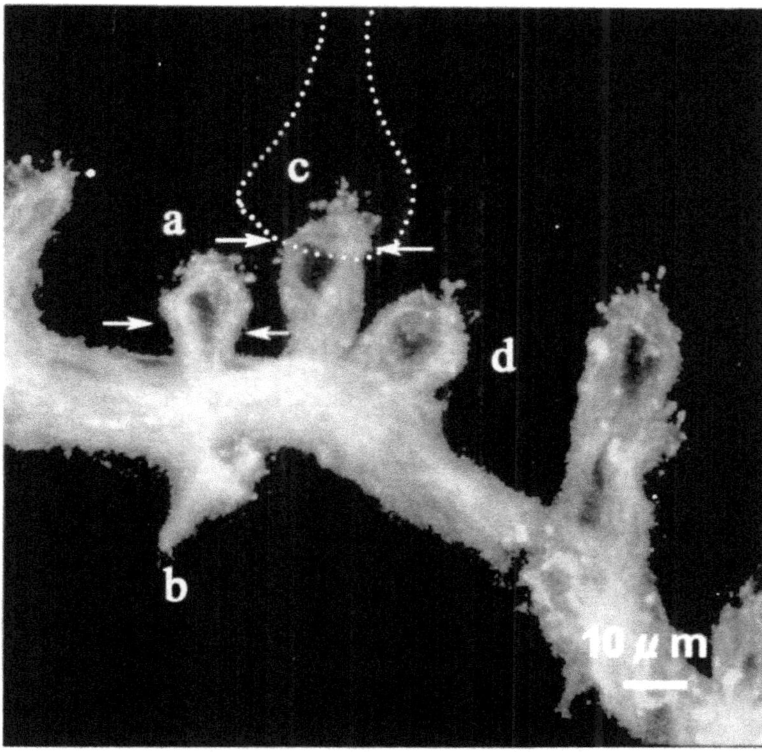

Fig. 8.9 Three-dimensional reconstruction showing pairs of nerve strands innervating taste buds and their subdivision into substrands in *Plotosus*. In this picture, two pairs (a and b, c and d) are clearly visible. Nerve strands (a and c) are seen to divide into two substrands (arrows) before entering a taste bud. Each substrand enters the same taste bud through one of the basolateral sides to form a plexus with the other substrand. Note some intragemmal fibers further enter the taste bud and end in a swelling. This micrograph was obtained from 150 confocal pictures taken at 0.2 μm intervals. A taste bud is outlined by dots. Reproduced from Sakata et al. (2001).

innervate the base of taste buds. Occasionally, elongate cells of the taste buds were labeled by transcellular diffusion of DiI, as observed in *Plotosus*, but usually only the nerve plexus at the base of the taste bud was labeled (Fig. 8.10). In goatfish, the nerve plexus forms a disk embracing the basal ends of the taste cells. When viewed from the surface, the taste buds are arranged in clusters, each cluster receiving its own branch of the main barbel nerve trunks (Fig. 8.11A,B). Each cluster comprises two hemiclusters, usually containing 7 taste buds each (range 5-8) (Fig. 8.10A-C). The taste buds in a cluster cover an oval area approximately 500 × 200 μm. Innervation of taste buds in the barbel follows an

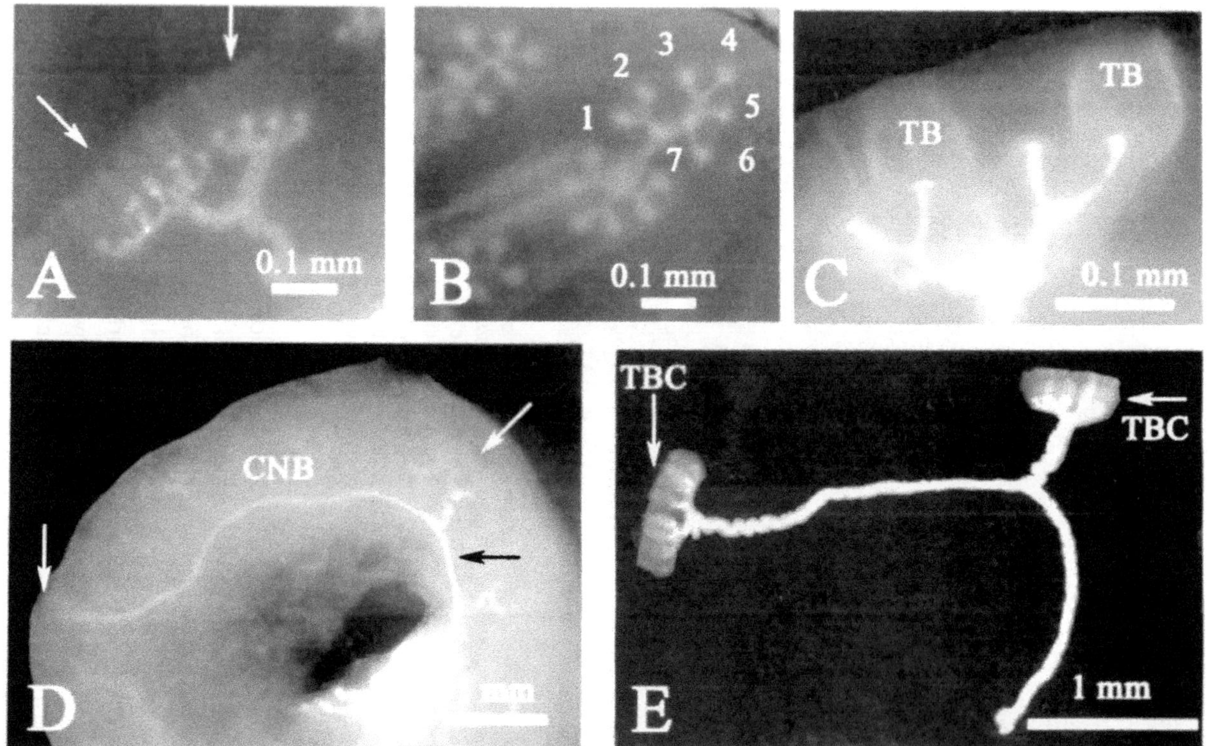

Fig. 8.10 Innervation of taste buds in the barbels of goatfish. The nerves are filled with the fluorescent tracer DiI. **A.** Lateral view of one cluster of nerve terminals. Note one cluster consists of two hemiclusters (arrows), which are innervated by the circumferential branch. **B.** Surface view of a cluster showing each hemicluster consists of 7 terminal arbors, one beneath each taste bud of the cluster. **C.** Side view of a hemicluster. **D.** Transversely cut barbel showing one circumferential nerve bundle (CNB; black arrow) which originated from a longitudinal bundle at the right side of the barbel to run around the central rod cartilage, which was removed in this piece of barbel. This branch ends in two terminal branches (white arrows). **E.** One CNB was dissected out from the left preparation (D). Note this branch innervates two taste bud clusters (TBC; white arrows). Reproduced from Kiyohara et al. (2002).

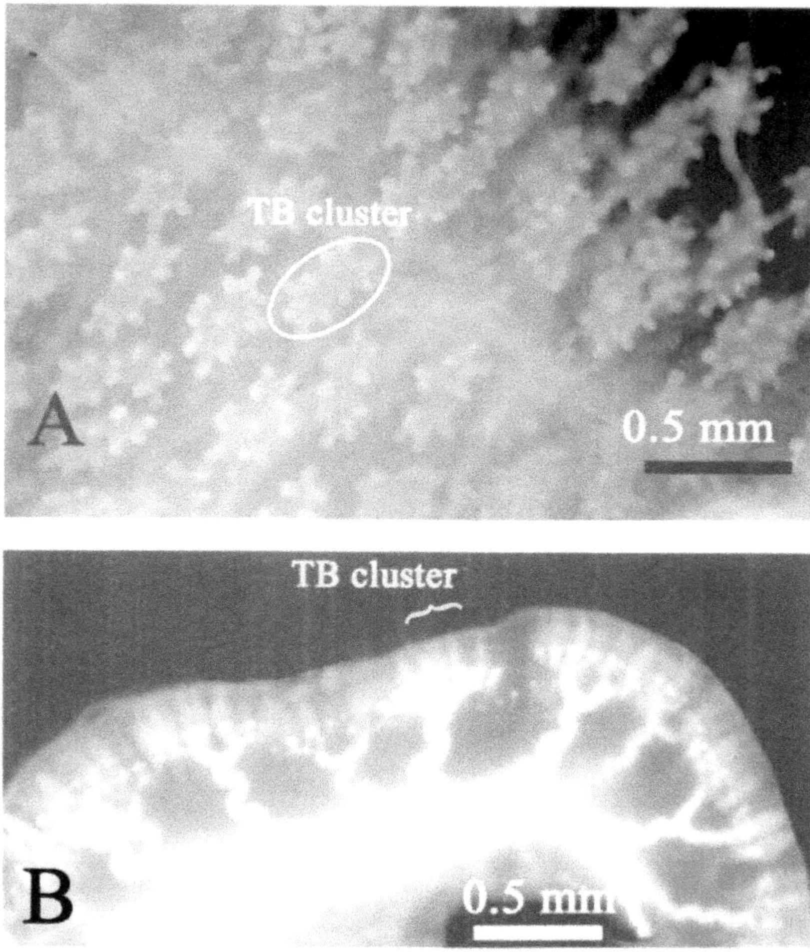

Fig. 8.11 **A.** Surface view of a goatfish barbel in which the nerve processes have been stained with the fluorescent tracer, DiI. The labeled nerve fibers branch to form stereotypically organized nerve terminals beneath the taste bud clusters (TBC) in the epithelium. **B.** Cross section through a DiI-labeled barbel showing each TBC innervated by a single radial nerve bundle emanating from the CNBs. Reproduced from Kiyohara et al. (2002).

orthogonal system. The principal barbel nerve runs longitudinally along the lateral edge of the cartilaginous core of the barbel. Fascicles leave the main trunk along the posterior margin of the core to form distally directed longitudinal bundles, which extend a short distance along the length of the barbel (Fig. 8.12). Each longitudinal bundle gives rise to paired circumferential branches extending medially and laterally around the

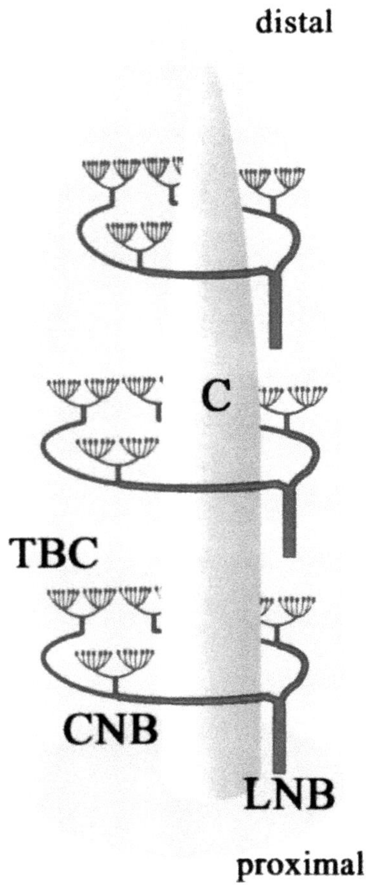

Fig. 8.12 Schematic representation of the pattern of innervation of TBs in the barbel of goatfish. Longitudinally running nerve branch (LNB) originating from the main nerve trunk sends a pair of circumferential branches (CNB) medially and laterally around the margins of the barbel to innervate 4 TBCs at each proximal-distal level. Each cluster consists of 14 TBs. Each longitudinal branch, which consists of approximately 90 fibers innervating 56 TBs, is a functional unit. Each 1-mm length of barbel contains approximately 15 CNBs. Reproduced from Kiyohara et al. (2002).

margins of the barbel to innervate the taste bud clusters at that proximodistal level (Figs. 8.10D,E, and 8.12). The axons of each circumferential branch innervate two clusters of taste buds at that particular transverse level; so, when DiI is placed into a single taste bud, labeled fibers are present in all taste buds along the same circumferential branch.

Each longitudinally running nerve branch (LNB) can be thought to be a functional unit, originating from the main trunk and dividing into two circumferential nerve bundles (CNB) extending respectively medially and laterally around the barbel (Fig. 8.12). Each CNB innervates two taste bud clusters in the epithelium. Since each taste bud cluster contains approximately 14 taste buds, each CNB innervates 28 taste buds. Thus each longitudinal bundle innervates 56 taste buds located at a defined transverse level of the barbel. There are on average 15 CNB per 1-mm length of barbel. Using these values, we can estimate the number of LNB, CNB, and taste buds in the barbel. For example, a 20-cm specimen of *Parupeneus trifasciatus* possesses a pair of 4 cm long barbel. In this barbel, there would be approximately 600 CNB and 33,600 taste buds.

Transmission electron microscopy of the nerve bundles at various regions of the barbel reveals that the LNB and the CNB each contain similar numbers of axons, suggesting that fibers in a LNB bifurcate to travel in both CNBs. Most of the LNBs consist of approximately 90 fibers with diameter less than 2 μm while some LNBs contain some coarse fibers greater than 3 μm in diameter. These results are supported by experiments involving DiI application to a single taste bud (not shown). Apparently, each nerve fiber of the circumferential branch innervates all taste buds innervated by the branch as a whole.

5.3 Comparison of Innervation of Taste Buds Between Catfish and Goatfish

The catfish and goatfish differ in the neural organization of their barbels. In the barbel of catfish, a taste fiber bundle or a functional unit carries information received from some longitudinal area of the barbel surface, to which the bundle is distributed to form networks. In the goatfish barbel however, one longitudinally running nerve bundle or a functional unit carries information originating from CNB fibers at a certain level of longitudinal extent of the barbel. Since a sharply defined somatotopical map is present in the FL of both *Plotosus* and goatfish, as is discussed below, these functional units of barbel nerves must determine receptive fields characteristic in the FL.

The pattern of innervation of a single taste bud is also different between *Plotosus* and goatfish. Each taste bud in *Plotosus* receives fibers through two substrands, each of which enters the bud basolaterally (Fig. 8.9). In the barbels of goatfish, a single strand enters the central portion of the basal region of a taste bud (Fig. 8.10C). In the pelvic fin of the gadid fish, *Ciliata mustela*, each taste bud is innervated by more than two fascicles which originate from different regions of the underlying bundle (Kotrschal et al., 1993). Therefore, the number of strands supplying a taste bud and their pattern of divergence from the parent bundle vary among species, or even depend on the location of taste buds in the same species.

It is of particular interest that the barbel taste buds of *Plotosus* are located in pairs on the underlying network (Fig. 8.8). Similar paired innervation of taste buds has also been found in the gill arches of *Gnathopogon biwae* (Iwai, 1964). Since a pair of nerve strands originates from the same bundle, their innervated taste buds might be functionally homogeneous. This may also be true for the 56 taste buds innervated by a single LNB of goatfish.

Reutter (1992) studied the ultrastructure of taste buds in the barbels of *Plotosus* and found synapse-like structures between fibers and t-cells (light cell) or f-cells (dark cell), or between fibers and basal cells. Intragemmal fibers often reach above the taste bud nerve fiber plexus and terminate as varicosities (Fig. 8.9). Since taste fibers are found to swell at synaptic regions in the taste buds of the gadid fish (Crisp et al., 1975), these varicosities may represent synapses on the basal processes of either t-cells or f-cells.

5.4 Fiber Components of Barbel Nerves and their Function

Barbels of both catfish and goatfish are sensitive to chemical and mechanical stimulation. However, the constituents of barbel nerves responsible for mechanosensitivity are not well understood. Barbel nerves of catfish contain sensory fibers of trigeminal and facial nerves. Kotrschal et al. (1993) selectively applied DiI to the stumps of the facial or spinal nerve supplying pelvic fins in the gadid fish *Ciliata mustela*. They observed that intragemmal and perigemmal fibers of taste buds were labeled but no free epithelial endings were labeled, after application to the facial nerve stumps. The spinal sensory fibers are thought to be functionally equivalent to trigeminal fibers (Dodd and Kelly, 1991). In addition, nerve tracing experiments in the puffer *Fugu pardalis* showed that taste information from the lips, which are innervated by both the trigeminal and facial nerves, is transmitted only by the facial nerve root (Kiyohara et al., 1975). Our results suggest that the intragemmal and perigemmal fibers in the sea catfish belong to the facial nerve while extragemmal fibers are trigeminal. The facial nerve of fishes can carry tactile as well as taste information (Davenport and Caprio, 1982; Kiyohara et al., 1985a). The origin of tactile sensitivity of the facial fibers is thought to arise from the perigemmal fibers. However, another origin is possible-connection of Merkel-like basal cells of the taste bud to intragemmal fibers (Toyoshima et al., 1984; Kiyohara et al., 1985a). Since the relative number of perigemmal fibers in fishes is quite few compared to intragemmal fibers, and tactile sensitivity is significant in facial nerve recordings (Kiyohara et al., 1985a), intragemmal fibers are likely to respond also to tactile stimulation. If this is the case, taste buds in *Plotosus* may function as compound sensory organs containing chemosensory and mechanosensory cells, as suggested by Reutter (1971, 1986) and Finger (1997).

The barbels of *Plotosus* and goatfish contain no intrinsic musculature and no retrograde labeling was found in the brain stem after application of horseradish peroxidase to the central stumps of barbel nerves. Therefore, motor fibers from the trigeminal or facial nerve do not contribute to the innervation of the barbels. However, other efferent components of barbel nerves, e.g. sympathetic fibers relating to the wall of the blood vessels, remain unknown (Harder, 1975).

6. GENERAL MORPHOLOGICAL FEATURES OF THE PRIMARY TASTE CENTER IN *PLOTOSUS* AND GOATFISH

The primary taste center of *Plotosus* is located in two bilateral pairs of extraordinary protrusions on the dorsal surface of the medulla oblongata, the FL and VL(Fig. 8.13A). Histologically, the FL is subdivided by fascicles of nerve fibers into 5 distinct lobules constituting 5 longitudinal columns (Fig. 8.13B). They are more clearly distinguished in *Plotosus* than in other previously studied catfishes such as *Ictalurus punctatus* (Hayama and Caprio, 1989) and *Arius felis* (Kiyohara and Caprio, 1996). The four lobules that receive afferents from the barbels are round in transverse section. A three-dimensional representation of these lobules shows that the four barbel lobules occupy approximately the anterior two-thirds of the FL (Fig. 8.13C). They are termed (from medial to lateral) : the medial mandibular barbel lobule (MML), lateral mandibular barbel lobule (LML), maxillary barbel lobule (MXL), and nasal barbel lobule (NBL). This is done on the basis of both histological and electrophysiological evidence

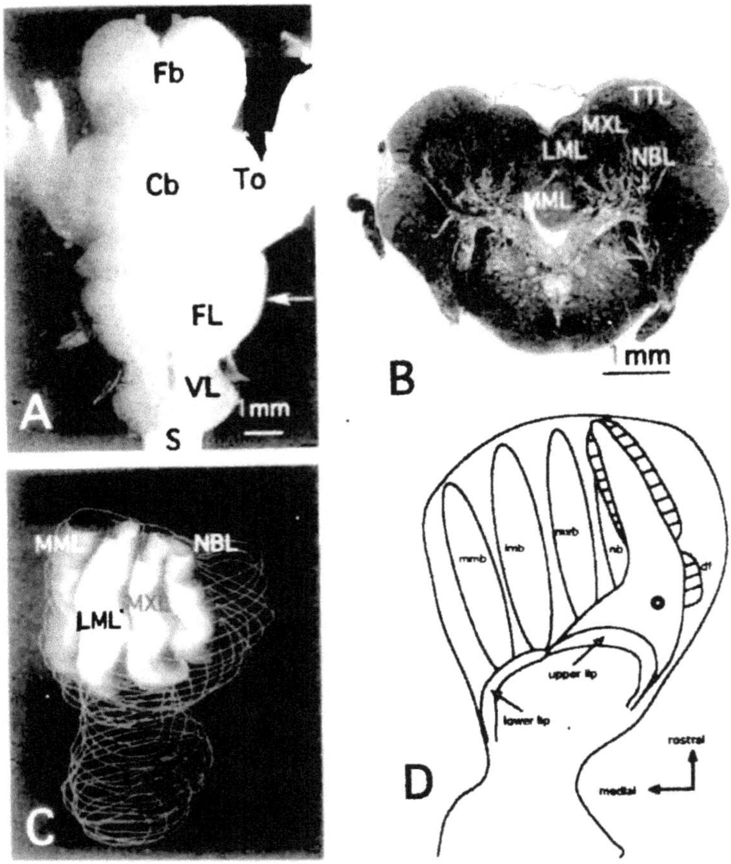

Fig. 8.13 Primary taste center of *Plotosus* and somatotopic organization in the facial lobe. **A.** Dorsal view of brain. **B.** Photomicrograph of transverse section through the anterior facial lobe (FL) (indicated by arrow in A). **C.** Dorsal view of a three-dimensional reconstruction showing the rostrocaudal extension of the four barbel lobules in the left side of the medulla oblongata. Solid lines show the outline of the FL and the vagal lobe (VL). **D.** Schematic representation of somatotopic map in the FL. Fb, forebrain; To, optic tectum; Cb, cerebellum; S, spinal cord; MML, medial mandibular barbel lobule; LML, lateral mandibular barbel lobule; MXL, maxillary barbel lobule; NBL, nasal barbel lobule. Reproduced from Kiyohara et al. (1996).

for the topographical projection of peripheral fibers to each respective lobule (Marui et al., 1988; Kiyohara et al., 1996). The fifth lobule is located dorsolateral to the barbel lobules and is dorsoventrally flattened. This lobule is termed the trunk-tail lobule (TTL) and extends rostrocaudally as do the other lobules. These lobules become less distinct caudally and no lobules are recognizable in the caudal one-third of the FL.

The lengths of the MML, LML, MXL, NBL, and TTL are approximately equivalent. Another columnar nucleus, the intermediate nucleus of the FL (NIF), is also found in the medioventral region of the FL. This nucleus appears ventral to the LML or MML, approximately at the level of the anterior one-third of the FL, and continues through the caudal FL.

Neurons of the 5 lobules of the FL can be classified by size as either medium or large (Kiyohara et al., 1996). The somata of medium neurons are oval and 7–9 μm in diameter, with little cytoplasm. They are located throughout the lobules, frequently in clusters of 20 to 50 cells. The medium neurons have slender dendrites which give off numerous secondary dendrites with sparse spines. Tertiary dendrites are also frequently seen. Medium cells have a 60 μm dendritic field around them. The somata of large neurons are polygonal or fusiform in shape, and 15–20 × 12–15 μm in major and minor diameter. They possess abundant cytoplasm and obvious Nissl bodies, and are found mainly in the circumferential portions of the lobules. The large neurons are subdivided into two types according to the morphology of their dendrites. One has relatively slender dendrites which project in all directions and often ramify extensively. The ramified dendrites of this type bear spines. The other type has a few thick, smooth dendrites without spines. These dendrites often run rostrocaudally in the periphery of the barbel lobules. The dendrites of both types of large neurons extend over 160 μm, producing a total dendritic field of 300–350 μm for each neuron. Since the minor and major diameters of a barbel lobule measured in transverse sections are 500–600 and 700–800 μm respectively, the dendritic fields of large cells cover substantial areas of each lobule. No neurons were found to extend dendrites from one lobule to another. Application of DiI to the cut surface of the ascending secondary gustatory tract retrogradely labeled only large neurons, suggesting that large and medium neurons are projection and intrinsic cells respectively.

In goatfish, the rostral medulla exhibits an elaborate FL protruding dorsally from the floor of the fourth ventricle and extending beneath the caudally deflected cerebellum. When the cerebellum is removed, it is apparent that the dorsal FL is not smooth but marked by numerous tubercles (Fig. 8.14A-C). Sections of the medulla reveal three parallel taste columns extending rostrocaudally as the dorsal facial lobe (FL), ventral FL and vagal lobe (VL) (Fig. 8.14C). The dorsal FL is remarkably enlarged and exhibits a tubercular appearance (Fig. 8.14C). This enlargement can be related to the enormous number of taste buds as well

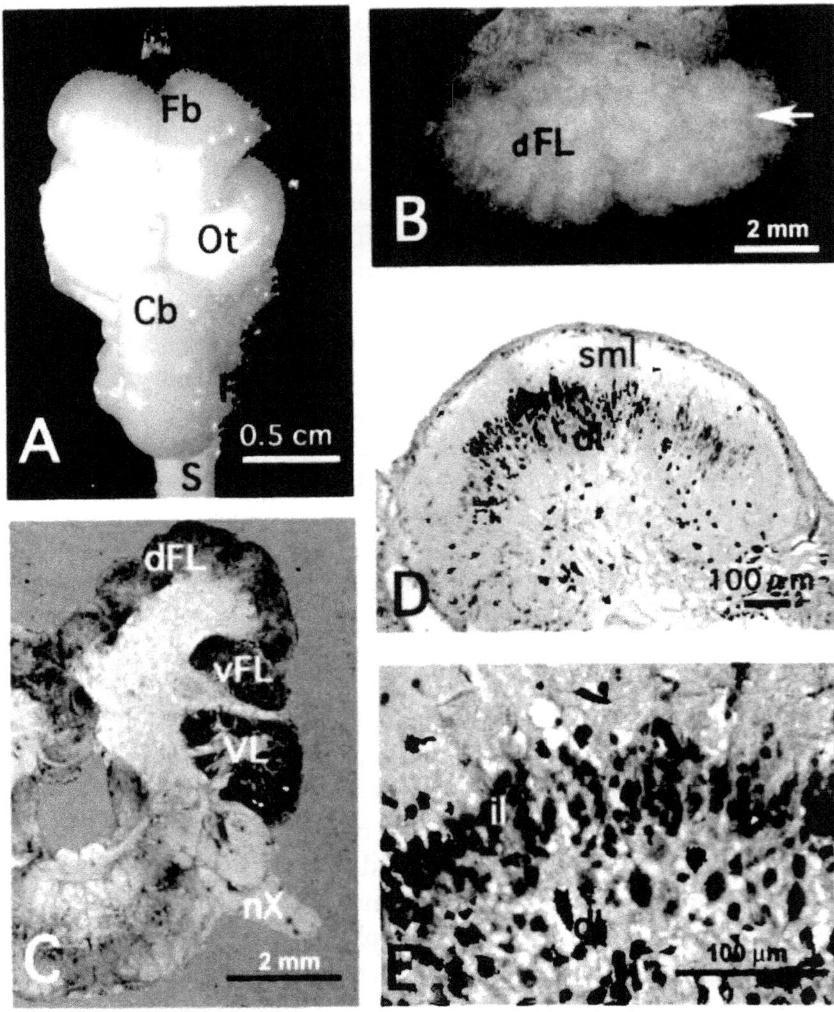

Fig. 8.14 Primary taste center of goatfish and cytoarchitecture of the facial lobe.

A. Dorsal view of the brain of *Parupeneus pleurotaeni*. Anterior is up. Fb, forebrain; Ot, optic tectum; Cb, cerebellum; FL, facial lobe; VL, vagal lobe; S, spinal cord. **B.** Dorsal view of facial lobe. Membranes covering facial lobe were removed to show the cauliflower-like surface of the dorsal division facial lobe (dFL). **C.** Transverse section at the indicated level (arrow) in B. vFL, ventral division of facial lobe; VL, vagal lobe; nX, vagal nerve. **D.** Longitudinal section of a tubercle of the dorsal facial lobe showing lamination with a superficial molecular layer (sml), an intermediate layer of densely packed medium-size neurons (il) and a deeper layer of elongate, larger neurons (dl). **E.** Higher magnification of cytoarchitecture of intermediate and deeper layers. Reproduced from Kiyohara et al. (2002).

as their large size. The dorsal FL seems to develop within a limited space between the medulla and cerebellum. As a result, the dorsal FL has become highly convoluted to accommodate the increased volume of brain tissue in a limited space. Histologically, the tubercles appear coarsely laminated with a superficial molecular layer, an intermediate layer of densely packed medium-size neurons, and a deeper layer of elongate, larger neurons (Fig. 8.14D, E). These latter two types of neurons are thought to correspond to the medium and large neurons found in the FL of *Plotosus*.

7. REPRESENTATION OF BARBELS AS LOBULES IN THE FACIAL LOBE OF *PLOTOSUS* AND GOATFISH

Anatomical (Finger, 1976; Kiyohara et al., 1985b; Morita and Finger, 1985b; von Bartheld and Meyer, 1985; Kanwal and Caprio, 1987; Puzdrowski, 1987; Kiyohara, 1988; Kotrschal and Whitear, 1988; Fukusako et al., 1993; Kiyohara and Kitoh, 1994) and physiological (Peterson, 1972; Marui, 1977; Marui and Caprio, 1982; Kanwal and Caprio, 1988; Marui et al., 1988; Hamaya and Caprio, 1989) studies in various species of fish clearly indicate that the primary medullary taste complex is topographically organized with vagal nerve-innervated oropharyngeal fields being represented viscerotopically in a caudal VL, and facial nerve-innervated taste buds being represented somatotopically in a rostral FL. This topographical organization is highly developed in the FL of both catfishes and goatfishes. As previously described, the regions representing catfish barbels are sharply defined and extraordinarily enlarged as different lobules extending rostrocaudally in the FL. In goatfish, the sensory inputs from the barbel terminate in a derived dorsal FL which has a highly convoluted surface forming a multitude of tubercles. The apparent tubercles of the goatfish dorsal FL are actually recurved flexures in a convoluted continuous columnar representation of the barbel. The results of *Plotosus* and goatfish are described in detail below.

7.1 *Plotosus*

The central projections of the four barbel nerves on the FL were traced by means of horseradish peroxidase neurohistochemistry (Kiyohara et al., 1996). When HRP was applied to only one of four barbel branches, only the corresponding barbel lobule was labeled (Fig. 8.13B, C). Labeling begins at the rostral tip of each barbel lobule and extends caudally to the point where the lobule becomes obscure. In *Arius felis*, the rostral and caudal surfaces of each barbel are separately represented in each barbel lobule (Kiyohara and Caprio, 1996). The maxillary barbel contains facial nerve fibers in large and small branches, corresponding to the rostral and caudal surfaces of the barbel respectively. The large branch terminates throughout the entire ipsilateral MXL while the terminal fields of the small branch progressively decrease caudally toward the dorsal portion of the ipsilateral MXL. These diffuse topographic projections of the two maxillary barbel branches indicate that representation of the rostral surface of the maxillary barbel is more extensive than that of the caudal surface in the MXL, and that the rostrocaudal axis of the maxillary barbel in the MXL is most likely represented in the ventrodorsal axis of the MXL.

The other branches also project to distinct areas of the ipsilateral FL. The terminal field of the recurrent branch fibers is limited to the TTL (Fig. 8.13B). HRP labeling from the recurrent nerve was heavy in the anterior two-thirds of TTL and became sparse caudally. The other branches such as the upper lip or lower lip also projected topographically to corresponding areas of the FL. A more complete somatotopy of the FL was revealed by electrophysiological mapping in *Plotosus* (Marui et al., 1988). This study showed that both chemosensitive and mechanosensitive neurons are topographically organized on superimposed maps in the FL. The tip to base axis of each barbel is represented in the rostrocaudal axis of each of the four barbel lobules. The anteroposterior body axis is represented posteroanteriorly in the TTL (Fig. 8.13D). However, this axis may be twisted and bent as suggested in *Arius felis*.

In other species of catfish, the somatotopic maps of the FL are also elucidated. The number and relative lengths of the barbel lobules correlate directly with the number and relative lengths of the barbels which a particular species of catfish possesses. The channel catfish *Ictalurus punctatus* has four pairs of barbels like *Plotosus*, but their lengths differ in the order of maxillary barbel > lateral mandibular barbel > medial mandibular barbel > nasal barbel. The lengths of the four lobules in *Ictalurus* reflect the same order (Hayama and Caprio, 1989). This relationship was also found for *Arius felis*, which possesses three pair of barbels of different lengths (Kiyohara and Caprio, 1996), and for *Silurus asotus*, which has long maxillary and short mandibular barbels (Kiyohara and Kitoh, 1994).

7.2 Goatfish

Labeling of various peripheral branches of the facial nerve showed that the barbel is represented in the dorsal, tubercular portion of the FL while nerves innervating the rest of the face and head terminate in the ventral subdivision of the lobe. Within the dorsal portion of the FL, the primary afferent terminals end within the superficial molecular layer of the lobe. In order to better understand the nature of the tubercles of the dorsal FL, microelectrodes were also utilized to electrophysiologically map the receptive fields of neurons situated in various areas of the FL. A recording electrode was driven vertically through the FL in a systematic grid of points projected onto the dorsal surface of the lobe. Because of the convoluted nature of tubercles of the dorsal FL, in any single dorsoventral electrode penetration, the electrode tip was likely to pass from one tubercle into another. Thus the observed discontinuities in the receptive fields of adjacent areas were not unexpected. Penetrations through the rostromedial portion of the FL yielded receptive fields near the base of the barbel, whereas penetrations along the lateral edge of the rostral part of the lobe revealed fields near the distal tip of the barbel (Fig. 8.15). There was not a smooth continuity of receptive fields between these areas however. For example, moving the electrode approximately 500 μm laterally, from position 1B to position 1C at middle levels (Fig. 8.15A) showed receptive fields moving from the base of the barbel to midway along its length with no intermediate representation. Such intermediately situated receptive fields were present, however, in more caudal levels of the lobe (Level 3, Fig. 8.15A). When recordings from the whole FL were taken into account, despite apparent discontinuities at any particular anteroposterior level, a continuous, albeit convoluted representation of the barbel was present in the dFL (Fig. 8.15B). The organization was as if a dangled strand of spaghetti were allowed to coil haphazardly upon itself when lowered onto a platter (Kiyohara et al., 2002).

8. PROJECTIONS OF TRIGEMINAL NERVE FIBERS TO THE FACIAL LOBE

Another striking difference in the organization of the FL between *Plotosus* and goatfish is the trigeminal projections onto the FL in each species. Previous anatomical studies on a variety of catfishes have established that the trigeminal nerve, as well as the facial nerve, projects somatotopically to the FL (*Arius felis*, Kiyohara and Caprio, 1996; *Plotosus lineatus*,

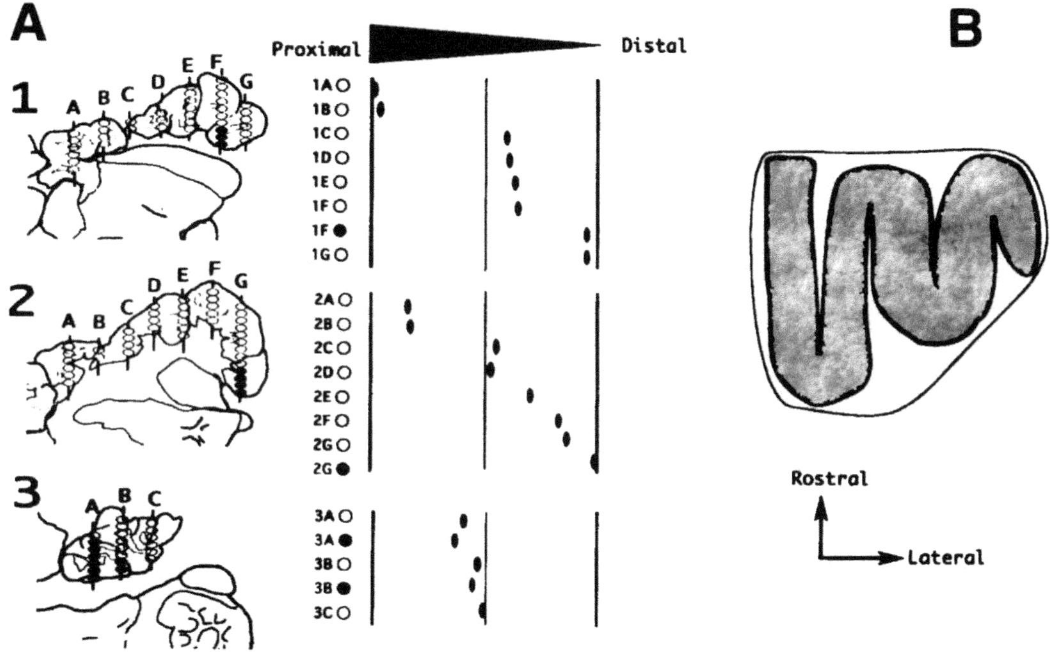

Fig. 8.15 Somatotopic representation of the barbel in the dorsal facial lobe of goatfish. Multiunit activity in response to mechanical stimulation was recorded in 6 transverse planes with 500 μm intervals. **A.** Examples of tactile receptive fields (RFs) for recording sites at three different transverse levels from rostral (1) to caudal (3). Intervals are 540 μm between 1 and 2 and 1,080 μm between 2 and 3. RFs for the recording sites are shown on the right as solid ovals located in the relative position on the barbel (shown at the top). In 1, for example, the RF was located on the base of the barbel for recording sites along the electrode track A. The RFs did not change as the electrode penetrated deeper in this track but the response changed considerably in magnitude. The maximum response was always obtained when the electrode was positioned in the intermediate or deeper layer of the tubercle. In some tracks, such as 1F or 2G, RFs changed after the electrode reached a certain depth in the dorsal FL. These changed sites are shown as solid circles in the sections on the left. **B.** Dorsal view of barbel representation in the right side of the dorsal FL. Reproduced from Kiyohara et al. (2002).

Kiyohara et al., 1996; *Silurus asotus*, Kiyohara and Kitoh, 1994). In the goatfish, this trigeminal projection to the FL is extremely sparse, similar to other species of fishes with less developed FLs (Kiyohara et al., 1998). This difference may be related to the fact that many free nerve endings are located in the barbel epithelium of the catfish (Sakata et al., 2001) compared to that of goatfish (Kiyohara et al., 2002).

Because of the intimate mixing of the trigeminal and facial roots and ganglia in the aforesaid species of catfish, it is not possible to determine the distribution of the trigeminal fibers within the FL independent of the facial nerve contribution. The advent of post-mortem tracing by application of the carbocyanine dye DiI to partially dissected specimens (Godement et al., 1987) permits such an analysis of this system. In the channel catfish (Kiyohara et al., 1999), the trigeminal projections onto the FL were examined by applying DiI to the central cut stump of the trigeminal root in isolated fixed brains. The course of trigeminal nerve fibers to the FL and their mode of termination in this species are described below.

The trigeminal motor nucleus and the principal sensory nucleus lie near the level of entrance of the trigeminal nerve (Fig. 8.16). The majority of primary trigeminal fibers however, sweep caudally after entering the brain to form the descending root (RDV). Fibers in the descending trigeminal root descend through the medulla to reach the medial funicular nucleus and subsequent regions in the spinal cord. Throughout the length of the RDV, fibers terminate in the spinal trigeminal nucleus lying just dorsomedial to the tract itself. At the level of the caudal FL, the RDV moves somewhat more dorsally in the medulla to approach the ventral margin of the FL (Fig. 8.16 levels 2,000–3,100). At this level, dorsomedially directed collaterals leave the RDV to enter the FL proper. Labeled fibers originate from various regions of the RDV and leave the RDV running dorsomedially through three or four bundles (Fig. 8.16 levels 2,200–2,700, Fig. 8.4A,B). The labeled trigeminal fibers are frequently branched in the RDV where the trigeminal bundles projecting to the FL arise, indicating the presence of axon collaterals in the trigeminal fibers projecting to the FL. The trigeminal fibers are coarser than the facial nerve fibers which terminate within the same structure.

The trigeminal bundles projecting to the FL comprise two groups. One bundle turns medially to terminate directly in the intermediate nucleus underlying the FL proper. Another fascicle turns rostrally immediately after reaching the ventral portion of the FL and ascends as

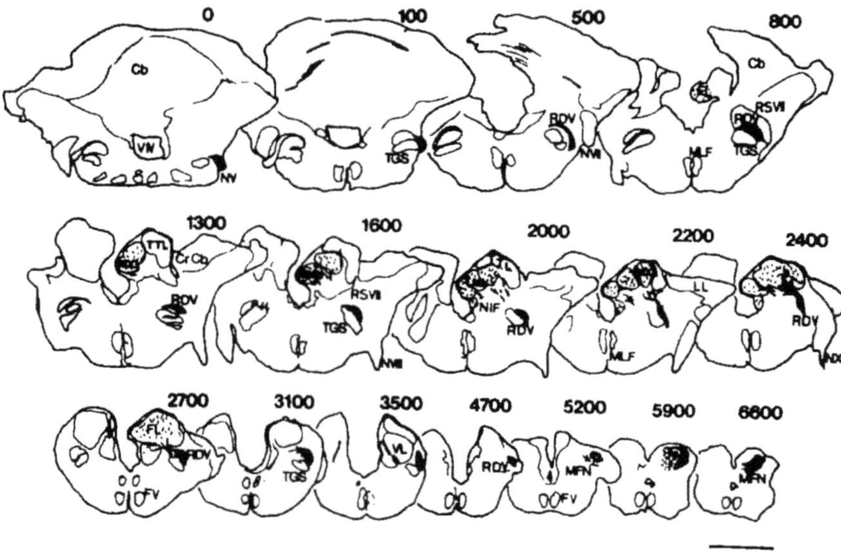

Fig. 8.16 Line drawing illustrating distribution of the trigeminal fibers to the FL as well as to the brain stem and the spinal cord in the channel catfish. Number above the right of each transverse section indicates the distance (µm) from the first section from which the trigeminal nerve enters the brain. The solid black indicates the labeled descending trigeminal root. DiI labeled fibers and fiber bundles projecting to the FL are shown by stippled lines and solid lines in the 800 – 2,700 sections, respectively. The stippled areas in 5,900 and 6,600 sections indicate the projections to the medial funicular region. Scale bar: 2 mm. Cb, cerebellum; CrCb, cerebellar crest; FL, facial lobe; FV, ventral fasciculus; LL, lateral line lobe; LML, lateral mandibular barbel lobule; MML, medial mandibular barbel lobule; MXL, maxillary barbel lobule; MFN, medial funicular nucleus; MLF, medial longitudinal fasciculus; NIF, intermediate nucleus of the facial lobe; NVII, facial nerve; NV, trigeminal nerve; NVIII, statoacoustic nerve; NX, vagal nerve; RDV, descending trigeminal root; RSVII, sensory root of facial nerve; TGS, ascending secondary gustatory tract; TTL, trunk-tail lobule; V IV, fourth ventricle; VL, vagal lobe. Reproduced from Kiyohara et al. (1999).

three or four fascicles situated in the ventral portion of the FL. During the rostral course of these fascicles, labeled fibers emerge dorsomedially and terminate in all anterior and intermediate portions of the FL, with the exception of the dorsolateral region (TTL) of the anterior FL. Other groups of trigeminal fibers descend through a few fascicles and terminate in posterior portions of the FL. The trigeminal axons ramify extensively within the innervated lobules. Numerous fine fibers with varicosities occur throughout the neuropil of the lobules providing extensive terminal networks (Kiyohara et al., 1999).

Thus the trigeminal fibers terminate throughout the FL except for the TTL which contains the representation of taste buds innervated by the recurrent branch of the facial nerve, i.e., those over the trunk and tail of the animal. In catfish, the trigeminal input to the primary gustatory complex is restricted to those portions of the nucleus receiving chemosensory inputs from the face and barbels, i.e. the trigeminally innervated sensory fields. The lack of trigeminal contribution to the trunk-tail lobule is not unreasonable considering the pattern of peripheral innervation of these nerves. While the trigeminal and facial nerves overlap in their cutaneous distributions over the face, barbels, and lips (Herrick, 1901; Finger, 1976; Kiyohara et al., 1985b), only the facial nerve has a recurrent branch reaching the flank and tail. Thus the extent of overlap in central termini for the trigeminal and facial nerves matches the extent of their overlap in the periphery. The trunk-tail lobule may be devoid of trigeminal cutaneous input, but it has unique connections with the spinal cord consonant with its receiving spinal cutaneous input via a relay in the dorsal horn (Finger, 1978; Kanwal and Finger, 1997). Thus all regions of the FL may receive corresponding somatotopically mapped information from both taste buds (via the facial nerve) and skin (via the trigeminal nerve or indirectly from spinal dorsal roots).

Since the facial nerve components convey mechanosensory as well as chemosensory information (Davenport and Caprio, 1982), it is not obvious why two types of presumed mechanosensory fibers, trigeminal and facial, should project to the FL of catfishes. It is possible that presumed trigeminal fibers projecting to the FL belong to the facial nerve (Northcutt et al., 2000); however, this is not likely since most of the trigeminal fibers terminating in the FL appear to be collaterals originating from the main trigeminal axons and are thicker in diameter than the facial fibers (Kiyohara et al., 1986; 1999). Unfortunately, only limited data are available regarding the nature of information conveyed by these nerves in other fishes. In puffer fish *Fugu pardalis*, the mechanosensory fibers of the facial nerve consist of only one type, which elicits phasic and tonic impulse trains in response to long-lasting mechanical stimulation (Kiyohara et al., 1985a). Likewise, mechanosensory fibers in the recurrent branch of the facial nerve appear to be responsive to tactile stimuli (Davenport and Caprio, 1982). It is possible that mechanosensory facial nerve fibers may convey only one quality of cutaneous stimulation. In contrast, trigeminal fibers respond in various ways to mechanical stimulation as shown in the sea lamprey (Matthews and Wickelgren, 1978) and mammals (Kandel and

Jessell, 1991). They show either rapid or slow adaptive responses to non-nocioceptive mechanical stimulation, and may also respond only to nocioceptive or proprioceptive stimuli. In fact, proprioceptive responses, produced only by directional movement of the barbels, were recorded from the trigeminofacial complex nerve of *Plotosus lineatus* (Konishi et al., 1966) and from the anterior ganglion of bullhead catfish, *I. nebulosus* (Biedenbach, 1971). Therefore, it is likely that trigeminal and mechanosensory facial fibers convey different qualities or modalities of cutaneous stimuli to the FL, contributing to discrimination of different qualities of mechanical simulation.

Acknowledgements

Major parts of goatfish study were carried out at the Tropical Biosphere Research Center, University of Ryukyus Sesoko Station. We thank K. Takano, M. Nakamura, A. Takemura, and S. Nakamura for generous support of these experiments and help in collecting the goatfish. We also thank C. Lamb IV for critical reading of the manuscript and valuable suggestions, and S. Ishida for assistance in catfish collection. This study was supported by grant-in-aids (Nos. 13460087 and 16380137) from the Ministry of Education, Science, Sports, and Culture of Japan.

References

Ariens-Kappers, C.U., G.C. Huber and E.C. Crosby. 1936. *The Comparative Anatomy of the Nervous System of Vertebrates, Including Man*.Hafner New York. (reprinted 1965).

Atema, J. 1971. Structures and functions of the sense of taste in the catfish (*Ictalurus natalis*). *Brain Behav. Evol.* 4: 273–294.

Biedenbach, M.A. 1971. Functional properties of barbel mechanoreceptors in catfish. *Brain Res.* 27: 360–364.

Crisp, M., G.A. Lowe and M.S. Laverack. 1975. On the ultrastructure and permeability of taste buds of the marine teleost *Ciliata mustela*. *Tissue & Cell* 7: 191–202.

Davenport, C.J. and J. Caprio. 1982. Taste and tactile recordings from the ramus recurrens facialis innervating flank taste buds in the catfish. *J. Comp. Physiol.* 147: 217–229.

Dodd, J. and J.P. Kelly. 1991. Trigeminal system. In: *Principles of Neural Science*, E.R. Kandel, J.H. Schwartz, T.M. Jessell (Eds). Elsevier, Amsterdam, pp. 367– 384.

Finger, T.E. 1976. Gustatory pathways in the bullhead catfish, I. Connections of the anterior ganglion. *J. Comp. Neurol.* 165: 513–526.

Finger, T.E. 1978. Gustatory pathways in the bullhead catfish. II. Facial lobe connections. *J. Comp. Neurol.* 180: 691–705.

Finger, T.E. 1997. Evolution of taste and solitary chemoreceptor cell systems. *Brain Behav. Evol.* 50: 234–243.

Finger, T.E. and B. Böttger. 1990. Transcellular labeling of taste bud cells by carbocyanine dye (DiI) applied to peripheral nerves in the barbels of the catfish, *Ictalurus punctatus*. *J. Comp. Neurol.* 302: 884–892.

Fukusako, H., H. Maeno and S. Kiyohara. 1993. Topographical projection of facial recurrent fibers to the medullary facial lobe of the carp *Cyprinus carpio*. *Nippon Suisan Gakkaishi* 59: 29–33.

Godement, P., J. Vanselow, S. Thanos and F. Bonhoeffer. 1987. A study in developing visual systems with a new method of staining neurons and their processes in fixed tissue. *Development* 101: 697–713.

Grover-Johnson, N. and A.J. Farbman. 1976. Fine structure of taste buds in the barbel of the catfish, *Ictalurus punctatus*. *Cell Tissue Res.* 169: 395–403.

Harder, W. 1975. *Anatomy of fishes*. E.Schweizerbart'sche Verlagsbuchhandlung (Naegele u. Obermiller), Stuttgart.

Hayama, T. and J. Caprio. 1989. Lobule structure and somatotopic organization of the medullary facial lobe in the channel catfish *Ictalurus punctatus*. *J. Comp. Neurol.* 285: 9–17.

Herrick, C.J. 1901. The cranial nerves and cutaneous sense organs of the North American siluroid fishes. *J. Comp. Neurol.* 11: 177–249.

Herrick, C.J. 1905. The central gustatory paths in the brains of bony fishes. *J. Comp. Neurol.* 15: 375–456.

Iwai, T. 1964. A comparative study of the taste buds in gill rakers and gill arches of teleostean fishes. *Bull. Misaki Mar. Bio. Inst. Kyoto Univ* No. 7: 19–34.

Joyce, E.C. and G.B. Chapman. 1978. Fine structure of the nasal barbel of the channel catfish, *Ictalurus punctatus*. *J. Morphol.* 158: 109–154.

Kandel, E.R. and T.M. Jessell. 1991. Touch, In: *Principles of Neural Science*, E.R. Kandel, J.H. Schwartz, T.M. Jessell (Eds). Elsevier, Amsterdam, pp. 367–384.

Kanwal, J.S. and J. Caprio. 1987. Central projections of the glossopharyngeal and vagal nerves in the channel catfish, *Ictalurus punctatus*: clues to different processing of visceral inputs. *J. Comp. Neurol.* 264: 216–230.

Kanwal, J.S. and J. Caprio. 1988. Overlapping taste and tactile maps of the oropharynx in the vagal lobe of the channel catfish, *Ictalurus punctatus*. *J. Neurobiol.* 19: 211–222.

Kanwal, J.S. and T.E. Finger. 1997. Parallel medullary gustatospinal pathways in a catfish: possible neural substrates for taste-mediated food search. *J. Neurosci.* 17: 4873–4885.

Kapoor, B.G. and T.E. Finger. 2003. Taste and solitary chemoreceptor cells. In: *Catfishes*, G. Arratia, B.G. Kapoor, M. Chardon and R. Diogo (Eds). Science Publishers Inc., Enfield (NH), USA and Plymouth, U.K., vol. 2, pp. 753–769.

Kitoh, J., S. Kiyohara and S. Yamashita. 1987. Fine structure of taste buds in the minnow. *Nippon Suisan Gakkaishi* 53: 1943–1950.

Kiyohara, S. 1988. Anatomical studies of the facial taste system in teleost fish. In: *Beidler Symposium on Taste and Smell:* A Festschrift to L.M. Beidler, I.J. Miller (Ed.). Winston-Salem, NC, Book Services Assoc., pp. 127–136.

Kiyohara, S. and J. Kitoh. 1994. Somatotopic representation of the medullary facial lobe of catfish *Silurus asotus* as revealed by transganglionic transport of HRP. *Fisheries Science* 60: 393–398.

Kiyohara, S. and J. Caprio. 1996. Somatotopic organization of the facial lobe of the sea catfish *Arius felis* studied by transganglionic transport of horseradish peroxidase. *J. Comp. Neurol.* 368: 121–135.

Kiyohara, S., I. Hidaka and T. Tamura. 1975. The anterior cranial gustatory pathway in fish. *Experientia* 31: 1051–1053.

Kiyohara, S., S. Yamashita and J. Kitoh. 1980. Distribution of taste buds on the lips and inside the mouth in the minnow, *Pseudorasbora parva*. *Physiology & Behavior* 24: 1143–1147.

Kiyohara, S., T. Shiratani and S. Yamashita. 1985a. Peripheral and central distribution of major branches of the facial taste nerve in the carp. *Brain Res.* 325: 57–69.

Kiyohara, S., I. Hidaka, J. Kitoh and S. Yamashita. 1985a. Mechanical sensitivity of the facial nerve fibers innervating the anterior palate of the puffer *Fugu pardalis* and their central projection to the primary taste center. *J. Comp. Physiol.* A 157: 705–716.

Kiyohara, S., H. Houman, S. Yamashita, J. Caprio and T. Marui. 1986. Morphological evidence for a direct projection of trigeminal nerve fibers to the primary gustatory center in the sea catfish *Plotosus anguillaris*. *Brain Res.* 379: 353–357.

Kiyohara, S., J. Kitoh, A. Shito and S. Yamashita. 1996. Anatomical studies of the medullary facial lobe in the sea catfish *Plotosus lineatus*. *Fisheries Science* 62: 511–519.

Kiyohara, S., K. Shintomo and S.Yamashita. 1998. The projections of trigeminal nerve fibers to the medullary taste center in some teleosts. *Fisheries Science* 64: 276–281.

Kiyohara, S., S.Yamashita, C.F. Lamb and T.E. Finger. 1999. Distribution of trigeminal fibers in the primary facial gustatory center of channel catfish, *Ictalurus punctatus*. *Brain Res.* 841: 93–100.

Kiyohara, S., Y. Sakata, T. Yoshitomi and J. Tsukahara. 2002. The "goatee" of goatfish: innervation of taste buds in the barbels and their representation in the brain. *Proc. Roy. Soc. London Biol. Sci.* B 269: 1773–1780.

Konishi, J., M. Uchida and Y. Mori. 1966. Gustatory fibers in the sea catfish. *Jpn. J. Physiol.* 16: 194–204.

Kotrschal, K. and M. Whitear. 1988. Chemosensory anterior dorsal fin in rocklings (*Gaidropsarus* and *Ciliata*, Teleostei, Gadidae): Somatotopic representation of the ramus recurrens facialis as revealed by transganglionic transport of HRP. *J. Comp. Neurol.* 268: 109–120.

Kotrschal, K., M. Whitear and T.E. Finger. 1993. Spinal and facial innervation of the skin in the gadid fish *Ciliata mustela*. *J. Comp. Neurol.* 331: 407–417

Marui, T. 1977. Taste responses in the facial lobe of the carp, *Cyprinus carpio* L. *Brain Res.* 130: 287–298.

Marui, T. and J. Caprio. 1982. Electrophysiological evidence for the topographical arrangement of taste and tactile neurons in the facial lobe of the channel catfish. *Brain Res.* 231: 185–190.

Marui, T., J. Caprio, S. Kiyohara and Y. Kasahara. 1988. Topographical organization of taste and tactile neurons in the facial lobe of the sea catfish *Plotosus anguillaris*. *Brain Res.* 446: 178–182.

Matthews, G. and W.O. Wickelgren. 1978. Trigeminal sensory neurons of the sea lamprey. *J. Comp. Physiol.* 123: 329–333.

Morita, Y. and T.E. Finger. 1985a. Reflex connections of the facial and vagal gustatory systems in the brainstem of the bullhead catfish, *Ictalurus nebulosus*. *J. Comp. Neurol.*231: 547–558.

Morita, Y. and T.E. Finger. 1985b. Topographic and laminar organization of the vagal gustatory system in the goldfish, *Carassius auratus*. *J. Comp. Neurol.* 238: 187–201.

Northcutt, R.G., P.H. Holmes and J.S. Albert. 2000. Distribution and innervation of lateral line organs in the channel catfish. *J. Comp. Neurol.* 421: 570–592.

Peterson, R.H. 1972. Tactile responses of the goldfish (*Carassius auratus* L.). *Copeia* 1972: 816–819.

Puzdrowski, R.L. 1987. The peripheral distribution and central projections of the sensory rami of the facial nerve in goldfish *Carassius auratus*. *J. Comp. Neurol.* 259: 382–392.

Reutter, K. 1971. Die Geschmacksknospen des Zwergwelses *Amiurus nebulosus* (Lesueur). Morphologische und histochemische Untersuchungen. *Z. Zellforch.* 120: 280–308.

Reutter, K. 1978. Taste organ in the bullhead (Teleostei). *Adv. Anat. Embryol. Cell Biol* 55: 1–98.

Reutter, K. 1986. Chemoreceptors. In: *Biology of the Integument*. Vertebrates. J. Bereiter-Hahn, A.G. Matoltsy and K.S. Richards (Eds). Springer-Verlag, Berlin vol. 2, pp. 586–604.

Reutter, K. 1992. Structure of the peripheral gustatory organ, represented by the siluroid fish *Plotosus lineatus* (Thunberg). In: *Fish Chemoreception*, T.J. Hara (Ed.). Chapman & Hall, London, New York, pp. 60–78.

Reutter, K. and A. Hansen. 2005. Subtypes of light and dark elongated taste bud cells in fish. In: *Fish Chemosenses*, K. Reutter and B.G. Kapoor (Eds). Science Publishers, Enfield, NH (USA) and Plymouth, UK, pp. 211-230 (this volume).

Royer, S.N. and J.C. Kinnamon. 1996. Comparison of high-pressure freezing/freeze substitution and chemical fixation of catfish barbel taste buds. *Micros. Res. Tech.* 35: 385–412.

Sakata, Y., J. Tsukahara and S. Kiyohara. 2001. Distribution of nerve fibers in the barbels of sea catfish *Plotosus lineatus*. *Fisheries Science* 67: 1136–1144.

Satô, M. 1937a. On the barbels of a Japanese sea catfish, *Plotosus anguillaris* (Lacepede). *Sci. Rep. Tôhoku Imp. Univ.* (Sendai, Japan). *Biol.* 11: 323–332.

Satô, M. 1937b. Histological observations on the barbels of fishes. *Sci. Rep. Tôhoku Imp. Univ.* (Sendai, Japan) *Biol.* 11: 265–276.

Satô, M. 1938. The sensibility of the barbel of *Upeneus spilurus* Bleeker, with some notes on the schooling. *Sci.Rep. Tôhoku Imp. Univ.* (Sendai, Japan) *Biol.* 12: 489–500.

Sibbing, F.A. 1982. Pharyngeal mastication and food transport in the carp (*Cyprinus carpio* L): a cineradiographic and electromyographic study. *J. Morphol.* 172: 223–258.

Toyoshima, K., O. Nada and A. Shimamura. 1984. Fine structure of monoamine-containing basal cells in the taste buds on the barbels of three species of teleosts. *Cell Tissue Res.* 235: 479–484.

von Bartheld, C.S. and D.L. Meyer. 1985. Trigeminal and facial innervation of cirri in three teleost species. *Cell Tissue Res.* 241: 615–622.

CHAPTER

9

Subtypes of Light and Dark Elongated Taste Bud Cells in Fish

Klaus Reutter[1] and Anne Hansen[2]

ABSTRACT

In taste buds of fish, as in other vertebrates, the elongated cells apically bear microvilli which together form the taste bud receptor area. As seen in the scanning electron microscope (SEM), the receptor areas of most species of fish contain two kinds of microvilli, large and small. In some fish however, three or four different types of microvillar structures have been observed. Transmission electron microscopic (TEM) studies revealed that all microvilli (receptor villi) belong to two main types of elongated cells, the electron lucent light cells and the electron denser dark cells. Here we demonstrate that in distinct species of fish both types of cells may have several subtypes. One subtype of light cell always ends in one large apical microvillus, the other subtype(s) show different numbers of distinctly shaped microvilli. The most common dark cell subtype normally bears small and undivided microvilli; other dark cell subtypes have microvilli of variable shapes. The number of

Address for Correspondence: Klaus Reutter, [1]Anatomisches Institut, Universität Tübingen, 72074 Tübingen, Germany. E-mail: klaus.reutter@web.de

[2]Department of Cell and Developmental Biology, University of Colorado Health Sciences centre at Fitzsimons, Aurora, CO 80045 USA.

elongated taste bud cells and the respective number of different types of microvilli may vary in fish of different systematic position and even between fish belonging to the same genus. We therefore postulate that in fish, taste bud micromorphology is species specific. It is likely that fish taste buds evolved while fish took possession of their own distinct ecological niche.

Key Words: Receptor area; Taste bud; Microvilli; Receptor villi; Ultrastructure; Evolution; Electron microscopy.

1. INTRODUCTION

Besides the olfactory system and the system of solitary chemosensory cells, the sense of taste is a further chemosensory system that allows a fish to detect chemical substances in its close environment, to find food or prey and to test it before swallowing. Peripherally, the gustatory sense is represented by taste buds (TBs) that occur in large numbers inside the mouth and oropharyngeal cavity. They may also occur in the body integument, especially on the head and its appendages, e.g. the barbels (Hansen and Reutter, 2004).

In most fish, TBs are ovoid or oval organs that rest atop a dermal papilla. TBs are oriented vertically in the epithelium. They reach the epithelium's surface and their apices are in contact with the environmental waters. A TB consists of several types of cells, the ultrastructure of which is more or less well known (reviews: Kapoor et al., 1975; Reutter, 1978, 1982, 1986, 1992; Jakubowski and Whitear, 1990; Reutter and Witt, 1993; Witt, 1996; Sorensen and Caprio, 1998; Finger and Simon, 2000; Jakubowski and Żuwała, 2000; Tagliafierro and Zaccone, 2001; Kapoor and Finger, 2003; Witt et al., 2003; Hansen and Reutter, 2004): The (electron) light and dark cells form the TB sensory epithelium proper. Apically they end with microvillar structures that build the organ's receptor area. The bases of the elongated cells are in synaptic contact with the organ's nerve fiber plexus located between the lobed lower parts of the elongated cells and the most basally situated basal cells. The interface of the bud and the regular surrounding epithelial cells is marked by the less specialized marginal cells.

In view of the TB function, the elongated cells of the sensory epithelium are of special interest. These cells are of epithelial origin (Hansen et al., 2002), and are polarized cells with respect to the apical receptor villi and the basal synaptic contacts, i.e., the elongated cells are secondary sensory cells (Hansen and Reutter, 2004). It is still debated whether the dark elongated cells also serve as sensory cells. Since dark cells can also be synaptically connected to the nerve fibers of the bud's plexus, we postulate that the dark cells are also sensory cells (Reutter, 1978, 1992; Reutter and Witt, 1993); see Discussion below.

As a rule, the apical microvillar structures of light and dark cells differ markedly from each other: the light cells mostly terminate in one conical large villus, while the dark cells bear several small and sometimes divided microvilli. This situation was observed for several fish species in the scanning (SEM) and transmission (TEM) electron microscope and has been described and reviewed in numerous publications (see Discussion). However, during the last few years it became obvious that the receptor areas of several fish species contain more than two different types of microvilli, as seen in the SEM and TEM in zebrafish TBs (Hansen et al., 2002) and as seen in TEM, *Astyanax* TBs contain a third type of elongated cell, the dense cored vesicles cell (Boudriot and Reutter, 2001). Since it is obvious that the elongated TB cells not only comprise the well-known light and dark cells, we assumed that additional cell types or subtypes of cells also exist in the TBs of fish other than *Danio* and *Astyanax*. We therefore scrutinized the relevant data of the literature and reexamined our own SEM and TEM micrographs of several species of fish. In about 10 species of different taxa we found subtypes of light and dark TB cells. These limited data do not allow speculation about fish TB phylogeny, but it is obvious that fish TBs vary within closely related groups and seem to be species specific.

2. MATERIALS AND METHODS

Our SEM and TEM investigations on TBs of fish of different systematic position (see below) were executed during the last two decades. As far as possible, we tried to treat the tissue samples with the same standard methods in order to ensure comparability of the results. The regulations of the laboratories involved and those published in the Declaration of Helsinki (1995) were respected.

After 3-aminobenzoic acid ethyl ester (Tricaine, MS 222; Sigma – Aldrich) anaesthetization of the fish, the tissues were excised and fixed in 0.1 M phosphate buffer (pH 7.2) containing 2.5% glutaraldehyde or 2% paraformaldehyde (TEM) or in 5% glutaraldehyde in 0.05M sodium phosphate buffer (pH 7.2; SEM). For SEM examination, tissues were rinsed in phosphate buffer, dehydrated via acetone and isoamyl acetate, transferred to liquid carbon dioxide and critical point dried. The specimens were then gold coated and viewed in a CamScan DV 4 SEM. For TEM examination, the probes were postfixed with 1% osmium tetroxide (2 h), dehydrated in ethanol and propylene oxide and embedded in Araldite (Serva) or, in the case of *Danio*, in glycid ether 100 (Serva). Ultrathin sections were contrasted with 1% uranyl acetate and 1% lead citrate and finally examined with a Philips EM 300, a Philips EM 420, or a LEO EM 12 OMEGA.

The following species of fish belonging to various taxa (Nelson, 1994) were investigated:

Elasmobranchii: *Scyliorhinus canicula* (TEM)

Dipnoi: *Lepidosiren paradoxa* (TEM), *Protopterus annectens* (SEM, TEM), *Neoceratodus forsteri* (TEM)

Chondrostei:

Polypteriformes: *Polypterus senegalus* (SEM)

Acipenseriformes: *Acipenser baeri* (SEM), *Scaphirhynchus platorynchus* (TEM)

Neopterygii:

Semionotiformes: *Lepisosteus oculatus* (SEM, TEM)

Amiiformes: *Amia calva* (TEM)

Teleostei:

Anguillidae: *Anguilla anguilla* (SEM)

Bothidae: *Scophthalmus maximus* (SEM, TEM)

Cichlidae: *Archocentrus nigrofasciatus* (SEM)

Characidae: *Astyanax ("Anoptichthys") mexicanus* (TEM)

Clupeidae: *Clupea (Harengus) harengus* (SEM), *Pantodon buchholzi* (SEM)

Cyprinidae: *Barbus barbus* (SEM), *Phreatichthys andruzzi* (SEM), *Danio rerio* (SEM, TEM)

Eleotridae: *Eleotris sandwicensis* (SEM)

Gobiidae: *Awaous guamensis* (SEM), *Lentipes concolor* (SEM), *Sicyopterus stimpsoni* (SEM)

Mormyridae: *Pollimyrus castelnaui* (SEM)

Poeciliidae: *Poecilia reticulata* (SEM), *Xiphophorus helleri* (SEM)

Siluridae: *Amiurus (Ictalurus) nebulosus* (SEM, TEM), *Ictalurus punctatus* (SEM), *Silurus glanis* (SEM, TEM)

3. RESULTS

3.1 SEM Data

TBs are situated either within more or less elevated epidermal hillocks or in the flat epithelium, with no elevation. This enabled a simple classification of them: type I TBs, elevated; type II TBs, slightly elevated and type III TBs, not elevated (Reutter, 1974; Reutter et al., 1974). The distal endings of the TB elongated cells are the only structures that form the TB receptor area and are in contact with the environmental waters and prospective tastants. In some cases the receptor area is slightly sunken so the surrounding superficial epidermal (marginal) cells form a ringlike ditch around the receptor area that is similar to the mammalian TB pore (Fig. 9.1G). In most cases the receptor area protrudes above the epithelial surface and the microvilli are clearly visible, especially when viewed laterally.

In almost all cases the receptor area is composed of at least two different microvillar structures, large receptor villi and small ones. As a rule, a large villus belongs to a light elongated cell, while a dark elongated cell always terminates in up to 20 small villi (TEM). After careful examination of the receptor areas of various species of fish we found some microvillar structures that could not be classified as large or small villi (Figs. 9.1 and 9.2). Moreover, the receptor areas of *Lepisosteus oculatus* TBs seem to contain only one type of small microvilli (Fig. 9.1 D).

Consonant with the data in the literature, we found two types of microvilli, large and small, in the receptor areas of the following species: *Anguilla anguilla*, *Amiurus nebulosus*, *Silurus glanis*, *Xiphophorus helleri*, *Archocentrus nigrofasciatus*, and *Protopterus annectens* (Fig. 9.2 C,D). *Pantodon buchholzi* and *Ictalurus punctatus* receptor areas seem to have more than two different microvilli. We regularly found three types of microvilli in the TB receptor areas of *Acipenser baeri*, (Fig. 9.1 E-G), *Harengus harengus*, *Barbus barbus*, *Phreatichthys andruzzi*, *Danio rerio* (Fig. 9.2 A,B), *Poecilia reticulata*, *Awaous guamensis*, *Lentipes concolor*, *Eleotris sandwicensis*, and *Scophthalmus maximus*. The receptor areas of *Polypterus senegalus* comprise even four types of differently shaped receptor villi: The most common are the small villi, followed by several large ones, a few bunches or brushes of short microvilli and, less abundantly, bunches of slender, relatively long and arborized villi. As only SEM micrographs were available, it remains open as to which kind of elongated cells the various microvillar structures belong.

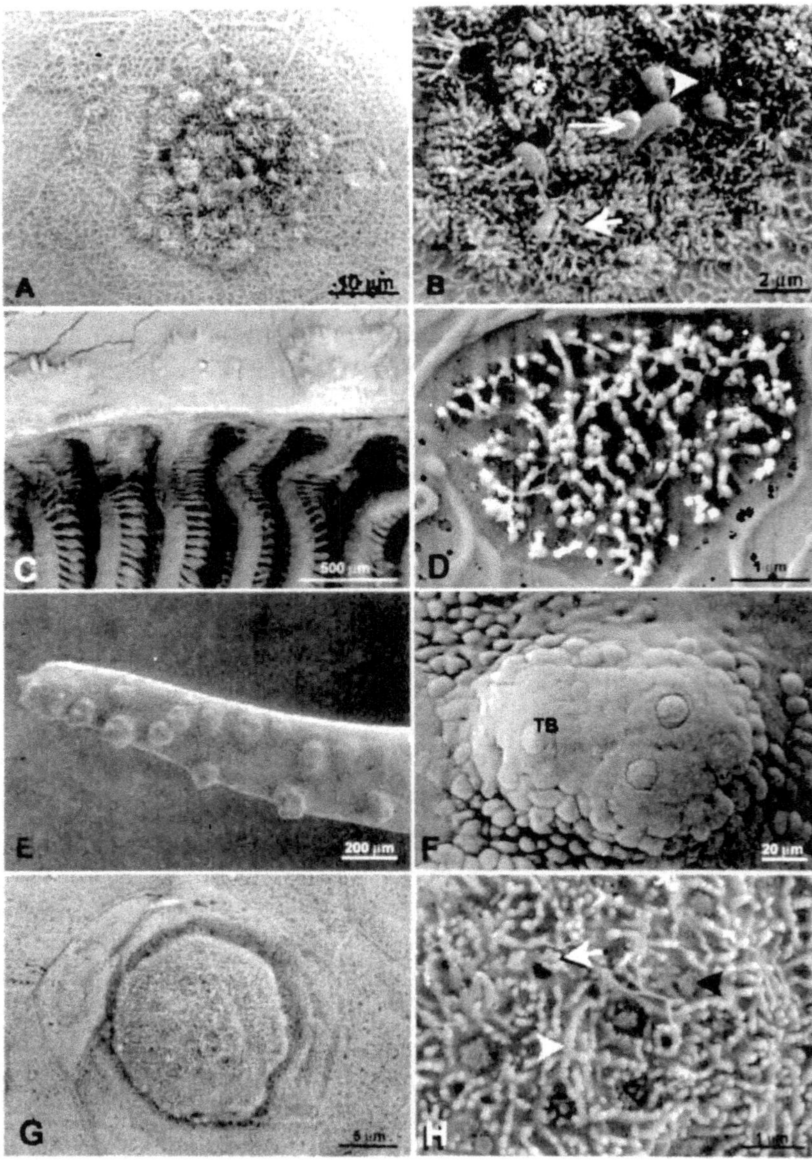

Fig. 9.1 A-H SEMs of taste buds of *Polypterus senegalus* **(A,B)**, *Lepisosteus oculatus* **(C,D)** and *Acipenser baeri* **(E-H)**. **A.** Receptor area of a tongue taste bud. **B.** Higher magnification of A. Receptor area contains single large receptor villi (long arrow), numerous small receptor villi (arrowhead), clustered slender villi (short arrow) and long brush arranged villi (asterisk). **C.** Transversely oriented gill arch with tooth pads (upper half) and region of gill filaments (lower half). TBs occur randomly in the gill arch region (not resolved due to low magnification; see D). **D.** Receptor area of TB of gill arch. Only small microvilli evident; fine mucous layer covers area. **E.** Tip of sturgeon barbel. Each epidermal hillock contains several taste buds. **F.** Higher magnification of E. Three receptor areas of TB clustered in one epidermal hillock. **G.** Receptor area with surrounding porelike ditch formed by neighboring epithelial (marginal) cells. **H.** Central part of receptor area with different microvillar profiles: single large receptor villi (black arrowhead), clustered small receptor villi (white arrow) and lobed or "winged" receptor villi (white arrowhead).

3.2 TEM Data

Longitudinal sections of a TB clearly show that light and dark elongated cells, in contrast to the rare developing or degenerating cells of the sensory epithelium, apically bear well-shaped receptor villi. As a rule, the apical parts of the light cells are roundish and terminate in one large conical microvillus, up to 2–3 μm long and 0.3–1.0 μm wide. The apical parts of the dark cells mostly have lobed processes that ensheath the light cells. Dark cells apically end with several small microvilli. These are about 1 μm long and 0.3 μm wide.

Both our TEM and SEM data as well as data in the literature prove that more than two types of elongated TB cells exist that contribute to a fish's TB receptor area. It is obvious that in some species of fish belonging to different taxa, the light and dark cells have distinct subtypes of cells that apically terminate in microvillar structures other than the above-mentioned large and small microvilli. Further, some of these elongated cells show an intermediate electron density, contain some "additional" organelles, and apically end with a specific type of microvillus. On comparing SEM and TEM data, we sometimes found discrepancies, e.g. fishes that showed two types of microvilli in the SEM revealed only one type in the TEM, and vice versa.

A TB of *Lepisosteus oculatus* shows one type of villus in the SEM, but in the TEM there are three (Figs 9.1D and 9.3). *Protopterus annectens* TBs show two types of villi in the SEM and four in the TEM (Figs 9.2 C, D and 9.6). Further, most of the species showing three types of villi in SEM (Figs 9.2C, D and 9.6) micrographs actually possess four different kinds of microvilli. These discrepancies are due to the fact that in the SEM microvillar structures with similar morphologies might be the apical endings of either a light or a dark cell (sub) type. It is difficult to differentiate in the SEM between longer and shorter villi when they are almost equal in diameter. Also, the cell borders of the villi-bearing cells cannot be detected within a receptor area by means of SEM. Moreover, differences seem to exist between cell type distribution depending on the location of the TBs. For instance, in TBs of *Archocentrus nigrofasciatus* several cell types were present on the lips, while TBs with only one cell type were located deep in the oropharygeal cavity.

As seen in the TEM, some species of fish possess more than one type of elongated light cell in their TBs (*Lepisosteus oculatus*, Fig. 9.3; *Scaphirhynchus platorynchus*, Fig. 9.4; *Danio rerio*, Fig. 9.5; *Protopterus*

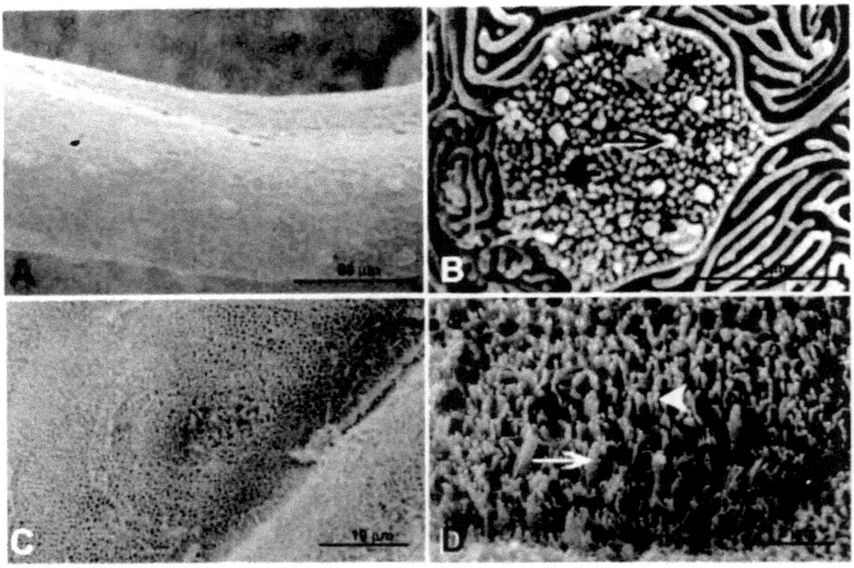

Fig. 9.2 A-D SEMs of TBs of *Danio rerio* (**A, B**) and *Protopterus annectens* (**C, D**). **A.** Middle part of zebrafish barbel with several TB receptor areas. **B.** Higher magnification of A. Receptor area contains three different types of receptor villi: Single large villi (long arrow), brush arranged villi (black arrowhead) and small villi (short black arrow). **C.** TB receptor area from extremely elongated and thin pectoral fin. **D.** Higher magnification of central part of receptor area in C. Single large receptor villi (long arrow) and aggregated small villi (arrowhead) visible.

annectens, Fig. 9.6). Apically, these different light cells contribute to the receptor area with distinct and relatively large microvillar structures. Contrarily, dark elongated cells seem to bear mostly one uniform type of small microvillus, with the exception of *Protopterus annectens*; in addition to the two subtypes of light cells we found two subtypes of dark cells, one of which bears a bunch of short and divided villi, while the other consists of straight, slender, and somewhat longer microvilli (Fig. 9.6A). In *Danio rerio* TBs, the brushlike microvillar structures (Fig. 9.2 B) belong to elongated cells of light to intermediate electron density. Their characteristic villi are somewhat longer and wider than the small microvilli of the dark cells and, at least in young animals, this cell type occurs mostly close to the lateral sides of the receptor area (Fig. 9.5 C, D).

Table 9.1 presents an overview of the different morphologies of the upper part of elongated TB cells and their apical microvillar structures, as occurring in different species of fish that (mostly) belong to different

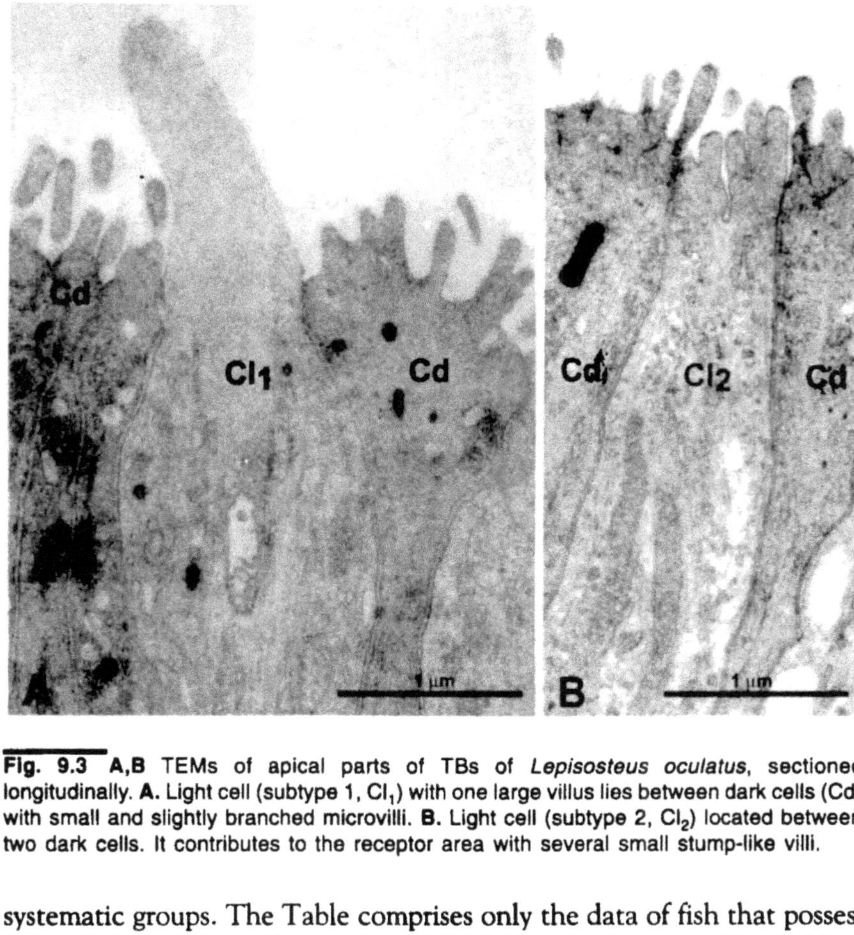

Fig. 9.3 A,B TEMs of apical parts of TBs of *Lepisosteus oculatus*, sectioned longitudinally. **A.** Light cell (subtype 1, Cl_1) with one large villus lies between dark cells (Cd) with small and slightly branched microvilli. **B.** Light cell (subtype 2, Cl_2) located between two dark cells. It contributes to the receptor area with several small stump-like villi.

systematic groups. The Table comprises only the data of fish that possess more than two types of receptor villi, with three exceptions. The Table is based on published data and this study. It shows that at least 9 different types of microvillar receptor structures exist based on the morphology of their apical endings; most of them belong to subtypes of the TB's light cells. Table 9.1 also shows that villus morphology varies even in fish that belong to the same taxon, and this is true even for members of the same genus, cf. *Amiurus*.

4. DISCUSSION

A comparison of TBs of various fish revealed that a variety of apical endings of TB cells exists in all the taxa examined to date.

Table 9.1 Synopsis of the different morphologies of apical microvillar structures of fish elongated TB cells when more than two different microvillar types were found per species in the TEM (exceptions: *Amiurus nebulosus*, *A. punctatus*, and *Lepidosiren paradoxa*) +, villus type common; (+), villus seldom occurs in this species.

Main group of fish	*Species*	*Microvillar Structures*									*References*
		of light cells						*of dark cells*			
Elasmobranchii	*Scyliorhinus canienla*	+*	+					+	+		Reutter (1994), Whitear and Moate (1994a)
Dipnoi	*Protopterus annectens*	+	+					+	+		this study
	Lepidosiren paradoxa	+						+			this study
	Neoceratodus forsteri	+	+	+				+			Reutter and Witt (1999) this study
Neopterygii											
Semionotiformes	*Lepisosteus oculatus*	+			+				+		Reutter et al. (2000)
Amiiformes	*Amia calva*	+					+	+		+	Reutter et al. (2000)
Chondrostei	*Scaphirhynchus platorynchus*	+	+	+				+			this study
Teleostei	*Danio rerio*	+	+			+		+			Hansen et al. (2000) this study
	Astyanax mexicanus	+					+	+			Boudriot and Reutter (2004)
	Amiurus (Ictalurus)										
	–nebulosus	+						+			Reutter (1971, 1978)
	–melas	+	+*					+			Desgranges (1965),* 3 subtypes
	–punctatus	+						+			Grover-Johnson and Farbman 1976), Royer and Kinnamon (1966)
	Silurus glanis	+	(+)				+	+	(+)		Reutter and Witt (1999)
	Scophthalmus maximus	+	+	+			+	+			this study

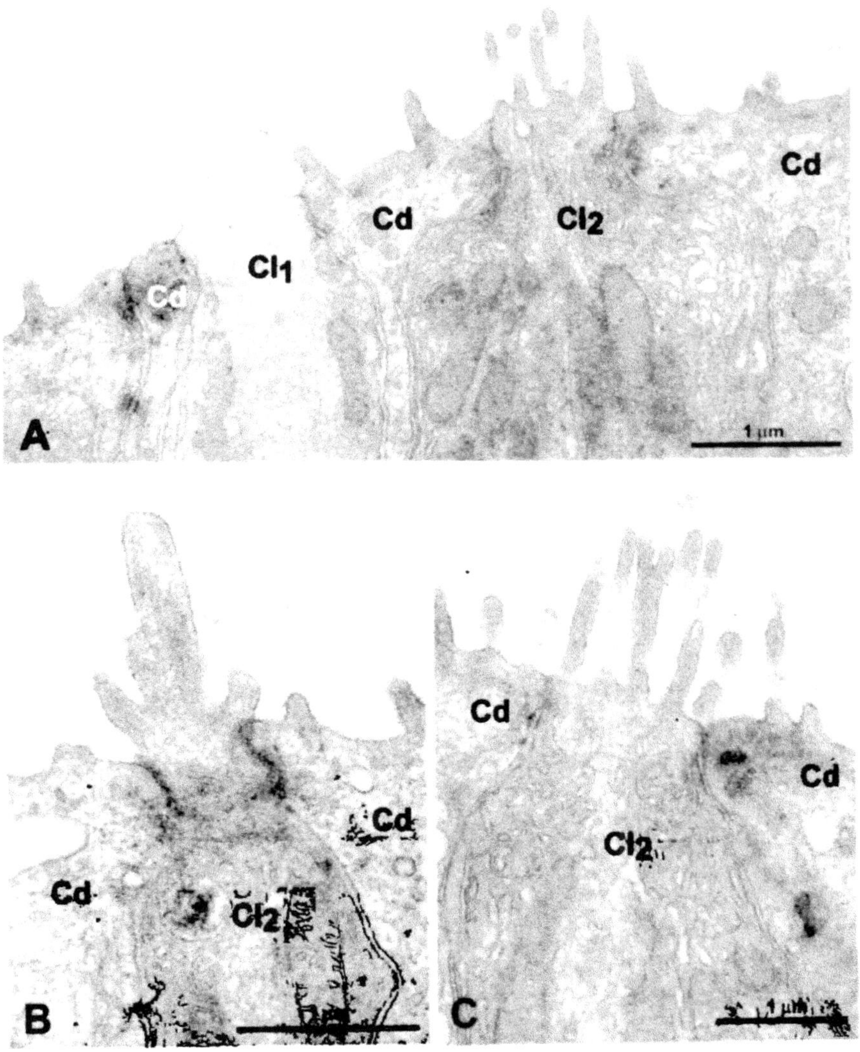

Fig. 9.4 A-C TEMs of longitudinally cut apical regions of taste buds in *Scaphirhynchus platorynchus*. Dark cells with small microvilli and light cells with differently shaped microvilli. **A.** Light cell (Cl_1) on left side bears one large conical villus, right one (Cl_2) several thin and long microvilli. **B.** Light cell (Cl_2) terminates in one large arborized or lobed microvillus. **C.** Light cell (Cl_2) projects to receptor area with bunch of long slender and, in part, ramified microvilli.

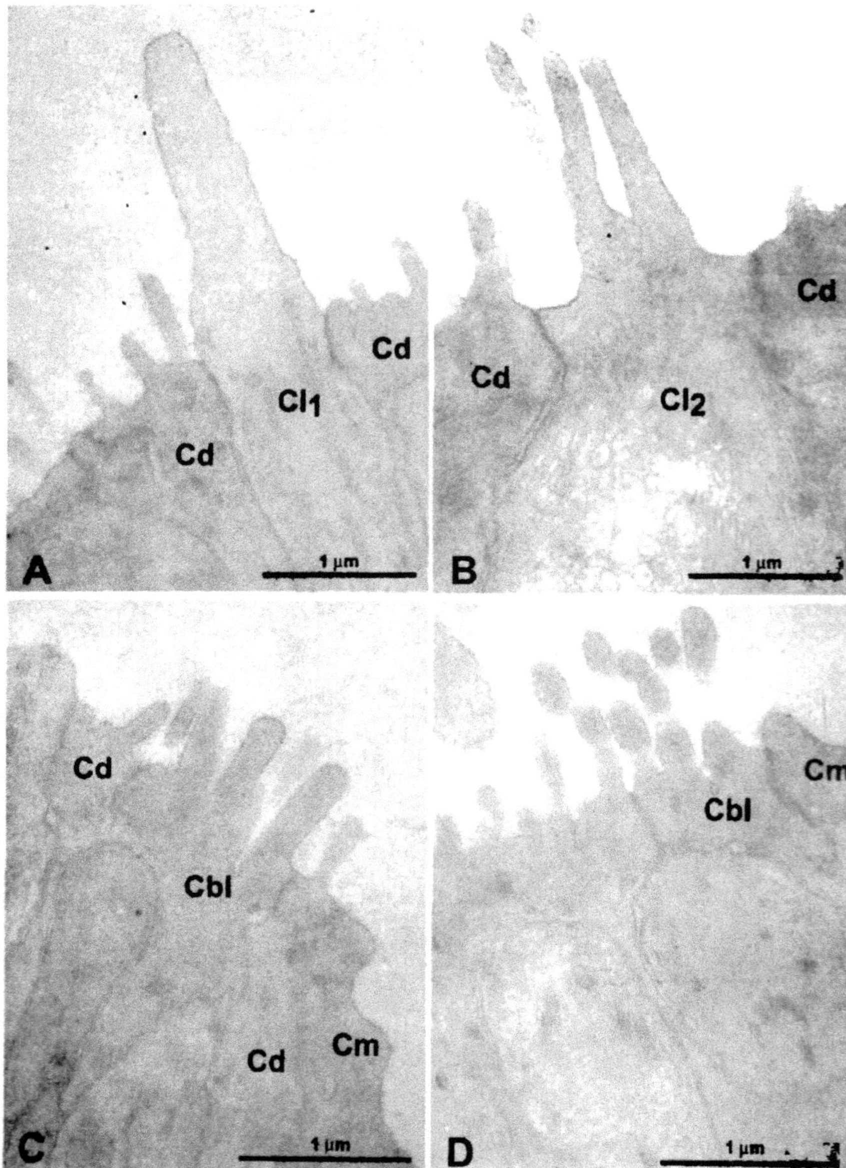

Fig. 9.5 A-D TEMs of apical parts of *Danio rerio* TBs sectioned longitudinally. All dark cells terminate in small microvilli, while apices of light cells differ. **A.** Light cell (Cl_1) with large conical microvillus. Dark cells (Cd) show several small microvilli. **B.** Light cell (Cl_2) bears some long and slender microvilli. **C, D**. Cell with brushy ending (Cbl) consists of relatively short, straight or "stiff" microvilli. These are somewhat longer and broader than those of dark cells. **D.** Microvilli of cell with a brush ending obliquely sectioned; "brush" may consist of about 10 microvilli.

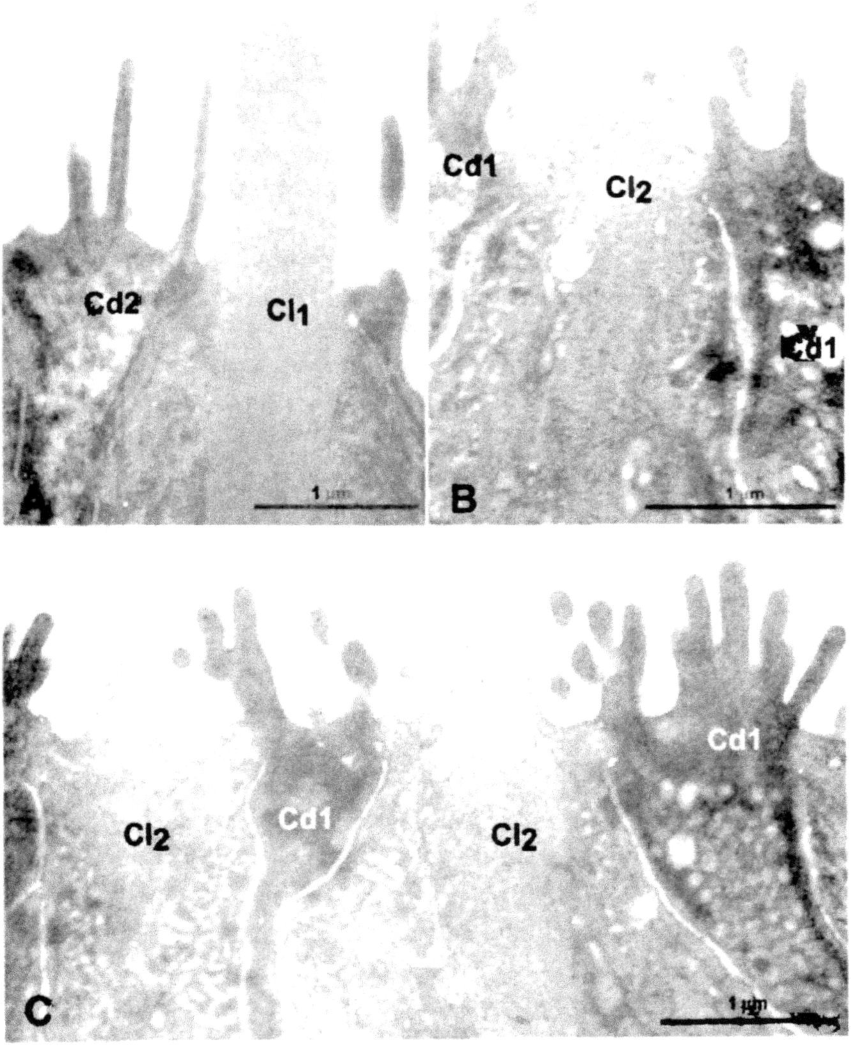

Fig. 9.6 A-C TEMs of longitudinally cut apical parts of taste buds of *Protopterus annectens*. Light cells as well as dark cells have two subtypes, each of which shows different receptor microvilli. **A.** Light cell (Cl_1) projects to the receptor area with one large conical microvillus. Dark cell (Cd_2) terminates with long thin "stiff" microvilli rich in microfilaments. **B, C**. Light cells (Cl_2) end with relatively large, multiple arborized or even irregularely lobed microvilli. Cd_1 dark cells are more common than Cd_2 cells. They bear tufts of short, branched or lobed microvilli.

It appears likely that in most species investigated thus far, the large receptor villi (SEM, TEM) belong to light cells (TEM) and the small receptor villi (SEM, TEM) to dark cells, as already described: SEM: *Amiurus nebulosus* (Breipohl et al., 1974; Reutter and Breipohl, 1975); *Xiphophorus helleri* (Reutter et al., 1974; Reutter and Breipohl, 1975); *Corydoras arcuatua* (Ovalle and Shinn, 1977); *Cyprinus carpio* (Kawakata et al., 1978); *Protopterus amphibius, Phoxinus phoxinus* (Lane and Whitear, 1982); *Gadus morhua* (Harvey and Batty, 1998). SEM and TEM: *Amiurus nebulosus* (Reutter, 1978); six genera of loricariid catfish (Ono, 1980); *Corydoras paleatus* (Fujimoto and Yamamoto, 1980); *Pimephales promelas* (Walker et al., 1981); *Salmo gairdneri* (Ezeasor, 1982); *Cobitis taenia* (Jakubowski, 1983); *Fundulus heteroclitus* (Hossler and Marchant, 1983); *Dicentrarchus labrax* (Connes et al., 1988); *Tinca tinca* (Żuwała and Jakubowski, 1993); *Scyliorhinus canicula* (Whitear and Moate, 1994a). TEM: *Cyprinus carpio, Parasilurus asotus, Cobitis taenia* (Hirata, 1966); *Clarias batrachus, Kryptopterus bicirrhis* (Welsch and Storch, 1969); *Amiurus nebulosus* (Reutter, 1971); *Pomatoschistus (Gobius) minutus, Trigla lucerna, Gasterosteus aculeatus, Phoxinus phoxinus* (Whitear, 1971); *Corydoras paleatus* (Schulte and Holl, 1971); *Blennius tentacularis* (Schulte and Holl, 1972); *Ciliata mustela* (Crisp and Laverack, 1975); *Ictalurus punctatus* (Grover-Johnson and Farbman, 1976; Royer and Kinnamon, 1996); *Anguilla anguilla* (Pevzner, 1978); *Acipenser güldenstädti, A. ruthenus, A. stellatus, Hucho hucho* (Pevzner, 1981); *Cyprinus carpio* (Toyoshima et al., 1984); *Pseudorasbora parva* (Kitoh and Kiyohara, 1987); *Plotosus lineatus* (Reutter, 1992). Reviews: (Kapoor et al., 1975; Reutter, 1986; Jakubowski and Whitear, 1990; Reutter and Witt, 1993; Jakubowski and Żuwała, 2000; Zaccone et al., 2001; Hansen and Reutter, 2004).

Parallel to the light cell-dark cell nomenclature for the TBs elongated cells, other nomenclatures are in use describing the same cells: The light cells are also termed gustatory cells (see Jakubowski and Whitear, 1990; Whitear, 1993) or, because of their richness in tubular profiles of the endoplasmic reticulum, t-cells (Crisp et al., 1975; Kiyohara et al., 1980; Royer and Kinnamon, 1996; Kiyohara and Tsukahara 2005, this volume). The dark cells are synonymous with supporting or sustentacular cells (see Jakubowski and Whitear, 1990 and above) and to f-cells so termed because of their richness in intermediate filaments (see Crisp et al., 1975; Kiyohara et al., 1980; Kiyohara and Tsukahara, 2005, this volume; Royer and Kinnamon, 1996). Because the light cells have and the dark cells may have synaptic contacts to nerve fibers (Reutter, 1971, 1978; Reutter and

Witt, 1993; Hansen and Reutter, 2004), we avoid the terms gustatory and sustentacular, etc., favoring the light cell-dark cell nomenclature. In any case, these discrepancies in nomenclature are irrelevant with respect to the different types of elongated cells and their polymorph microvillar structures.

It is of special interest that in several species of fish light cells and, to a lesser extent, dark cells occur in different subtypes (Table 9.1). As shown for several species light cell subtypes can differ from each other in three ways: a) with respect to shape of their receptor villi (e.g. *Scaphirhynchus*, *Scophthalmus*, *Neoceratodus*), or b) with respect to their electron density and receptor villi ("brush-like cell," *Danio*), or c) with respect to electron density, cell organelles, and receptor villi ("dense cored vesicles cell," *Astyanax*). Subtypes of dark cells occur only in a few species (e.g. *Scyliorhinus*, *Amia*). The different morphologies of the receptor villi are well pronounced in both light and dark cells and therefore it is most likely that the different microvilli are not merely transient structures of one and the same light or dark cell.

The functional significance of the different micromorphologies of the microvillar structures of elongated TB cells remains unclear. In principle, the plasmalemmata of the receptor villi are thought to be the site of the primary events of chemoreception. Up to now, this has been proved only once, for *Ictalurus punctatus* TBs in which the amino acid arginine specifically binds to the plasmalemmata of the large receptor microvilli (Finger et al., 1996). Possibly the different morphologies of microvillar structures observed are a sign of their chemoreceptive capacities. Most of the species listed in Table 9.1 are not laboratory animals and, in part, highly protected. Therefore the chances to use modern techniques (immunocytochemistry, molecular biology, biochemistry) to elucidate the function of the different types of villi are very low.

The heterogeneity of TB microvillar structures in different taxa and, to some extent, in different species (even of the same genus, as *Amiurus*), lets us believe that the receptor villi are species specific. It is obvious that the large and small microvilli are the most common and characterize the main types of light and dark elongated TB cells resp., since they occur in all species investigated. The less numerous subtypes of light and dark cells are mostly found in only one species and not in the other members of its taxonomic group investigated (Table 9.1). This was discussed earlier for the small taxon of neopterygian Semionotiformes and Amiiformes (holosteans; Reutter et al., 2000) as well as the main taxa of fish (Reutter and Witt, 1999): In fish TBs, the receptor villi of elongated cells vary in each group of fish and, to some extent, are also unique to a distinct species. Further, considering other criteria of TB morphology (TB size, location of the TB in the epithelium, TB sitting on a dermal papilla) and cytology (nerve fiber plexus, innervation, basal cells) it is obvious that the TBs of all vertebrate main taxa have TBs of their own types, as seen in fish, reptiles, birds and mammals (Reutter and Witt, 1993; Reutter, 1995).

The diversity of fish TBs, as demonstrated above, suggests that fish TBs did not evolve in a strict monophyletic way. It is more likely that fish TBs evolved polyphyletically in different directions, depending on the fish's individual ecological situation or niche. It is of special interest that in one main group of fish, the Elasmobranchii (Chondrichthyes), *Scyliorhinus canicula* does have well organized, morphologically distinct TBs (Reutter, 1994; Whitear and Moate, 1994a), whereas *Raja clavata* has no real TBs whatsoever (Whitear and Moate, 1994b). Further, as mentioned above, *Danio rerio* has a "brush-like ending cell" while goldfish, a close relative, lacks this cell type (Hansen et al., 2002). *Astyanax mexicanus* seems to exclusively possess the "dense cored vesicles cell" which up to now has not been found in other teleosts. The cave dwelling dark-adapted form of *Astyanax*, "*Anoptichthys*", has, compared to its epigean relative *Astyanax*, an enlarged nerve fiber plexus that possibly improves synaptic transmission of chemical stimuli (Boudriot and Reutter, 2001). We consider this enlargement an apomorphic derived character.

Comparing cell types of TBs in different species poses some technical problems. Sometimes we found discrepancies between the different microvillar structures as seen in the SEM for a distinct species of fish and the corresponding number of different microvilli as seen in the TEM. The reason is that microvilli of nearly the same morphology but belonging to two different types or subtypes of elongated TB cells, cannot be distinguished from each other in the SEM. Further, in some cases or even species the mucous surface coat atop the receptor area cannot be removed by standard methods. In this case it is likely that not all types of receptor villi penetrate the mucus and therefore are not detectable in the SEM (Reutter, 1980). Insofar as possible, SEM studies should precede and support a TEM investigation and not be the only technique applied when TB receptor areas are investigated. It remains an interesting question as to how many different types of elongated cells or subtypes of cells a *Polypterus* TB really contains: In the SEM we found 4 different kinds of microvillar structures but do not have the respective TEM results as yet. Moreover, in some species TBs seem to be equipped with a different set of cell types according to the location of the TB. For instance, in *Archocentrus nigrofasciatus* SEM micrographs we found TB cells with only short microvilli in the oral cavity. In this case the covering mucus cannot be the reason since the longer larger villi of the light cells would protrude further than the small villi of the dark cells.

In summarizing these data we postulate that no "common" type of TB exists in fish. It is more reasonable to assume that within the main taxa of fish and even within their genera, TBs vary considerably with respect to morphology and especially cell types and cellular subtypes.

Acknowledgement

This work was supported in part by National Institutes of Health Grant P30 DC 04657 to Diego Restrepo, University of Colorado Health Sciences Center, Denver, Colorado.

References

Boudriot, F. and K. Reutter. 2001. Ultrastructure of the taste buds in the blind cave fish *Astyanax jordani* ("*Anoptichthys*") and the sighted river fish *Astyanax mexicanus* (Teleostei, Characidae). *J. Comp. Neurol.* 434: 428–444.

Breipohl, W., G.J. Bijvank and G. Pfefferkorn. 1974. Scanning electron microscopy of various sensory receptor cells in different vertebrates. In: *Proc. Workshop Advances in Biomedical Applications of the SEM*, O. Johari and J. Gorvin (Eds). I.T.T. Research Institute, Chicago, IL (USA), pp. 557–564.

Connes, R., M. Granie-Prie, J.P. Diaz and J. Paris. 1988. Ultrastructure des bourgeons du goût du téléostéen marin *Dicentrarchus labrax* L. *Can. J. Zool.* 66: 2133–2142.

Crisp, M., G.A. Lowe and M.S. Laverack. 1975. On the ultrastructure and permeability of taste buds of the marine teleost *Ciliata mustela*. *Tissue & Cell* 7: 191–202.

Declaration of Helsinki. 1995. Recommendations from the Declaration of Helsinki. *Chem. Senses* 20: 181.

Desgranges, J.-C. 1965. Sur l'existence de plusieurs types de cellules sensorielles dans les bourgeons du goût des barbillons du poisson-chat. C. *R. Acad. Sci.* (*D*). Paris 261: 1095–1098.

Ezeasor, D.N. 1982. Distribution and ultrastructure of taste buds in the oropharyngeal cavity of the rainbow trout, *Salmo gairdneri* Richardson. *J. Fish Biol.* 20: 53–68.

Finger, T.E. and S.A. Simon. 2000. Cell biology of taste epithelium. In: *The Neurobiology of Taste and Smell*, T.E. Finger, W.L. Silver and D. Restrepo (Eds). Wiley-Liss Inc., New York, pp. 287–314.

Finger, T.E., B.P. Bryant, D.L. Kalinoski, J.H. Teeter and B. Böttger. 1996. Differential localization of putative amino acid receptors in taste buds of the channel catfish *Ictalurus punctatus*. *J. Comp. Neurol.* 373:129–138.

Fujimoto, S. and K. Yamamoto. 1980. Electron microscopy of terminal buds on the barbels of the silurid fish, *Corydoras paleatus*. *Anat. Rec.* 197: 133–141.

Grover-Johnson, N. and A.L. Farbman. 1976. Fine structure of taste buds in the barbel of the catfish, *Ictalurus punctatus*. *Cell Tissue Res.* 169: 395–403.

Hansen, A. and K. Reutter. 2004. Chemosensory systems in fish: Structural, functional and ecological aspects. In: *The Senses of Fish: Adaptations for the Reception of Natural Stimuli*, G. von der Emde, J. Mogdans and B.G. Kapoor (Eds). Narosa Publishing House, New Delhi, and Kluwer Academic Publishers, Dordrecht. pp. 55–89.

Hansen, A., K. Reutter and E. Zeiske. 2002. Taste bud development in the zebrafish, *Danio rerio*. Devel. Dyn. 223: 483–496.

Harvey, R. and R.S. Batty. 1998. Cutaneous taste buds in cod. *J. Fish Biol.* 53: 138–149.

Hirata, Y. 1966. Fine structure of the terminal buds on the barbels of some fish. *Arch. Histol. Jap.* 26: 507–523.

Hossler, F.E. and L.H. Merchant. 1983. Morphology of taste buds on the gill arches of the mullet *Mugil cephalus*, and the killifish *Fundulus heteroclitus*. *Amer. J. Anat.* 166: 299–312.

Jakubowski, M. 1983. New details of the ultrastructure (TEM, SEM) of the taste buds in fishes. *Z. mikrosk.- anat. Forsch.* 97: 849–862.

Jakubowski, M. and K. Żuwała. 2000. Taste organs in lower vertebrates: Morphology of the gustatory organs in fishes. In: *Vertebrate Functional Morphology*, H.M. Dutta and J.S. Datta Munshi (Eds). Science Publishers Inc., Enfield (NH), USA, pp. 161–174.

Jakubowski, M. and M. Whitear. 1990. Comparative morphology and cytology of taste buds in teleosts. *Z. mikrosk.- anat. Forsch.* 104: 529–560.

Kapoor, B.G. and T.E. Finger. 2003. Taste and solitary chemoreceptor cells. In: *Catfishes*, G. Arratia, B.G. Kapoor, M.Chardon and R. Diogo (Eds). Science Publishers Inc., Enfield (NH), USA, vol. 2, pp. 753–769.

Kapoor, B.G., H.E. Evans and R.A. Pevzner. 1975. The gustatory system in fish. *Adv. Mar. Biol.* 13: 53–108.

Kawakita, K., T. Marui and M. Funakoshi. 1978. Scanning electron microscopic observations on the taste buds of the carp (*Cyprinus carpio* L.). Jpn. *J. Oral Biol.* 20: 103–113. (In Japanese).

Kitoh, J., S. Kiyohara and S. Yamashita. 1987. Fine structure of taste buds in the minnow. *Nippon Suisan Gakkaishi* 53: 1943–1950.

Kiyohara, S. and J. Tsukahara. 2005. Barbel taste system in catfish and goatfish. In: *Fish Chemosenses*, K. Reutter and B.G. Kapoor (Eds). Science Publishers, Inc., Enfield, (NH), USA, and Plymouth, UK., pp. 175–209 this volume.

Kiyohara, S., S. Yamashita and J. Kitoh. 1980. Distribution of taste buds on the lips and inside the mouth in the minnow, *Pseudorasbora parva*. *Physiol Behav.* 24: 1143–1147.

Lane, E.B. and M. Whitear. 1982. Sensory structures at the surface of fish skin. I. Putative chemoreceptors. *Zool. J. Linn. Soc.* 75: 141–151.

Nelson, J.S. 1994. *Fishes of the World*. 3rd edition. John Wiley, New York, NY.

Ono, R.D. 1980. Fine structure and distribution of epidermal projections associated with taste buds on the oral papillae in some loricariid catfishes (Siluroidei: Loricariidae). *J. Morphol.* 164: 139–159.

Ovalle, W.K. and S.L. Shinn. 1977. Surface morphology of taste buds in catfish barbels. *Cell Tissue Res*. 178: 375–384.

Pevzner, R.A. 1978. Electron microscopic study of the taste buds of the eel, *Anguilla anguilla*. *Tsitologiya* 20: 1112–1118. (Russian, with English summary).

Pevzner, R.A. 1981. The fine structure of taste buds of the ganoid fishes. I. Adult Acipenseridae. *Tsitologiya* 23: 760–766. (Russian, with English summary).

Reutter, K. 1971. Die Geschmacksknospen des Zwergwelses *Amiurus nebulosus* (Lesueur). Morphologische und histochemische Untersuchungen. *Z. Zellforsch*. 120: 280–308.

Reutter, K. 1974. Typisierung der Geschmacksknospen von *Xiphophorus helleri* Heckel (Poeciliidae, Teleostei). *Verh. Anat. Ges*. 68: 851–854.

Reutter, K. 1978. Taste organ in the bullhead (Teleostei). *Adv. Anat. Embryol. Cell Biol*. 55: 1–98.

Reutter, K. 1980. SEM-study of the mucus layer on the receptor field of fish taste buds. In: *Olfaction and Taste* VII, H. van der Starre (Ed.). IRL Press, London, p. 107.

Reutter, K. 1982. Taste organ in the barbel of the bullhead. In: *Chemoreception in Fishes*, T.J. Hara (Ed.). Elsevier, Amsterdam, pp. 77–91.

Reutter, K. 1986. Chemoreceptors. In: *Biology of the Integument*. vol. 2. *Vertebrates*, J. Bereiter–Hahn, A.G. Matoltsy and K.S. Richards (Eds). Springer-Verlag, Berlin, pp. 586–604.

Reutter, K. 1992. Structure of the peripheral gustatory organ represented by the siluroid fish *Plotosus lineatus* (Thunberg). In: *Fish Chemoreception*, T.J. Hara (Ed.). Chapman & Hall, London, New York, pp. 60–78.

Reutter, K. 1994. Ultrastructure of taste buds in the spotted dogfish *Scyliorhinus canicula* (Selachii). In: *Olfaction and Taste* XII. K. Kurihara, N. Suzuki and H. Ogawa (Eds). Springer-Verlag, Tokyo, p. 754.

Reutter, K. 1995. Comparative aspects of vertebrate taste cell microvillar structures. In: *Chemical Signals in Vertebrates VII*, R. Apfelbach, D. Müller-Schwarze, K. Reutter and E.Weiler (Eds). Pergamon Press, Oxford, (*Adv. Biosci*. 93: 3–9, 1994)

Reutter, K. and W. Breipohl. 1975. Rasterelektronenmikroskopische Untersuchung der Geschmacksknospen von Fischen (*Amiurus nebulosus* and *Xiphophorus helleri*). *Verh. Anat. Ges*. 69: 879–884.

Reutter, K. and M. Witt. 1993. Morphology of vertebrate taste organs and their nerve supply. In: *Mechanisms of Taste Transduction*, S.A. Simon and S.D.Roper (Eds). CRC Press, Boca Raton, pp. 29–82.

Reutter, K. and M. Witt. 1999. Comparative aspects of fish taste buds ultrastructure. In: *Advances in Chemical Signals in Vertebrates*, R.E. Johnston, D. Müller–Schwarze and P.W. Sorensen (Eds). Kluwer Academic Publishers, New York, pp. 573–581.

Reutter, K., W. Breipohl and G.J. Bijvank. 1974. Taste bud types in fishes. II.Scanning electron microscopical investigations on *Xiphophorus helleri* Heckel (Poeciliidae, Cyprinodontiformes, Teleostei). *Cell Tissue Res*. 153: 151–165.

Reutter, K., F. Boudriot and M. Witt. 2000. Heterogeneity of fish taste bud ultrastructure as demonstrated in the holosteans *Amia calva* and *Lepisosteus oculatus*. *Phil. Trans. R. Soc. Lond*. B. 355: 1225–1228.

Royer, S.M. and J.C. Kinnamon. 1996. Comparison of high-pressure freezing/freeze substitution and chemical fixation of catfish barbel taste buds. *Microsc. Res. Techn.* 35: 385–412.

Schulte, E. and A. Holl. 1971. Untersuchungen an den Geschmacksknospen der Barteln von *Corydoras paleatus* Jenyns. I. Feinbau der Geschmacksknospen. *Z. Zellforsch.* 120: 450–462.

Schulte, E. and A. Holl. 1972. Feinbau der Kopftentakel und ihrer Sinnesorgane bei *Blennius tentacularis* (Pisces, Blenniiformes). *Mar. Biol.* 12: 67–80.

Sorensen, P.W. and J. Caprio. 1998. Chemoreception. In: The Physiology of Fishes, D.H. Evans (Ed.). 2nd edition. CRC Press, Boca Raton, pp. 375–405.

Tagliafierro, G. and G. Zaccone. 2001. Morphology and immunohistochemistry of taste buds in bony fishes. In: *Sensory Biology of Jawed Fishes*, B.G. Kapoor and T.J. Hara (Eds). Science Publishers Inc., Enfield (NH), USA, and Plymouth, UK. pp. 335–345.

Toyoshima, K., O. Nada and A. Shimamura. 1984. Fine structure of monoamine-containing basal cells in the taste buds on the barbels of three species of teleosts. *Cell Tissue Res.* 235: 479–484.

Walker, E.R., S.F. Fidler and D.E. Hinton. 1981. Morphology of the buccopharyngeal portion of the gill in the fathead minnow *Pimephales promelas* (Rafinesque). *Anat. Rec.* 200: 67–81.

Welsch, U. and Storch, V. 1969. Die Feinstruktur der Geschmacksknospen von Welsen [(*Clarias batrachus* (L.) und *Kryptopterus bicirrhis* (Cuvier et Valenciennes)]. *Z. Zellforsch.* 100: 552–559.

Whitear, M. 1971. Cell specialization and sensory function in fish epidermis. *J. Zool. Lond.* 163: 237–264.

Whitear, M. 1993. Epithelial sensory cells in fish. In: Advances in Fish Research, B.R. Singh (Ed.). Narenda Publishing House, Delhi, vol. 1, pp. 169–184.

Whitear, M. and R.M. Moate. 1994a. Microanatomy of taste buds in the dogfish, *Scyliorhinus canicula*. *J. Submicrosc. Cytol. Pathol.* 26: 257–367.

Whitear, M. and R.M. Moate. 1994b. Chemosensory cells in the oral epithelium of *Raja clavata* (Chondrichthyes). *J. Zool. Lond.* 232: 295–312.

Witt, M. 1996. Carbohydrate histochemistry of vertebrate taste organs. *Progr. Histochem. Cytochem.* 30/4: 1–172.

Witt, M., K. Reutter and I.J. Miller Jr. 2003. Morphology of the peripheral taste system. In: *Handbook of Olfaction and Gustation*, R.L. Doty (Ed.). 2nd edition. Marcel Dekker Inc., New York, pp. 651–677.

Zaccone, G., B.G. Kapoor, S. Fasulo and L. Ainis. 2001. Structural, histochemical and functional aspects of the epidermis of fishes. *Adv. Mar. Biol.* 40: 253–348.

Żuwała, K. and M. Jakubowski. 1993. Light and electron (SEM, TEM) microscopy of taste buds in the tench *Tinca tinca* (Pisces: Cyprinidae). *Acta Zool.* (Stockholm) 74: 277–282.

CHAPTER

10

Efferent Synapses in Fish Taste Buds

Klaus Reutter[1] and Martin Witt[2]

ABSTRACT

In fish, nerve fibers of taste buds are organized within the organ's nerve fiber plexus. The plexus is located between the sensory epithelium consisting of light and dark elongated cells and basal cells. It comprises the basal parts and processes of the light and dark cells that intermingle with nerve fibers, which are the dendritic endings of the taste sensory neurons belonging to the cranial nerves VII, IX or X. Most of the synapses at the plexus are afferent; they have synaptic vesicles on the light (or dark) cells side, which is presynaptic. In contrast, the presumed efferent synapses may be rich in synaptic vesicles on the nerve fibers (presynaptic) side, whereas the cells (postsynaptic) side may contain a subsynaptic cistern, a flat compartment of the smooth endoplasmic reticulum. This structure is regarded as a prerequisite of a typical efferent synapse, as occurring in cochlea hair cells. In fish taste buds, efferent synapses

Address for Correspondence: [1]Klaus Reutter, Anatomisches Institut, Universität Tübingen, 72074 Tübingen, Germany. E-mail: klaus.reutter@web.de

[2]Anatomisches Institut, Technische Universität Dresden, Dresden, Germany.

are rare and were found only occasionally in a few species belonging to different taxa. The significance of efferent synapses in fish taste buds is not well understood because efferent connections between the gustatory nuclei of the medulla with taste buds are not yet proven.

Key Words: Afferent synapse; Subsynaptic cistern; Membrane specialization; Ultrastructure.

1. INTRODUCTION

In the last decades numerous electron microscopic investigations were carried out on fish taste buds (TBs). The studied species belong to almost all fish taxa, such as Elasmobranchii (Chondrichthyes; Whitear and Moate, 1994 a,b; Reutter, 1994), Dipnoi (Reutter, 1991), neopterygian Semionotiformes and Amiiformes (Holostei; Reutter et al., 2000), Chondrostei (Pevzner, 1981), and Teleostei (for references, see Reutter and Hansen, 2005; this volume). The literature has been revised several times, most recently by Jakubowski and Żuwała (2000), Kapoor and Finger (2003), Hansen and Reutter (2004) and Reutter and Hansen (2005, this volume).

TBs are intraepithelial organs of ovoid shape. They are about 100 µm high and 40 µm broad and consist of electron light and dark elongated cells, basal cells, marginal cells and the nerve fiber plexus. In view of TB function, receptor villi at the apical end of the elongated cells (see Reutter and Hansen, 2005, this volume) and the nerve fiber plexus on the basal side of these cells are of special interest. Chemical substances, respectively gustatory stimuli, are recognized at the receptor villi (Finger et al., 1996); then, at the basal part of the cell, the generated signal is synaptically transmitted to an axon of the nerve fiber plexus and afferently conducted to the oblongate medulla of the brain, respectively to the facial nucleus, glossopharyngeal nucleus or vagal nucleus (see Finger 1976, 1978; Hansen and Reutter, 2004). The presence of synapses at their bases makes the elongated cells to secondary sensory cells; consequently, the afferent nerve fibers are the dendritic endings of the first neuron of the gustatory pathway (refs. as above).

Interest focuses here on the small region of the nerve fiber plexus, the place where synapses occur. The unmyelinated axons of the plexus are highly intermingled with the bases of the elongated (dark and light) cells.

Synapses were regularly found there and depicted, despite the fact that fish TB synapses are often poorly equipped with well-known structural details, as synaptic vesicles and synaptic membrane densities. Synapses are located between a light cell's part or, more regularly, at one of its basal processes, and an axon. These synapses are afferent because synaptic vesicles and their presynaptic densities (dense projections) are located on the light (or dark) cell's side. They occur in TBs of all vertebrates. On the other hand, there is some evidence of efferent (or efferent-like) synapses in TBs also. Such synapses typically occur in the organ of Corti and the crista ampullaris, at the bases of the hair cells. They typically contain synaptic vesicles on the nerve fiber's (presynaptic) side and a subsynaptic cistern (as part of the smooth endoplasmic reticulum) on the hair cells (postsynaptic) side, situated directly below the subsynaptic membrane (Engström and Sjöstrand, 1954; Engström, 1958; Saito, 1980; Emmerling et al., 1998; and others). Synapses of comparable structure were occasionally found in mammalian TBs (man: Graziadei, 1970; fetal monkey: Zahm and Munger, 1983 a, b; guinea pig: Yoshi et al., 1990; rabbit: Fujimoto and Murray, 1970; Murray and Murray, 1970; Toyoshima and Tandler, 1987; rat: Akisaka, 1980; Yoshie et al., 1996; mice: Takeda, 1976; Kondo, 1983; Kudoh, 1988; Royer and Kinnamon, 1988) and also birds (parrot: Suzuki and Takeda, 1976), reptiles (*Clemmys japonica*: Uchida, 1980), newts (*Necturus maculosus*: Delay and Roper, 1988), frogs (*Rana sp.*: Osculati and Sbarbati, 1995) and fish (*Amiurus melas*: Desgranges, 1966, 1972; *Phoxinus phoxinus*: Jakubowski and Whitear, 1990). In TBs, efferent synapses are rare through all taxa of vertebrates and were identified mostly by their subsynaptic cistern (lacking in *Necturus* TBs). In fish, synapses with a subsynaptic cistern have been found only twice (see above). The reasons might be that efferent synapses can be oversighted, or they are rare, or they are missing in the TBs of a distinct species or even in a distinct taxon of fish. Consequently, we took a closer look at our collection of electron micrographs and found some additional efferent synapses in the TBs of fish belonging to different systematic groups.

2. MATERIALS AND METHODS

The electron micrographs used in this study were obtained from several different species of fish investigated in our laboratories during the past decades. To provide a better comparability of the results tissues were fixed and processed in the same way, insofar as this was possible. The ethical regulations for animal care and treatment recommended in the Declaration of Helsinki (1995) were respected.

After 3-aminobenzoic acid ethyl ester (Tricaine, MS 222; Sigma – Aldrich) anaesthetization of the fish, the tissues were excised and fixed in 0.1 M phosphate buffer (pH 7.2) containing 2.5 % glutaraldehyde or, in a few cases, also 2% paraformaldehyde. After osmication, the tissues were dehydrated in ethanol and propylene oxide and embedded in Araldite or in glycerid ether 100 (*Danio rerio*: Hansen et al., 2002). Ultrathin sections were stained with uranyl acetate and lead citrate and examined with a Philips EM 300 or a LEO EM 912 OMEGA transmission electron microscope.

The following species of fish belonging to various taxa (Nelson, 1994) were investigated:

Elasmobranchii (Chondrichthyes): *Scyliorhinus canicula*; **Dipnoi**: *Protopterus annectens*, *Lepidosiren paradoxa*, *Neoceratodus forsteri*; **Neopterygii,** Semionotiformes: *Lepisosteus oculatus*; Amiiformes: *Amia calva*; **Chondrostei**: *Scaphirhynchus platorynchus*; **Teleostei,** Cyprinidae: *Danio rerio*; Siluridae: *Silurus glanis*, *Amiurus* (*Ictalurus*) *nebulosus*; Bothidae: *Scophthalmus maximus*.

3. RESULTS

At first glance, fish TBs seem to be organized uniformly, but they may vary considerably (Reutter and Witt, 1993, 1999; Hansen and Reutter, 2005, this volume). Nevertheless, the basic cellular components are more or less the same and using the light cell-dark cell nomenclature (see Reutter and Hansen, this volume), a fish TB consists of electron-lucent light cells and electron-dense dark cells. The two cells types are elongated and form the TB's sensory epithelium proper. Apically, they terminate with microvilli (receptor villi) in the TB receptor area. At the TB base, the basal cells are located directly on the basal lamina. At the TB basolateral border lie the marginal cells. Between the bases of the elongated cells including their processes and the basal cells the organ's nerve fiber plexus is located. The TB is completely embedded in a stratified squamous epithelium and rests atop a dermal papilla (Fig. 10.1).

The **nerve fiber plexus** is the place where the organ is innervated and therefore synapses occur. Synapses are never found in great numbers and only in a few cases are they typical asymmetric Gray type I synapses. TB synapses often have only poor synaptic features, as synaptic vesicles and, at their active zone, less membrane specializations. Nevertheless, afferent synapses were found in TBs of nearly all fish investigated so far. An

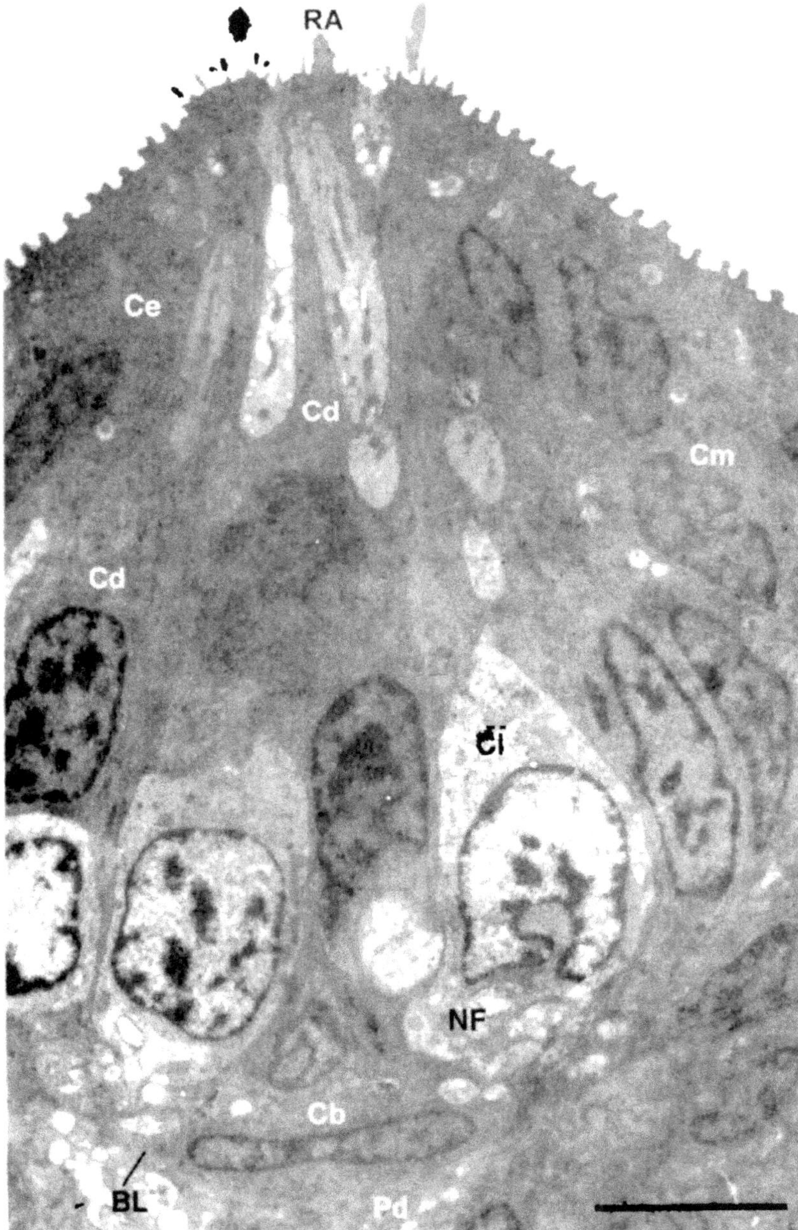

Fig. 10.1 Low power TEM of a slightly elevated taste bud of a 40-day-old young zebrafish, *Danio rerio*, in longitudinal section. The main structural details are indicated: Light cell (Cl), dark cell (Cd), receptor area (RA), basal cell (Cb), marginal cell (Cm), epidermal cell (Ce), dermal papilla (Pd), nerve fiber plexus (NF), basal lamina (BL). Bar: 10 μm.

example of an **afferent** TB synapse is shown in Figure 10.2a. At this synapse the light cell side is the presynaptic side and the nerve fiber (or axon) side the postsynaptic. Presynaptically, numerous small and mostly clear vesicles (20–30 nm) are located close to the presynaptic membrane. The active zone of the synapse is marked by the presynaptic densities (dense projections) of the presynaptic membrane. Both the pre- and sub(post)synaptic membranes limit the synaptic cleft. The latter is of overall constant width (16–20 nm). In Figure 10.1a the postsynaptic density is mostly lacking. The postsynaptic axon (atypically) contains some vesicles, too.

In contrast to the afferent, **efferent** synapses are sparse and obviously do not occur in every species. Their distinct organization cannot be confused with that of afferent synapses. As an efferent synapse transmits a signal from the central nervous system to the periphery, in this case to a TB cell, the axon is on the synapses presynaptic side, whereas the light (or, very seldomly dark) cell is at the postsynaptic side. Consequently, synaptic vesicles are seen in the axon but not in each case: In Figure 10.2b the nerve fiber is mostly empty of vesicles, but not the axons seen in Figure 10.2c. Pre- and subsynaptic membranes are again parallel to each other and the synaptic cleft of constant width. The cells (postsynaptic) side is more or less empty of vesicles. Membrane specializations are rare (Figs. 10.2b, c and 10.4a) and undistinctive (Fig. 10.3a, with presynaptic density). The most striking structural detail of an efferent synapse is its subsynaptic cistern. This flat and sometimes discoid compartment of the smooth endoplasmic reticulum consists of two parallel running membranes (in continuation with each other) that include a lumen, the cistern. It is located directly below the subsynaptic membrane. This means that at the active zone of the synapse, four membranes are running parallel to each other. As the pre- and postsynaptic membranes are mostly symmetric to each other, this efferent synapse is a Gray type II synapse.

We could identify such typical efferent synapses at the nerve fiber plexes of *Lepisosteus oculatus* (Fig. 10.2c), *Scaphirhynchus platorynchus* (Fig. 10.2b), *Danio rerio* (Fig. 10.3a), and *Neoceratodus forsteri* (Fig. 10.4a). The efferent synapses of *Neoceratodus* TBs are particularly well developed and impressive, given by their extended subsynaptic cisterns.

Besides these typical efferent synapses, two other efferent synapse types also exist in fish TBs. In a *Lepidosiren paradoxa* TB we found a junction that possesses a bulk of synaptic vesicles on the axon side but no

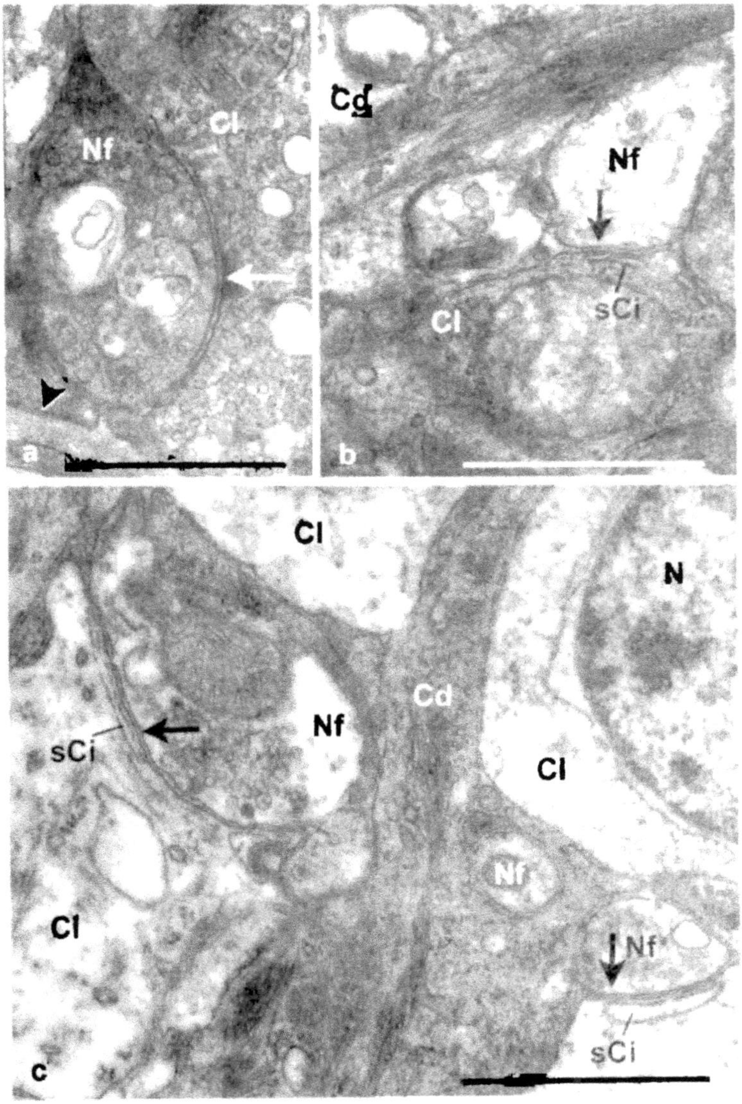

Fig. 10.2 a-c Synaptic contacts of different morphologies from the basal parts of *Scaphirhynchus* (a, b) and *Lepisosteus* (c) taste buds. **a.** Example for an afferent synapse of a fish taste bud. Near the taste buds underlying basal lamina (arrowhead), the basal process of a light cell (Cl) is in afferent synaptic contact (white arrow) to an axon (nerve fiber, NF) of the taste bud's nerve fiber plexus. Numerous small clear synaptic vesicles occur on the light cell side (presynaptic side); clear and a few dense-cored vesicles are seen on the axon (postsynaptic) side. Presynaptic and poor postsynaptic densities are given. **b.** Efferent synapses. An axon (NF) of the TB nerve fiber plexus in synaptic contact to a light cell (Cl): Axon side presynaptic while the light cell side contains a subsynaptic cistern (sCi), situated near the subsynaptic membrane, and therefore the cell is postsynaptic (black arrow). Dark cell (Cd). **c.** Two efferent synapses (black arrows) situated between the axons (NF) and the light cells (Cl): Presynaptic axons contain synaptic vesicles while light cell sides are characterized by subsynaptic cisterns (sCi). Dark cell (Cd). Scale bars in a - c: 1 µm.

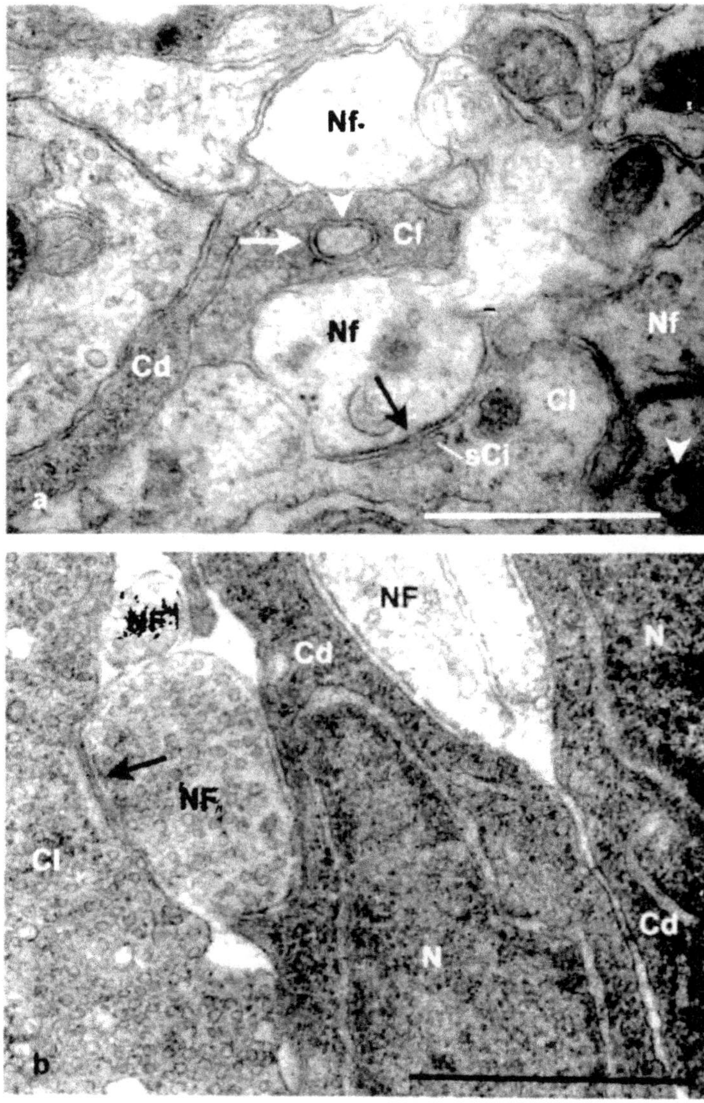

Fig. 10.3 a-b Synapses and synaptic contacts of *Danio* (a) and *Scophthalmus* (b) taste buds. **a.** Region of nerve fiber plexus of a 25-day-old zebrafish. A small process of a light cell (Cl) is in afferent synaptic contact to a cross-sectioned spinelike process of an axon (arrowhead; center of Figure). At the lower right side, a longitudinally cut axonal spine penetrates a light cell; another spine is cut transversely (arrowhead). In the lower middle, a presumably efferent synapse is located: Presynaptic axon (Nf) empty of vesicles; on light cell side lies a subsynaptic cistern (sCi). Dark cell (Cd). **b.** Medial part of a developing TB from a 14-day-old halibut larva. Axons (Nf) filled with vesicles are in close contact to a light cell (Cl) and a dark cell (Cd). Also the light cell is rich in vesicles. The arrow points to a presumed efferent synapse; a subsurface cistern is missing. Nucleus (N). Scale bars in a, b: 1 μm.

subsynaptic cistern on the – in this case – dark cell side (Fig. 10.4c). We could find no efferent synapses with a subsynaptic cistern in *Lepidosiren*. A third type of synapse with a presumably efferent function occurs in the developing TBs of halibut larvae, *Scophthalmus maximus* (Fig. 10.3b). In the center of the young sensory epithelium (not at its base) of an advanced developed TB anlage, the nerve fibers, densely filled with synaptic vesicles, are in close contact to (prospective) light cells (also rich in vesicles) and to dark cells. At these contacts typical subsynaptic cisterns are missing and the membrane specializations are poor.

The data concerned with efferent synapses in fish TBs from the literature and the results described here are summarized in Table 10.1. The Table shows that efferent synapses occur in the TBs of most, but not all, the main taxa of fish. In only six species (*Lepisosteus oculatus, Scaphirhynchus platorynchus, Ictalurus melas* (Desgranges, 1966, 1972), *Phoxinus phoxinus* (Jakubowski and Whitear, 1990), *Danio rerio* and *Neoceratodus forsteri*) efferent synapses with typical subsynaptic cisterns were found. All the typical efferent synapses are located at the light cells.

4. DISCUSSION

In principle there are two possible ways by which the central nervous system may control perception of sensory impulses. The first is activation of short inhibitory interneurons, for example in the visual or olfactory systems. In the olfactory bulb, there are local synaptic circuits between granule cells (interneurons) and mitral cells, which constitute the second olfactory neuron. The reciprocal synaptic circuit between mitral cells and granule cells (dendrodendritic inhibition) contributes to olfactory processing along with lateral inhibition of mitral cells (Isaacson and Vitten, 2003). Efferent pathways are usually established between secondary olfactory structures and the interneurons (periglomerular cells) in the olfactory bulb, but these axons do not reach the olfactory neuron (= olfactory receptor cell of the olfactory epithelium) directly.

The second way uses anatomically distinct tracts, which connect central sensory nuclei directly with secondary sensory cells, as observed in the vestibulocochlear system.

Afferent and efferent synapses are located in the organ of Corti and have been well studied (for Refs., see Introduction). Both types of synapses are located near each other at the hair cells base. The structures are well developed and, in the case of efferent synapses, serve as a prototype, which regularly shows synaptic vesicles at the axon side and

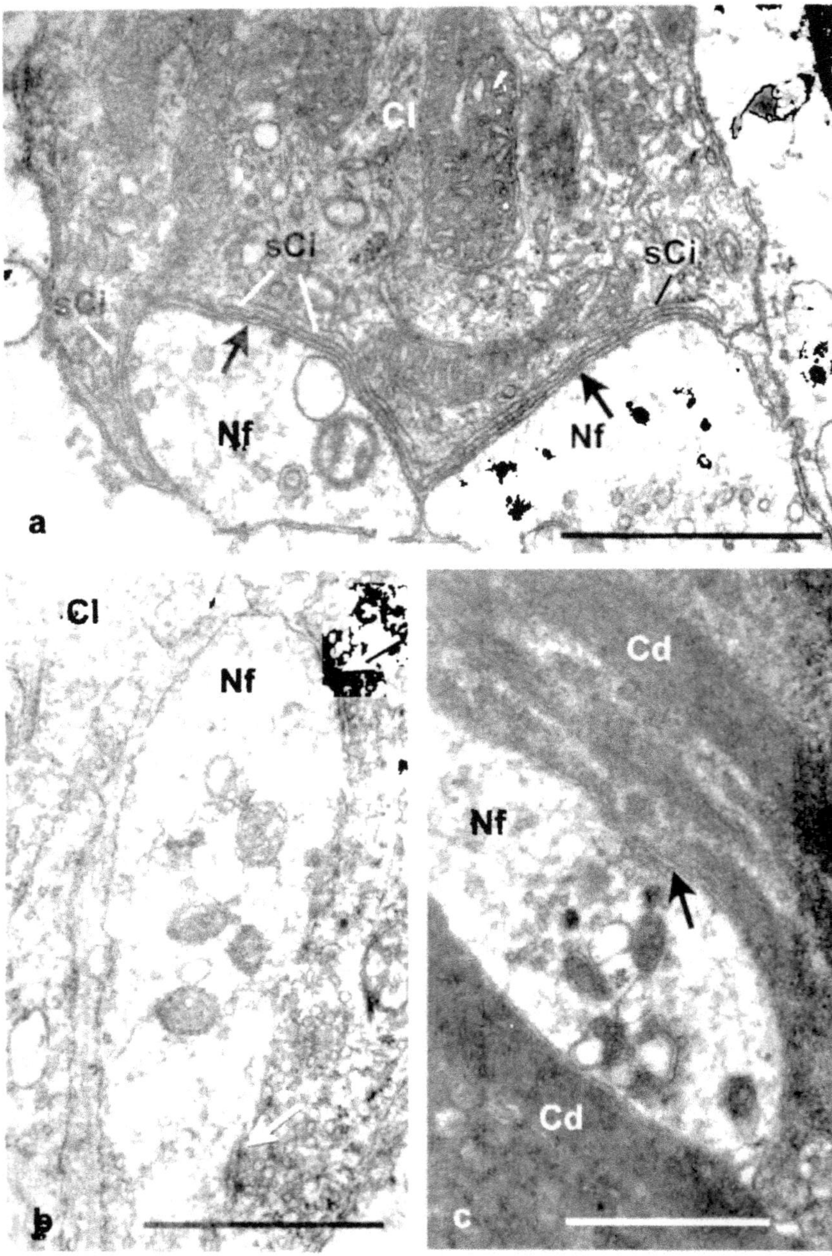

Fig. 10.4 a–c Efferent and afferent synapses in *Neoceratodus* (a, b) and *Lepidosiren* (c) taste buds. **a.** Basal part of a longitudinally cut light cell (Cl). Two axons (Nf) of the nerve fiber plexus are in efferent synaptic contact (black arrows) to the base of the light cell. The latter shows well-formed subsynaptic cisterns (sCi). On the axon presynaptic sides, vesicles are mostly lacking. **b.** A light cell (Cl) synapses twice afferently (white arrows) to a longitudinally cut axon (Nf). **c.** Supranuclear region of a longitudinally cut *Lepidosiren* taste bud. An axon (Nf), situated between two dark cells (Cd), synapses to the dark cell (black arrow). On the axon side synaptic vesicles are located, on the dark cell side a subsynaptic cistern is lacking. Scale bars in a–c: 1 µm.

Table 10.1 Overview of data on efferent synapses or efferent synapse-like structures in fish taste buds (+)

Main fish group	Species	*Efferent Synapses*				*References*
		with subsynaptic cistern	*without subsynaptic cistern*	*situated between*		
				nerve fiber and light cell	*nerve fiber and dark cell*	
Dipnoi	*Lepidosiren paradoxa*		+		+	this study
	Neoceratodus forsteri	+	+	+		this study
Chondrostei	*Scaphirhynchus platorynchus*	+		+		this study
Neopterygii: *Semionotiformes*	*Lepisosteus oculatus*	+		+		this study
Teleostei	*Ictalurus melas*	+		+		Desgranges (1966, 1972)
	Phoxinus phoxinus	+		+		Jakubowski and Whitear (1990)
	Danio rerio	+		+		this study
	*Scophthalmus maximus**		+	+	+	this study

*Developmental stage: 14 days old

subsynaptic cisterns on the hair cell side. The functional significance of these efferent synapses is predicted by the olivocochlear tract that runs efferently from the medulla (its upper oliva) to the cochlea, in detail to the hair cells of the organ of Corti. The fibers of this tract may carry impulses that downregulate the firing rate of hair cells, especially during long-lasting noises (adaptation). On the other hand, in the gustatory system of all vertebrates including mammals no neuronal tract is known that connects central gustatory nuclei with peripheral TBs efferently. So efferent synapses were not expected when TBs were studied in the TEM. Neither, as far as we know, is there any physiological proof of an efferent pathway in the gustatory system. All TEM reports on efferent synapses in TBs therefore tiptoe around the issue with authors speaking of "efferent-like" or "presumed efferent" synapses. The criteria for efferent synapses in TBs are the morphological similarities they have in common with efferent synapses of hair cells. What we need is proof of an efferent gustatory pathway.

Undeniably, "efferent-like" synapses have been found in TBs of all classes of vertebrates (see above). For fish, Desgranges (1966, 1972) described the "double innervation" of *Ictalurus* TBs and depicted synapses possessing typical subsynaptic cisterns. Another subsynaptic cistern was published for a *Phoxinus* TB (Jakubowski and Whitear, 1990). Furthermore, in these cases the postsynaptic cell was a light cell (Table 10.1). In view of the relatively great number of TEM works on fish TBs (for references, see Reutter and Hansen, 2005, this volume), these two findings indicate that efferent synapses in fish TBs are really rare. This is true for their propagation within systematic groups and within individual TBs, too. Detection of efferent synapses needs screening of TEM pictures in high numbers and even then it is possible to oversight them. Though we searched for efferent synapses in 11 species of fish; we found them, equipped with typical subsynaptic cisterns, in only four species.

The developing TBs of *Scophthalmus maximus* have an interesting synapse-like formation that we propose to be efferent in function. The axon terminal as the presumed presynaptic side is filled with synaptic vesicles; membrane specializations are poor and a subsynaptic cistern is lacking. Zahm and Munger (1983a, b) reported such formations, equipped with and without subsynaptic cisterns, in developing macaque TBs also. They discussed the possibility that these axon terminals might deliver a trophic substance to the adjacent differentiating cells. The early TB cells often contain numerous synaptic vesicles of different size and contents.

This might be a sign that these cells may soon be ready for function and then synapse afferently to nerve fibers. Well-developed afferent synapses occur only in further developed fish larvae (Reutter et al., 1995; Hansen et al., 2002).

To date there are no physiological studies that give evidence for efferent nerve fibers or efferent synapses in either mammalian or fish TBs. Therefore, it is difficult to name their function appropriately. Even in competent and comprehensive work on TBs "efferent" synapses are only mentioned and not functionally interpreted (see Finger and Simon, 2000). What is more, the notion that efferent synapses might control TB cells lacks certitude. So we agree with Roper (1989): "Although there are several examples in the literature that antidromic impulses in gustatory afferent axons affect taste transduction..., physiological evidence for *bona fide* efferent synaptic control of taste cells is scanty but suggestive".

References

Akisaka, T. 1980. Morphological and functional aspect of the rat taste bud by means of electron microscopy. *J. Osaka Dent. Univ.* 14: 1–28.

Declaration of Helsinki. 1995. Recommendations from the Declaration of Helsinki. Chem. *Senses* 20: 181.

Delay, R.J. and S.D. Roper. 1988. Ultrastructure of taste cells and synapses in the mudpuppy *Necturus maculosus*. *J. Comp. Neurol.* 277: 268–280.

Desgranges, J.-C. 1966. Sur la double innervation des cellules sensorielles des bourgeons du goût des barbillons du Poisson-chat. C. *R. Acad. Sc. Paris, Série D.* 263: 1103–1106.

Desgranges, J.-C. 1972. Sur les bourgeons du goût du Poisson-chat *Ictalurus melas*: ultrastructure des cellules basales. C. *R. Acad. Sc. Paris, Série D.* 274: 1814–1817.

Emmerling, M.R., H.M. Sobkowicz, C.V. Levenick, et al., 1990. Biochemical and morphological differentiation of acetylcholinesterase-positive efferent fibers in the mouse cochlea. *J. Electron Microsc. Techn.* 15: 123–143.

Engström, H. 1958: On the double innervation of the sensory epithelia of the inner ear. Acta *Otolaryngol.* 49: 109–118.

Engström, H. and F.S. Sjöstrand. 1954. The structure and innervation of the cochlear hair cells. *Acta Otolaryngol.* 44: 490–501.

Finger, T.E. 1976. Gustatory pathways in the bullhead catfish. I. Connections of the anterior ganglion. *J. Comp. Neurol.* 165: 513–526.

Finger, T.E. 1978. Gustatory pathways in the bullhead catfish. II. Facial lobe connections. *J. Comp. Neurol.* 180: 691–705.

Finger, T.E. and S.A. Simon. 2000. Cell biology of taste epithelium. In: *The Neurobiology of Taste and Smell*, T.E. Finger, W.L. Silver and D. Restrepo (Eds). Wiley-Liss Inc., New York, pp. 287–314.

Finger, T.E., B.P. Bryant, D.L. Kalinoski, J.H. Teeter, and B. Böttger, 1996. Differential localization of putative amino acid receptors in taste buds of the channel catfish *Ictalurus punctatus*. *J. Comp. Neurol.* 373:129–138.

Fujimoto, S. and R.G. Murray. 1970. Fine structure of degeneration and regeneracion in denervated rabbit vallate taste buds. *Anat. Rec.* 168: 399–413.

Graziadei, P.P.C. 1970. The ultrastructure of taste buds in mammals. In: *2nd Symp. Oral Sensation and Perception*, J.F. Bosma (Ed.). Thomas Publ., Springfield, Jll. pp. 5–35.

Hansen, A. and K. Reutter. 2004. Chemosensory systems in fish: Structural, functional and ecological aspects. In: *The Senses of Fish: Adaptations for the Reception of Natural Stimuli*, G. von der Emde, J. Mogdans and B.G. Kapoor (Eds). Narosa Publishing House, New Delhi, and Kluwer Academic Publisher, Dordrecht, The Netherlands. pp. 55–89.

Hansen, A., K. Reutter and E. Zeiske. 2002. Taste bud development in the zebrafish, *Danio rerio*. *Devel. Dyn.* 223: 483–496.

Isaacson, J.S. and H. Vitten. 2003. GABA(B) receptors inhibit dendrodendritic transmission in the rat olfactory bulb. *J. Neurosci.* 23: 2032–2039.

Jakubowski, M. and M. Whitear. 1990. Comparative morphology and cytology of taste buds in teleosts. *Z. mikrosk.- anat. Forsch.* 104: 529–560.

Jakubowski, M. and K. Żuwała. 2000. Taste organs in lower vertebrates: Morphology of the gustatory organs in fishes. In: *Vertebrate Functional Morphology*, H.M. Dutta and J.S Datta Munshi (Eds). Science Publishers Inc., Enfield, (NH), USA and Plymouth, UK. pp. 159–172.

Kapoor, B.G. and T.E. Finger. 2003. Taste and solitary chemoreceptor cells. In: *Catfishes*, G. Arratia, B.G. Kapoor, M.Chardon and R. Diogo (Eds). Science Publishers Inc., Enfield (NH), USA and Plymouth, UK. *vol. 2*, pp. 753–769.

Kondo, I. 1983. A histochemical study on degeneration and regeneration of mouse circumvallate taste buds. *Jpn. J. Oral Biol.* 25: 745–762.

Kudoh, M. 1988. Ultrastructural and histochemical localization of acetylcholinesterase in the taste bud of mouse vallate papilla. *Fukushima J. Med. Sci.* 24: 27–44.

Murray, R.G. and A. Murray. 1970. The anatomy and ultrastructure of taste buds. In: *Taste and Smell in Fishes*, G.E.W. Wolstenhome and J. Knight (Eds). Churchill, London, pp. 3–30.

Nelson, J.S. 1994. *Fishes of the World*. 3rd edition. John Wiley, New York.

Osculati, F. and A. Sbarbati. 1995. The frog taste disc: A prototype of the vertebrate gustatory organ. *Progr. Neurobiol.* 46: 351–399.

Pevzner, R.A. 1981. The fine structure of taste buds of the ganoid fishes. I. Adult Acipenseridae. *Tsitologiya* 23: 760–766. (Russian, with English summary).

Reutter, K. 1991. Ultrastructure of taste buds in the Australian lungfish, *Neoceratodus forsteri*. *Chem. Senses* 16: 404.

Reutter, K. 1994. Ultrastructure of taste buds in the spotted dogfish *Scyliorhinus canicula* (Selachii). In: *Olfaction and Taste*, K. Kurihara, N. Suzuki and H. Ogawa (Eds). Springer-Verlag, Tokyo, *Vol. XI*, p. 754.

Reutter, K. and A. Hansen. 2005. Subtypes of light and dark elongated taste bud cells in fish. In: *Fish Chemosenses*, K. Reutter and B.G. Kapoor (Eds). Science Publishers, Enfield, (NH), USA and Plymouth, UK pp. 211–230, this volume.

Reutter, K. and M. Witt. 1993. Morphology of vertebrate taste organs and their nerve supply. In: *Mechanisms of Taste Transduction*, S.A. Simon and S.D.Roper (Eds). CRC Press, Boca Raton, pp. 29–82.

Reutter, K. and M. Witt. 1999. Comparative aspects of fish taste bud ultrastructure. In: *Advances in Chemical Signals in Vertebrates*, R. E. Johnston, D. Müller-Schwarze and P. W. Sorensen (Eds) Kluwer/Plenum, New York, pp. 573–581.

Reutter, K., F. Boudriot and M. Witt. 2000. Heterogeneity of fish taste bud ultrastructure as demonstrated in the holosteans *Amia calva* and *Lepisosteus oculatus*. *Phil. Trans. R. Soc. Lond.* B. 355: 1225–1228.

Reutter, K., M. Witt, J.A. Knutsen and K.B. Døving. 1995. Taste bud development in turbot larvae (Teleostei). *Chem. Senses* 20: 764–765.

Roper, S.D. 1989. The cell biology of vertebrate taste receptors. *Ann. Rev. Neurosci.* 12: 329–353.

Royer, S.M. and J.C. Kinnamon. 1988. Ultrastructure of mouse foliate taste buds: Synaptic and nonsynaptic interactions between taste cells and nerve fibers. *J. Comp. Neurol.* 270: 11–24.

Saito, K. 1980. Fine structure of the sensory epithelium of the guinea pig organ of Corti: Afferent and efferent synapses of hair cells. *J. Ultrastruct. Res.* 71: 222–232.

Suzuki, Y. and M. Takeda. 1984. Ultrastructure of taste buds in birds. *Jpn. J. Oral Biol.* 26: 669–678. (Japanese, with English summary).

Takeda, M. 1976. An electron microscopic study on the innervation in the taste buds of the mouse circumvallate papillae. *Arch. Histol. Jpn.* 37: 395–413.

Toyoshima, K. and B. Tandler. 1987. Modified endoplasmic reticulum in type II cells of rabbit taste buds. *J. Submicrosc. Cytol.* 19: 85–92.

Uchida, T. 1980. Ultrastructural and histochemical studies on the taste buds in some reptiles. *Arch. Histol. Jpn.* 43: 459–478.

Whitear, M. and R.M. Moate. 1994a. Microanatomy of taste buds in the dogfish, *Scyliorhinus canicula*. *J. Submicrosc. Cytol. Pathol.* 26: 257–367.

Whitear, M. and R.M. Moate. 1994b. Chemosensory cells in the oral epithelium of *Raja clavata* (Chondrichthyes). *J. Zool. Lond.* 232: 295–312.

Yoshie, S., H. Kanazawa and T. Fujita. 1996. A possibility of efferent innervation of the gustatory cell in the rat circumvallate taste bud. *Arch. Histol. Cytol.* 59: 479–484.

Yoshie, S., C. Wakasugi, Y. Teraki and T. Fujita. 1990. Fine structure of the taste bud in guinea pigs. I. Cell characterization and innervation patterns. *Arch. Histol. Cytol.* 53: 103–119.

Zahm, D.S. and B.L. Munger. 1883a. Fetal development of primate chemosensory corpuscles. I. Synaptic relationships in late gestation. *J. Comp. Neurol.* 213: 146–162.

Zahm, D.S. and B.L. Munger. 1983b. Fetal development of primate chemosensory corpuscles. II. Synaptic relationships in early gestation. *J. Comp. Neurol.* 219: 36–50.

CHAPTER

11

Comparison of Taste Bud Types and Their Distribution on the Lips and Oropharyngeal Cavity, as well as Dentition in Cichlid Fish (Cichlidae, Teleostei)

Lev Fishelson

ABSTRACT

Taste buds (TBs) on the lips, jaws, and oropharyngeal cavity of the mouth-brooding and substrate-brooding species of cichlid fishes from Africa, America, and Israel, were studied using LM and SEM, concomitant with observations on dentition. The cytological structure of the TBs was found to conform in all species to the three types (Types I, II and III) already recognized in teleost fishes.

However, the diameter of the receptor areas in some fishes was twice as great as in species from other fish groups. The total number of TBs on the lips

Address for Correspondence: Lev Fishelson, Department of Zoology, George S. Wise Faculty of Life Sciences, Tel Aviv University, Tel Aviv 69978, Israel. E-mail: fishelv@post.tau.ac.il and in the oropharyngeal cavity varied from about 3,500 in the largest specimens of *Labeotropheus trewavasae* to about 18,000 in *Tilapia zillii*. Differences were revealed in number and distribution of TBs in the oropharyngeal cavities of the various species. Calculated per mm^2, the highest number (440/mm^2) of TBs was found in *Dimidiochromis compressiceps* on the lower jaw, and the lowest (12 TB/mm^2) on the hypopharyngeal bone of *Tilapia zillii*. With fish growth the number and dimensions of the TBs increased, attesting to a constant novogenesis of these sense organs. According to the distribution of the majority of TBs, the fish studied could be divided into those such as *Dimidiochromis compressiceps* and *Tilapia zillii* in which most TBs are aggregated on the jaws in the front part of the mouth, and those such as *Astatotilapia flaviijosefii* and *Cichlasoma cyanoguttatum* in which the pharyngeal bones bear the highest density of TBs. The former group feed mainly on larger prey while the latter feed on smaller items. Regarding dentition, the various species differ in form and distribution of the monocuspid, bicuspid, and tricuspid teeth on the jaws and pharyngeal bones. *Tilapia zillii* features a special type of tricuspid and quadricuspid teeth on the epipharyngeal and hypopharyngeal bones. In these teeth the highest cusp is not situated centrally, as in all other cichlids observed, but is the most posterior one, with the other cusps forming a single file in front of it.

Key Words: Cichlid fish; Taste buds; Lips; Oropharyngeal cavity; Dentition.

1. INTRODUCTION

In fish, gustation takes place predominantly in the taste buds (TBs) situated around the mouth, in the oropharyngeal cavity, on the basal parts of the gills, and often also on the skin and its appendages also by means of sensory microvilli protruding above the epithelium (Whitear, 1971; Gomahr et al., 1992; Hansen and Reutter, 2004). The signals from these organs are transferred from the mouth to the brain via the facial (VII, cranial) and glossopharyngeal (IX, cranial) nerves, and from the middle and posterior part of the oral cavity by the vagus (X, cranial) nerve (for references, see Reutter and Witt, 1993). For instance, the cytology of TBs has been studied in carp (Hirata, 1966), catfish (Reutter, 1971, 1978; Grover-Johnson and Farbman, 1976; Kapoor and Finger, 2003), blenniid and gobiid fishes (Fishelson and Delarea, 2004a), cardinal fish (Fishelson et al., 2004b), flatfishes (Tsura and Omori, 1976), rainbow trout (Ezeasor, 1980), minnow (Kiyohara et al., 1980), poeciliids (Reutter, 1973; Reutter et al., 1974), lungfish (Reutter, 1991) and in holostean fishes (Reutter et al., 2000). Fish gustation has been summarized in articles by Atema (1971), Kapoor et al. (1975), Caprio (1984), Hara (1993), Zaccone et al.,

2001 and in books by Beidler (1971), Hara (1992), Finger et al. (2000) and Doty (2003). It is generally accepted that there are three basic types of TBs in fishes. These are up to 100 μm high and have a diameter of 50–80 μm at the base, and 3.5–5.0 μm at their exposed receptor area. Here they possess receptor microvilli that extend above the epithelium's surface. Type I and II TBs are raised on papillae, while the receptor areae of Type III TBs are in level with the epithelium (Reutter, 1973; Reutter et al., 1974; for references, see Hansen and Reutter, 2004).

Until now TBs were not studied intensely in the Cichlidae, a species-rich family of paleotropical and neotropical fish. Recently Fishelson (2004, online) described the histogenesis of TBs and relevant brain parts in two species of cichlid fishes. Cichlids are morphologically characterized by the presence of only one nare on either side of the snout, and the so-called pharyngeal or throat jaws, which are modifications of the basal ossicles of the gill arches. The "jaws" comprise the upper, epipharyngeal bones (EBs) and the lower, single hypopharyngeal bone (HB). The EBs are formed by the pharyngobranchials of the 2nd, 3rd, and 4th gill arches and consist of two bony plates attached to the basicranium. The HB is triangular and formed by the merger of the two ceratobranchials of the 5th gill arch (Liem, 1991). These modified pharyngeal bones are intensely used during maceration, laceration, or mastication of the food items on their passage into the esophagus (Huysseune, 1983; Greven, 2002). According to various authors (e.g. Ribbink, 1991; Liem, 1991; Stiassny and Meyer, 1999) they show a strong adaptive morphology to the nature and type of food consumed. Cichlid fishes occur in inland waters of most continents (except Australia) and have served for numerous studies, including taxonomic-evolutionary investigations previously based on the morphology of bone structures and associated muscles (Kaufman and Liem, 1982; Barel, 1983; Meyer, 1990; Greenwood 1991, and cited therein, Stiassny, 1991; Stiassny and Meyer, 1999; Verheyen et al., 2003).

Most cichlids are herbivores but some are carnivores specializing in piscivory and insectivory, including scale-eaters and eye-peckers, while others are planktonivores or benthivores. Paralleling these remarkable trophic diversifications are the adaptive modifications of the cichlid feeding apparatus (Liem and Osse, 1975; Liem, 1978, 1979, 1991; Meyer, 1990; Greven, 2002). Several recent studies of genetic material, such as mitochondrial and nuclear DNA of various species (Nagl et al., 2001; Klett and Meyer, 2002) have increased the differences between the traditional established morphological classification and the suggested genetic entities, and added a new dimension to the controversy over cichlid classification. Without going into the polemics of what is correct, in the present study I follow the established division of Cichlidae into Tilapinae (or Tilapinii), which include the commercially familiar genera *Tilapia, Oreochromis, Sarotherodon,* and related forms. These are mostly larger fish exploited for aquaculture. The African Pseudocrenilabrinae (Haplochrominii) and the American Cichlasomatinae, which predominantly comprise hundreds of smaller species, are widely used in the aquarium trade. The latter two subfamilies, in addition to genus *Haplochromis* (late sense) of the African species, and *Cichlasoma* of the American species, include tens of genera and hundreds of forms inhabiting the rivers and lakes of Africa, as well as North, Central and (partly) South America (Fryer and Iles, 1972; Ribbink et al., 1983; Mc Kay, 1984; Greenwood, 1991). Their members mostly form flocks of species endemic to specific lakes.

Numerous observations on these fishes, both in nature and captivity, have revealed an extremely high environmental adaptability as well as a broad spectrum of behavioral and social styles, especially of reproduction (Fishelson, 1966, 1983, 2002; Keenleyside, 1979, 1991; Mc Kay, 1984; Ribbink, 1991).

The best-known and basic behavioral division of species in all three subfamilies belongs to one of these two groups: a) Substrate-brooders in which females attach spawn capable of adhering to either exposed or concealed hard substrate. The males then swim above the spawn and ejaculate their sperm. The parents tend the spawn through hatching and until the young become free swimming, and then remain and protect them for 3-4 weeks. b) Mouth-brooders, in which the female, male, or both parents gather the eggs and sperm into their mouth during or following spawning. In most of these latter species fertilization occurs prior to this, when the males swim in displays to attract the females and ejaculate low over the spawning site (Fishelson, 1983). Further, as observed in *Oreochromis macrochir* (Wickler, 1965) and *Astatotilapia flaviijosefii* (pers. obs.), most of the fertilization occurs in the mouth cavity, as argued by Mrowka (1987). The mouth-brooders incubate the eggs and juveniles until metamorphosis is complete and then release them into the surrounding waters, taking them back into the mouth when encountering danger. The two types of relations between offspring and parents, especially concerning exposure of the young to environmental dangers, are manifested in the dynamics of larval and postlarval development not only in terms of external morphology, but also in the ontogeny of the various vital organs (Fishelson, 1966, 1995a, b, 2004).

The present study compares the distribution, type and number of TBs, as well as dentition, in the oropharyngeal cavity of 11 species of cichlid fish (Table 11.1), in order to better understand their ecomorphological adaptations. It does not include the TBs situated on the gill arches. Some of the species studied are substrate-brooders, e.g. *Tilapia zillii* (Israel), *Cichlasoma cyanoguttatum* (Central America) living in pairs or, like *Neolamprologus spilostetus* (East Africa), living in families with helpers, whereas others—*Astatotilapia flaviijosefii, Pseudotropheus fuellebornii*, and *Oreochromis aureus*—are mouth-brooders.

2. MATERIAL AND METHODS

The fish studied were 50–100 mm TL; for comparison, samples of both smaller and much larger fish were also studied. The fish were grown in aquaria of various dimensions at the Department of Zoology, Tel Aviv University, or purchased from fish dealers. They were kept in fresh water at 24 (± 1.5)°C and constantly aerated, under 12 L / 12 D lighting regime, and fed Tetraflecks and commercial fish pellets containing 30% protein. Prior to sacrifice most fish were placed in cold-water baths at 6-8°C in

Table 11.1 Cichlid species studied, their sites of origin, brooding style and total length (in mm)

Species and specimen number	*Site of origin*	*Brooding style*	*Dimension*	
			Min	*Max*
Tilapinae				
Tilapia zillii (6)	Israel	Substrate	50	300
Oreochromis aureus (16)[1]	Israel	Mouth	10	300
Sarotherodon galilaeus (6)	Israel	Mouth	60	220
Pseudocrenilabrinae				
Neolamprologus spilosetotus (5)	E. Africa	Substrate	50	60
Dimidiochromis compressiceps (4)	E. Africa	Mouth	55	150
Astatotilapia flaviijosefii (16)	Israel	Mouth	30	110
Labeotropheus trewavasae (6)	E. Africa	Mouth	60	130
Pseudotropheus fuelleburnii (4)	E. Africa	Mouth	60	90
Aulonocara nayasse (5)	Africa	Mouth	50	60
Cichlasomatinae				
Cichlasoma cyanoguttatum (12)	C. America	Substrate	30	80
Cichlasoma paraguyaensis (4)	S. America	Substrate	55	60

[1]Possibly hybrid of *O. aureus* x *O. niloticus*

order to prevent excessive secretion of mucus in the oral cavity, which might have coated the sensory organs. The fish were then killed with an overdose of MS 222. The heads were dissected and the oral parts separated.

For light microscopic (LM) histology the mouthparts were fixed in Bouin for 24 h, then washed in ethanol. Following this the samples were either decalcified or the soft mouth-covering skin separated, dehydrated along an ascending series of ethanol, and embedded in paraffin blocks. Serial sections 6 and 10 μm thick were prepared, stained in Ehrlich's hematoxylin-eosin or Massons' trichrome, and studied with a Zeiss Ultrascope. Photographs were taken with an automatic Nikon camera and a ChipEr DSP digital camera. For scanning electron microscopy (SEM), smaller parts of the same tissues were fixed in 3.5% glutaraldehyde buffered with 0.1 M cocadylate at pH 7.2. Then the fixed upper and lower parts of the oropharyngeal cavity were critical point dried, gold plated, and studied with a JEOL JSM 840. Some tissues were transferred to ethanol, air dried, then studied or photographed. For comparison, similar samples were fixed in 4% formaldehyde, transferred to ethanol and also air dried. For methodological reasons, as in the study of TBs in gobies and blennies (Fishelson and Delarea, 2004a), the shape and numerical distribution of TBs at the diverse parts of the oropharyngeal cavity are given separately: e.g. for the lips, jaw teeth bands, breathing valves, palate, epipharyngeal bones (EBs), tongue, hypopharyngeal bone (HB) (Fig. 11.1A, B). The total number of TBs on the various sites are presented in Table 11.2. The TBs of the lips and oropharyngeal cavity in the African mouth-brooding cichlid *Dimidiochromis (Haplochromis) compressiceps* are compared with those of the Israeli substrate-brooder *Tilapia zillii*, and both served as a basis against which the other species were compared.

3. RESULTS

3.1 Oropharyngeal cavity of *Dimidiochromis compressiceps*

Dimidiochromis compressiceps (80–180 mm TL) differs from most of the members of Pseudocrenilabrinae in its highly compressed head and body. Even in specimens of more than 180 mm TL therefore, the width of the upper jaw is only 7.3 mm. The upper lip is mostly smooth (Fig. 11.2A); only the slightly wider lateroposterior site possesses a group of 600

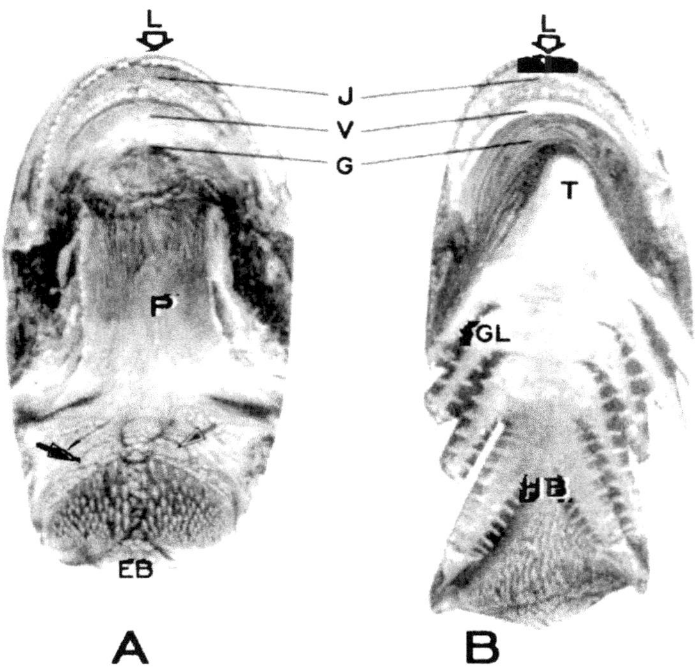

Fig. 11.1 A-B Upper (A) and lower (B) mouth surface of *Tilapia zillii* (1/2 natural size). EB, epipharyngeal bones; G, gular region in front of tongue; GL, gill bases; HB, hypopharyngeal bone; L, lips; J, jaws, P, palatinum; T, tongue; V, breathing valves; arrow, 2nd pharyngobranchiales.

papillae, each with a single Type I TB, their receptor area measuring 6.5 –7.5 µm in diameter, with around 40 thick and 180 slender receptor microvilli (Fig. 11.2B). The inner margin of the lip has 180 lobules, 30–130 µm broad, each bearing 2–4 papillae with Type I and II TBs. Receptor villi on Type II TBs are larger, 0.6–0.7 µm long, around 56 per bud, surrounded by about 260 smaller villi 0.2–0.3 µm long (Fig. 11.2D). The upper teeth band has monocuspid teeth, larger and strongly curved, caninelike in the front row and two rows of much smaller teeth (Figs. 11.2 A). The band bears over 3,600 lobules, 30–110 µm, each with 2–5 papillae with TBs, totaling 4,200 TBs, with about 320 TB per mm^2. The clavate ends of the receptor villi dominate these buds. The upper valve possesses a group of 1,600 lobules and papillae, 40–140 µm, each topped by Type II TBs (Figs. 11.2A, C). The palate in smaller fish is almost smooth; in larger fish it has 10 longitudinal folds, each with 30–40 papillae

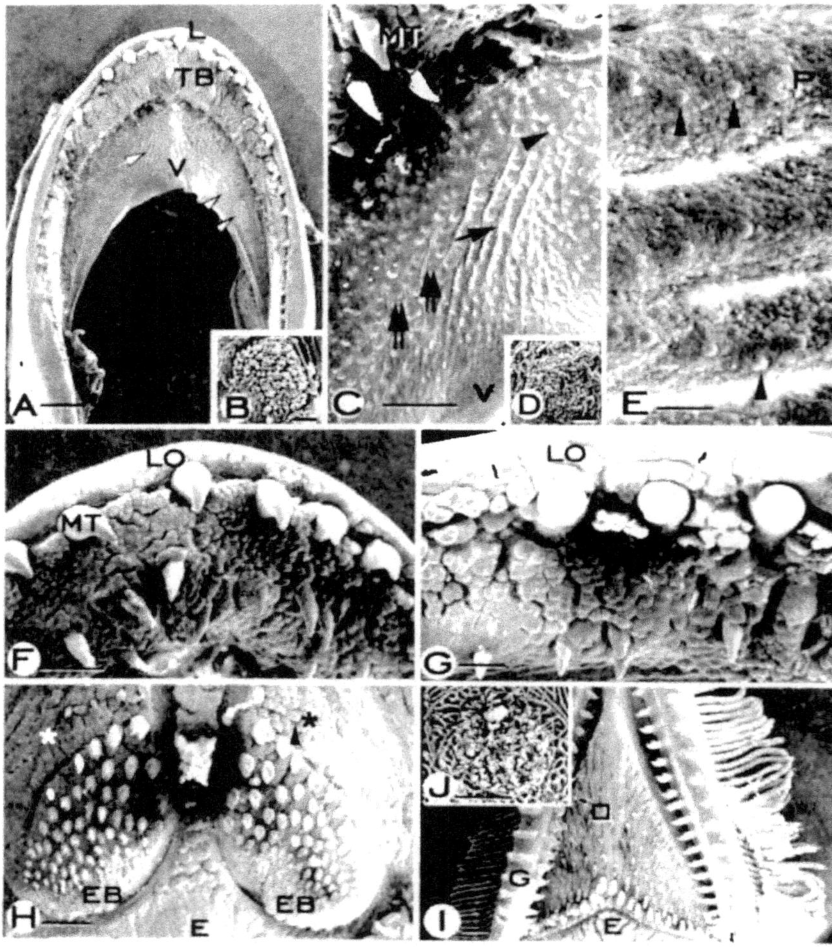

Fig. 11.2 A-J Oropharyngeal cavity of *Dimidiochromis compressiceps*. **A.** Upper jaw. L, lip; TB teeth band; V valve; arrow elongated lobules with TB; arrowhead, papillae. *Scale bar* 2 mm. **B.** Type I TB. *Scale bar* 2 µm. **C.** Upper valve (V) with elongated lobules (arrow) and separated papillae (arrowhead). Pale tops (double arrow), of TBs. *Scale bar* 200 µm. **D.** Type II TB; grouping of sensory villi. *Scale bar* 2 µm. **E.** Palate with plications (P), arrowhead papillae. *Scale bar* 50 µm. **F.** enlarged part of upper jaw. LO lobules of lip; MT, monocuspid teeth, numerous TB-bearing lobules. *Scale bar* 1 mm. **G.** Lower jaw. LO, lobules of lip; arrowhead, TB-bearing lobules. *Scale bar* 0.5 mm. **H.** Epipharyngeal bone (EB). E, open esophagus lamellae; asterisks, 2nd pharyngobranchials with lobules (arrowhead) and TB. *Scale bar* 1 mm. **I.** Hypopharyngeal bones. E, open esophagus; G, gill arches; □, lobule from where J is enlarged. *Scale bar* 0.5 mm. **J.** Type III TB. *Scale bar* 5 µm.

with Type II and III TBs (Fig. 11.2E). Each of the EBs (Fig. 11.2H), has 12 rows of teeth, with 3–4 thicker ones in the central row; 400 irregular lobules with Type III TBs occur between the teeth.

The lower jaw has a lobulated lip margin (Fig. 11.2G) with Type I TBs. The teeth band has monocuspid teeth; the front row of 26 larger and strong teeth is followed by irregular, posterior rows of smaller ones. Across the band are about 1,300 lobules with papillae, with 4,000 Type I and II TBs. The lower valve has two lateral pads of papillae, each with 200 Type II TBs.

The tongue is narrow, apically pointed, featuring 60 Type I TBs, then becomes smooth at the pharyngeal region, where 400 Type III TBs are situated. The HB (Fig. 11.2I) has 24 rows of compressed teeth, 6–7 per row, the central-posterior ones thicker; between them are rows of lobules with about 2,500 Type III TBs (Fig. 11.2J), 115 TB per mm^2. This predatory species possesses 12,000 TBs in the oropharyngeal cavity, 82% of which are situated in the frontal part of the mouth.

Tilapia zillii (120–290 mm TL) is the only Israeli substrate-brooding species of cichlid (Fishelson, 1983). In specimens of 120 mm TL, the upper jaw bears a 2.0 mm wide band of three rows of teeth (Fig. 11.3A); the front row is regular and uninterrupted, while the more posterior two rows contain dispersed, smaller teeth. Most of the teeth band is covered by 500 papillae of various size, each bearing 1–2 Type I TBs, 6.5–8.5 μm in diameter (Fig. 11.3A,B; Table 11.2). The upper valve is 1.2 mm wide, with 120 elongated lobules with papillae and TBs along its anterior portion. The EBs (Fig. 11.3C) have 13-15 rows of compressed tricuspid and partly quadricuspid teeth each, 4-5 in the median rows. However, unlike in the other species studied, the cusps do not form a row but a single file, with the posteriormost cusp the highest (Fig. 11.3D). Between the teeth of each EB are 18 rows of lobules with 300 TBs; 300 Type III TBs-carrying lobules are also found in front of and between the two bones. The lower jaw and the valve display numerous TB-bearing papillae (Fig. 11.3E); tongue almost smooth, with few TBs. The HB (Fig. 11.3F) bears irregular rows of flat teeth, slightly thicker in the median rows. Lines of papillae are interspaced between them, each with about 30 Type III TBs (Table 11.2).

The lips and oropharyngeal cavity of the larger, 300 mm TL *Tilapia zillii*, strongly differ from the previously described smaller specimens: The upper lip, strongly widening laterally bears around 300 Type II TBs on low

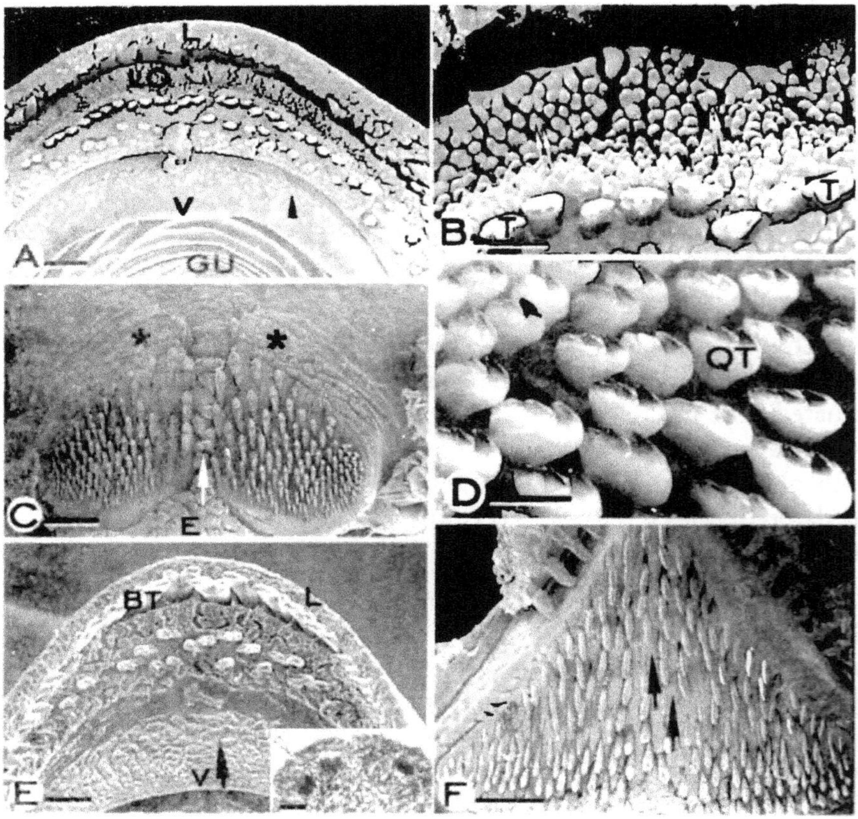

Fig. 11.3 **A**- F Oral cavity of *Tilapia zillii*. **A.** Upper jaw. GU, gular folds; L, lip; LO, lobules on teeth band; V, valve, arrowhead, papillae. *Scale bar* 1 mm. **B.** Enlarged part of upper jaw. T, tricuspid teeth, double arrow lobule with pale ends of TB. *Scale bar* 0.5 mm. **C.** Epipharyngeal bones. E. esophagus. Arrow, lobules with TBs, asterisks, 2nd pharyngobranchials. *Scale bar* 1 mm. **D.** Quadricuspid teeth (QT) on EB. *Scale bar* 0.3 mm. **E.** Lower jaw. BT, bicuspid teeth; L, lip; V, valve; arrow, elongated lobules. *Scale bar* 200 μm (inset LM section of a lobule with 3 TBs. *Scale bar* 100 μm). **F.** Hypopharyngeal bone; arrow, TB-bearing lobules. *Scale bar* 0.5 mm

papillae along its lobulated margin, each TB is 7.5–8.5 μm, with 42 larger and 110 smaller receptor villi. The upper jaw teeth band bears over 2,000 lobules and papillae, with 6,500 Type I and II TBs, 93 TBs/mm^2. The valve is almost entirely covered by several hundred lobules, some small with single TBs, others 0.5–1.0 mm long, irregular, each topped by 6–18 TBs, totaling about 1800 TBs. The palate displays about 10–12 long folds, each with Type III TBs. The EBs have 16 rows of teeth each, the central one with 6 teeth, the larger of which are compressed, quadricuspid and

Table 11.2 Site and maximal number of TBs on lips and in the oropharyngeal cavity of the largest cichlid fish studied**

Species	*TL mm*	*Lips*	*U-teeth band*	*U-valve*	*Pal*	*EB*	*L-teeth band*	*L-valve*	*Tongue*	*HB*	*Total TBs*
T. zillii	300	380	6,560	1,840	300	1,600	2,700	1,800	300	2,500	17,980
O. aureus[1]	300	450	1,200	800	400	1,200	1000	750	240	800	6,840
S. galilaeus	220	300	600	400	200	1,000	400	300	180	800	4,180
D. compressiceps	160	1,680	4,200	220	200	600	3,000	800	460	800	11,960
A. flaviijosefii	110	400	1,400	800	1,040	600	1050	300	60	1400	7,050
L. trewavasae	110	280	800	450	800	300	320	120	40	280	3,490
P. fuelleburnii	120	320	800	500	1,300	500	480	360	180	300	4,760
C. cyanoguttatum	90	380	800	480	60	1,400	400	250	140	400	4,500
C. paraguyaensis	70	300	1,200	360	300	1,100	700	340	150	300	4,760
Au. nyassae	80	500	1,150	180	200	960	800	160	320	700	5,240
N. spilosetotus	70	360	800	220	200	1,500	600	180	240	1,600	7,260

[1] This is possibly a natural hybrid of *O. aureus* x *O. niloticus*.

U = upper; L = lower; EB = epipharyngeal bones; HB = hypopharyngeal bone

** TBs on gill bases are not counted

tricuspid, forming a line (Fig. 11.3C, D). Between the teeth are lines of lobules with Type III TBs. The lower jaw lip bears 180 lobules with 540 Type I TBs 6.0–10.0 μm in diameter. The teeth band bears three rows of mainly tricuspid teeth. Behind the teeth are elongated lobules with TBs. The valve has 900 lobules, each with 1–4 papillae topped by Type I and II TBs, totaling almost 2,700 TBs. The central part of the valve bears 170 papillae each with one Type II TB. The gular epithelium has 12 transverse folds, the five anterior ones bearing 150 Type II TBs. The tongue has numerous transverse irregular folds with papillae and Type II and III TBs. The HB has 36 rows of compressed tricuspid and quadricuspid teeth, 11–12 in the central rows. Between the teeth are elongated lobules, each with 10–40 papillae and Type I TBs (Table 11.2). In the largest fish of this species, over 80% of the TBs were situated in the frontal part of the mouth.

3.2 Other Mouth-brooders Studied

Astatotilapia (Haplochromis) flaviijosefii (30–120 mm TL) is the only Israeli pseudocrenilabrid species. In 30 mm TL fish the upper jaw has a teeth band bearing a front row of 20 flat bicuspid teeth and an additional line of tricuspid teeth, of which the central cusp is sharp and higher than those flanking (Fig. 11.4A). Between the teeth are 60 lobules with papillae bearing about 140 Type I and II TBs. The upper valve bears about 30 irregular lobules with TBs. The EBs each possess 111 rows of slightly bicuspid teeth, five of which, in the central row, are molarlike (Fig. 11.4B); the higher cusp curves backward. Numerous lobules extend among the rows, each with Type III TBs. The lower jaw with all teeth embedded in a layer of lobules with TBs (Fig. 11.4C). The lower valve bears about 20 papillae with TBs. The HB (Fig. 11.4D) has 20 rows of bicuspid teeth in which the larger cusp is curved forward. Each median row has 7 teeth, of which the central and posterior ones are molarlike; lobules with Type II and III TBs are interspersed between the teeth.

Fishes 40 mm TL and 50 mm TL are characterized by lobulation of the lip margins and development of an additional, third irregular row of tricuspid teeth on the jaws, with an increased number of lobules between them, each with 1–3 papillae and corresponding number of Type I TBs. Both upper and lower valves possess 120–160 lobules, 50–300 μm respectively, each with 2–12 papillae with Type I and II TBs. The palate shows 12 longitudinal folds containing papillae with Type II and III TBs.

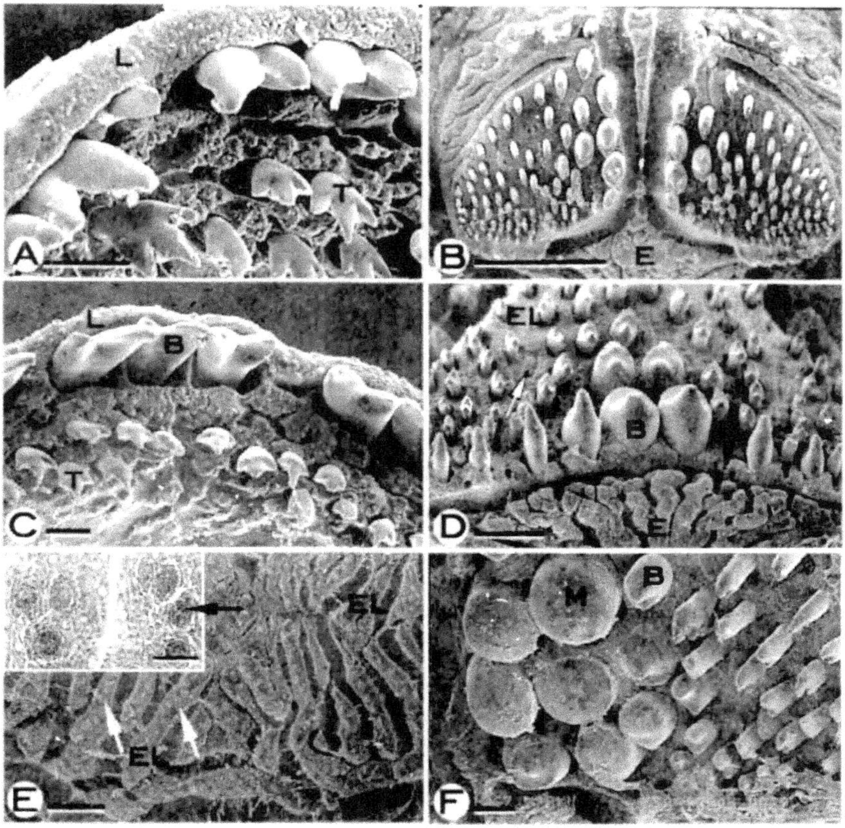

Fig. 11.4 **A - F** Oropharyngeal cavity of *Astatotilapia flaviijosefii*. **A.** Part of upper jaw. L, lip; T, tricuspid teeth. *Scale bar* 100 μm. **B.** Epipharyngeal bones. E, open esophagus, asterisks 2nd pharyngobranchials. *Scale bar* 1 mm. **C.** Lower jaw. B, bicuspid teeth; L, lip; T, tricuspidal teeth. *Scale bar* 100 μm. **D.** Hypopharyngeal bone. B, bicuspid teeth; E esophagus lamellae; EL, TB-bearing elongated lobules. *Scale bar* 150 μm. **E.** Valve of upper jaw. EL, elongated lobules; arrows, papillae with TB. *Scale bar* 200 μm (inset LM section of two lobules with cross sections of TBs; *Scale bar* 120 μm). **F.** Epipharyngeal bone. M, molar teeth; B, bicuspid teeth. *Scale bar* 100 μm

The EB have 12 rows of teeth each, of which the central 7–9 are molarlike, interspersed with rows of lobules with 400 Type II and III TBs. The lower jaws possess a front row of 20 flat, bicuspid teeth, with the inner cusp higher and sharper than the others; behind this are two irregular rows of tricuspid teeth, between which are lobules with Type I TBs. The HB has 22 rows of teeth, 9–10 of which are centrally posterior, molarlike, and interspersed with TB-bearing papillae.

In larger, adult fish the upper and lower lips are marginally lobulated, with about 400 Type I TBs on each. The upper jaw of a 70 mm TL fish is 8.2 mm wide, of an 80 mm fish 8.6 mm, and of a 100 mm fish 10.0 mm wide. The teeth band in the largest fish features about 700 lobules with 1–3 Type I TBs on each. The upper valve in the largest fish has 50 elongated lobules across it (Fig. 11.4E and inset), each bearing about 800 papillae and Type I and II TBs. The palate has 26 longitudinal folds, each with about 40 Type III TBs. The EBs have 15 rows of teeth each, with 5–6 teeth in the innermost row; the central group of teeth are molarlike, 0.5–0.6 mm thick (Fig. 11.4F), between which are rows of about 120 lobules with about 600 Type III TBs. The 2nd pharyngobranchial ossicles are attached anteriorly to the EBs, showing irregular papillae (with 3–4 teeth in between) with 200 Type III TBs. The lower jaw has a prominent front row of 14–16 loosely dispersed, larger and flat teeth, followed as in the upper jaw, by two rows of smaller, tricuspid teeth, the central cusp wider and higher than the flanking ones. Interspaced between the teeth are about 800 round or irregular lobules with about 1,050 Type I and II TBs, 300 TB per mm^2. The lower valve bears around 200 elongated or irregular lobules, 40–100 µm, each with papillae and 300 Type II TBs. The gular epithelium is ridged and reveals about 300 papillae with Type I and II TBs. The tongue is apically narrow, with 40 papillae and TBs, then widens, with 8 folds, each with papillae and TB. The HB has 16–20 rows of teeth, 5 teeth in the central rows; the more median are molarlike (Fig. 11.4F), while the lateral ones are bicuspid, resembling those found in *Aulonocara baenschi* (Greven, 2002). Between the rows of teeth are 24 rows of lobules with about 1,400 Type III TBs. In total, the oropharyngeal cavity of adult *Astatotilapia flaviijosefii* (360 mm^2) possesses over 7,000 TBs (Table 11.2), which yields about 20 TBs per mm^2. However, the density of TBs per mm^2 in the highest sensory areas, such as the jaws and HB, is 39 TBs and 233 TBs per mm^2 respectively. The TBs in this species are equally located on the frontal and pharyngeal part of the oropharyngeal cavity, with the lowest number found in the midregion, palate and tongue.

Pseudotropheus fuelleburnii, 50–120 mm TL. In adults of this species the upper jaw band is composed of four rows of tricuspid teeth, the largest in the anteriormost row (Fig. 11.5A). Sparsely distributed between the teeth are approximately 600 papillae, 25–30 µm diameter at the base, above which lie Type I TBs, 3.5–5.0 µm in diameter (Figs 11.5B, C). The narrow, postteeth region bears several epithelial folds, frequently divided

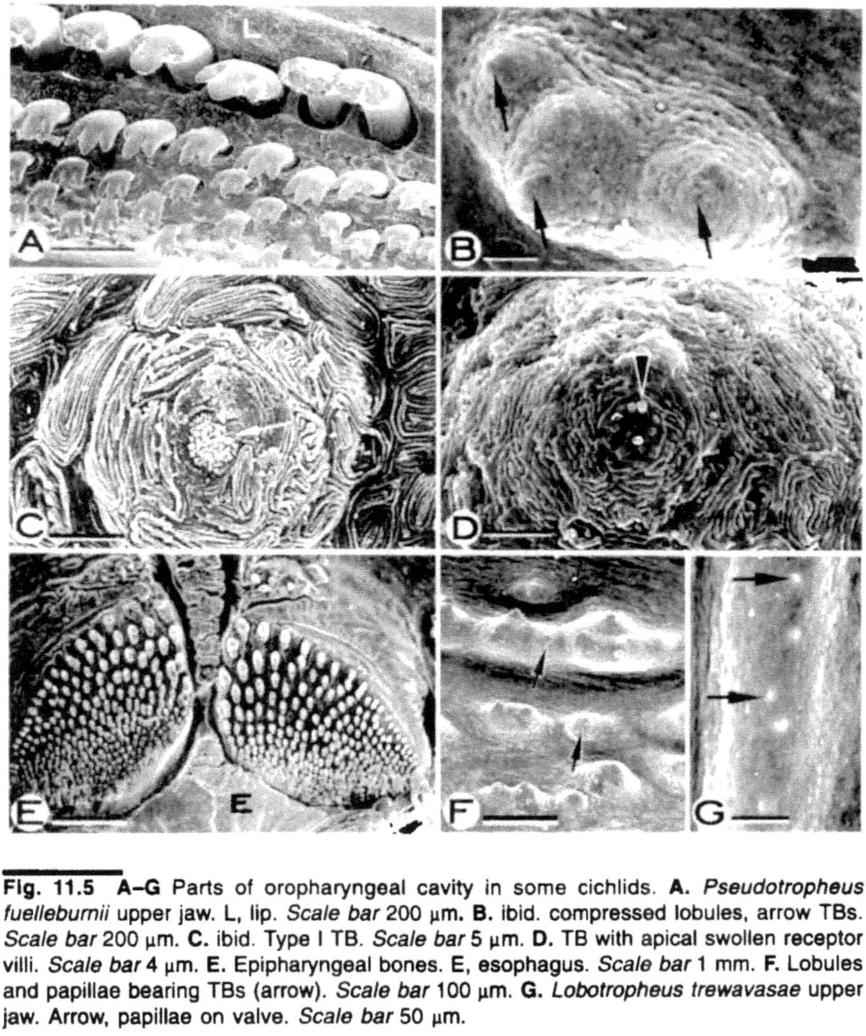

Fig. 11.5 **A–G** Parts of oropharyngeal cavity in some cichlids. **A.** *Pseudotropheus fuelleburnii* upper jaw. L, lip. *Scale bar* 200 µm. **B.** ibid. compressed lobules, arrow TBs. *Scale bar* 200 µm. **C.** ibid. Type I TB. *Scale bar* 5 µm. **D.** TB with apical swollen receptor villi. *Scale bar* 4 µm. **E.** Epipharyngeal bones. E, esophagus. *Scale bar* 1 mm. **F.** Lobules and papillae bearing TBs (arrow). *Scale bar* 100 µm. **G.** *Lobotropheus trewavasae* upper jaw. Arrow, papillae on valve. *Scale bar* 50 µm.

into round or elongated lobules 50–140 µm; each lobule bears 1–4 papillae, above which lie Type II TBs, 4.5–6.0 µm in diameter (Fig. 11.5E); each receptor area has 6–8 clavate like receptor villi (Fig. 11.5D). The upper breathing valve has a row of 25 large papillae on the basal part, each with 2–3 TBs; the more anterior part is covered by about 450 small papillae, each with one Type III TB, 3–5 µm in diameter, and 8–12 receptor villi in the receptor area. The palate shows 16–24 narrow longitudinal folds, each with a row of 60–70 papillae bearing Type II TBs.

Close to the EBs the palatal folds disappear and here the smooth palate bears about 150 small papillae with Type III TBs. The EBs are covered with bicuspid teeth that form 15–18 regular rows, 10 teeth per row in the central part (Fig. 11.5E), interspaced with small papillae and TBs. The epithelium between the EBs forms 28–30 larger lobules, 60–120 μm, each with several Type II and III TBs (Table 11.2).

The teeth band of the lower jaws resembles that on the upper jaw. The postteeth band site has around 100 papillae, scattered or forming lines (Fig. 11.5F), each with a TB. The valve has 140–160 papillae, 80–100 μm, each with Type I TBs. The gular epithelium below the valve and tongue has 13–16 longitudinal folds, each with 80–90 TBs.

The tongue is with 16–20 longitudinal folds, each with 12–14 papillae with TBs. The HB bears 26–50 longitudinal rows of bicuspid teeth, with a maximum of 15 teeth in the median rows. However, the number of these increases with fish growth, as observed in 60 mm TL and 120 mm TL fish. For example, the upper valve bears 300 and 450 papillae with TBs respectively; the palate has 12 and 24 folds with TBs respectively. In total, there are about 4,600 TBs in the oral cavity of *Pseudothtropheus fuelleburnii* (Table 11.2).

In *Labeotropheus trewavasae* (70 mm TL), the oropharyngeal cavity strongly resembles that of the previously described species. The upper jaw bears 4 rows of tricuspid teeth with 500 papillae and Type I TBs. On the postteeth site and extending along the valve are papillae with Type II TBs (Fig. 11.5G). Along the palate are 10 rows of papillae with 80 Type II and III TBs (Table 11.2). The lower jaw has 4 rows of teeth and lobules that partly extend across the valve. The HB each have 56 regular rows of teeth, 15 in the median rows.

In *Aulonocara nyassae* (60–80 mm TL) from Lake Victoria, the upper and lower jaws are armored with sharp bicuspid teeth in the front row and monocuspid teeth behind them, forming irregular rows (Fig. 11.6A). The upper lip has 300–350 Type I TBs; between the teeth are 150 lobules, some round, 40–60 μm, each with a single papilla and one TB; others are elongated, 100–120 μm, each with 4–5 Type I TBs (Fig. 11.6B). The upper valve has 160 dispersed papillae bearing Type II TBs; the palate has 10 longitudinal folds, each bearing 16 TBs. Each of the EBs (Fig. 11.6C, D) has 12 rows of bicuspid teeth, between which are numerous lobules with Type II and Type III TBs. The tongue apex reveals a group of 30 papillae with TBs; it widens with length and shows transversal folds featuring Type III TBs, 3.4–4.0 μm. The HB has 6–7 teeth in the central

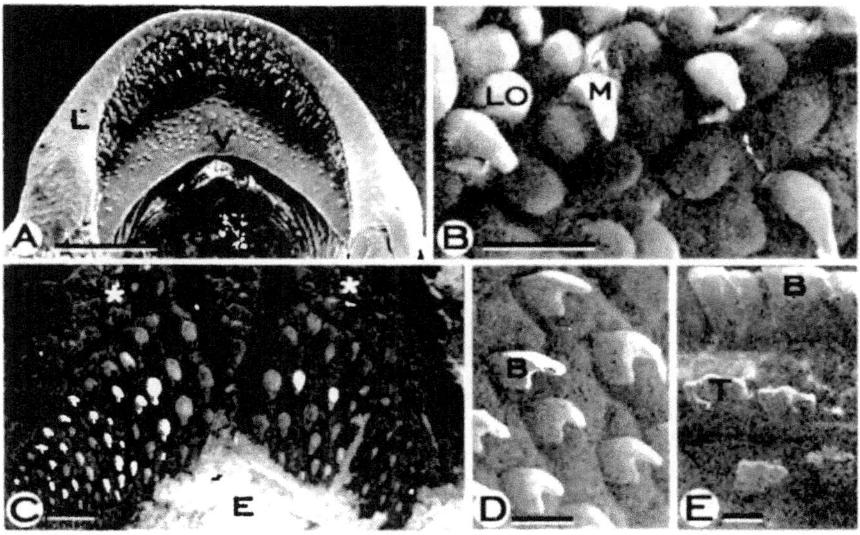

Fig. 11.6 **A–E** Parts of oropharyngeal cavity of some cichlids. **A.** *Aulonocara nyassae* upper jaw. L, lip; V, valve with papillae. *Scale bar* 1 mm. **B.** Ibid. teeth enlarged. LO, lobules; M, monocuspid teeth. *Scale bar* 100 µm. **C.** Ibid. epipharyngeal bones. E, esophagus; asterisc, 2nd pharyngobranchials. *Scale bar* 60 µm. **D.** Ibid. teeth enlarged. *Scale bar* 200 µm. **E.** *Oreochromis aureus* upper jaw. B, bicuspid and T, tricuspid teeth. *Scale bar* 100 µm.

row and numerous Type III TBs (Table 11.2). In this species 62% of TBs are found in the pharyngeal region.

Oreochromis aureus (120 mm–300 mm TL) represents the largest Israeli mouth-brooding *Tilapia*. The upper jaw has a narrow band of 2–3 rows of teeth, bicuspid flat in front row, tricuspid in two posterior ones (Fig. 11.6E), changing laterally to one line of monocuspid teeth. Between the teeth are numerous irregular lobules with Type I TBs. The upper valve bears 350–400 papillae with Type I and II TBs (Table 11.2). The palate has longitudinal lines of small papillae with TBs. The EBs each have 24 irregular rows of bicuspid compressed teeth, with a maximum of 15 teeth in the central rows (Figs. 11.7A, B). Each bone bears around 1,000 Type III TBs. About 500 Type II and III TBs are situated on lobules and papillae of various sizes between the two EBs. The lower jaw resembles the upper; teeth bands have strong frontal teeth and 3–4 rows of inner teeth, interspaced with numerous papillae and TBs. On the valves of smaller fish

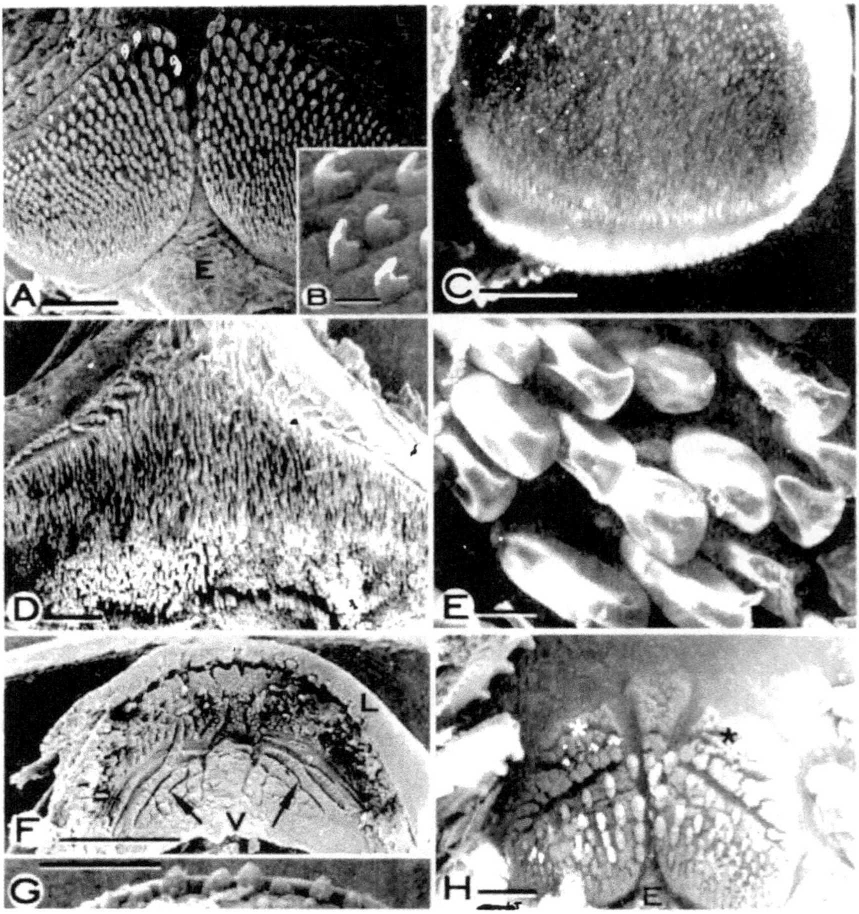

Fig. 11.7 A-H Mouth and pharyngeal bones of some cichlids. **A.** Epipharyngeal bones of *Oreochromis aureus*. E, esophagus. *Scale bar* 1 mm. **B.** Ibid. Teeth enlarged. *Scale bar* 100 μm. **C.** Epipharyngeal bone of *Sarotherodon galilaeus*. *Scale bar* 1 mm. **D.** Ibid. hypopharyngeal bone. *Scale bar* 1 mm. **E.** curved teeth. *Scale bar* 120 μm. **F.** *Cichlasoma cyanogutattum* upper jaw. L, lip; V, valve; arrow, elongate lobules with TBs. *Scale bar* 0.5 mm. **G.** Ibid. lower jaw. *Scale bar* 1 mm. **H.** Ibid. epipharyngeal bones. asterisk, TBs on 2nd pharyngobranchials. *Scale bar* 0.5 mm.

papillae with TBs are found only at the center of the valve, while in the largest fish the entire valve was covered with Type I and II TBs. The HB in the largest fish show 18 rows of strongly compressed, slightly bicuspid teeth interspaced with lobules with about 1,800 Type II and III TBs (Table 11.2). Of the oropharyngeal TBs, 61% are situated in the frontal region of the mouth.

Sarotherodon galilaeus is a mouth-brooding, biparental or monoparental tilapine (Fishelson, 2002), feeding predominantly on planktonic organisms, especially algae. In adults of this species (220 mm TL) the upper jaw has a teeth band with four irregular rows of slender teeth, the frontal bicuspid, and the more inner tricuspid, interspaced with numerous lobules and papillae with Type I TBs (Table 11.2). A few TBs are scattered on the upper valve and palate. The EBs of this species differ strongly from all other species studied: Delicate brushlike slender bicuspid teeth densely cover the bones, leaving only a wee bit of space for small lobules and TBs (Fig. 11.7C). The lower jaw also has a front row of bicuspid teeth, followed by five scattered rows of tricuspid ones. The tongue is wide, with dispersed papillae and TBs. The HB, like the EBs, has a very dense layer of slender bicuspid teeth in which the terminal cusp is sharply curved forward (Figs. 11.7D, E).

3.3 Other Substrate-brooders Studied

In *Cichlasoma cyanogutattum* (30–90 mm TL), a Central American cichlid, the oral cavity of fish of 25–30 mm TL already resembles that of larger fish. The teeth band of the upper jaw features a row of monocuspid flat teeth in front and irregular rows of delicate monocuspid teeth beyond, among which are numerous papillae with Type II TBs (Table 11.2). The longitudinal, radiating lobules of the postteeth epithelium extend over the upper valve and are covered with numerous TBs (Fig. 11.7F). The upper valve has a few dispersed TBs. The EBs have mostly bicuspid and monocuspid teeth, forming 7–9 longitudinal rows each with 4–5 teeth. They are thick in the median rows and delicate at the periphery. The 2nd pharyngobranchiales display 6 bicuspid teeth (Fig. 11.7H). Between the teeth are large lobules with papillae and 1–3 Type I TBs (Table 11.2). The lower jaw has a much thinner lip, exposing the front row of strong, chisel-like, monocuspid teeth (Fig. 11.7G). The following band of teeth is narrow; the teeth are monocuspid and irregularly dispersed among the longitudinal epithelial lobules that extend posteriorly on the narrow valve. Along these folds are rows of TBs. The HB bears 18 rows of teeth; the two central rows have 8 thick round teeth each.

In *Cichlasoma paraguyaensis* (60–80 mm TL) the oropharyngeal cavity resembles that of the previously described species: the teeth on the upper and lower jaws are monocuspid, round and sharp, forming 3–4 irregular rows (Fig. 11.8A). Between them are circa 200 irregular lobules, each with 1–2 Type I TBs. The upper valve shows parallel and obliquely arranged lobules bearing about 800 Type II TBs. The lower valve and tongue have

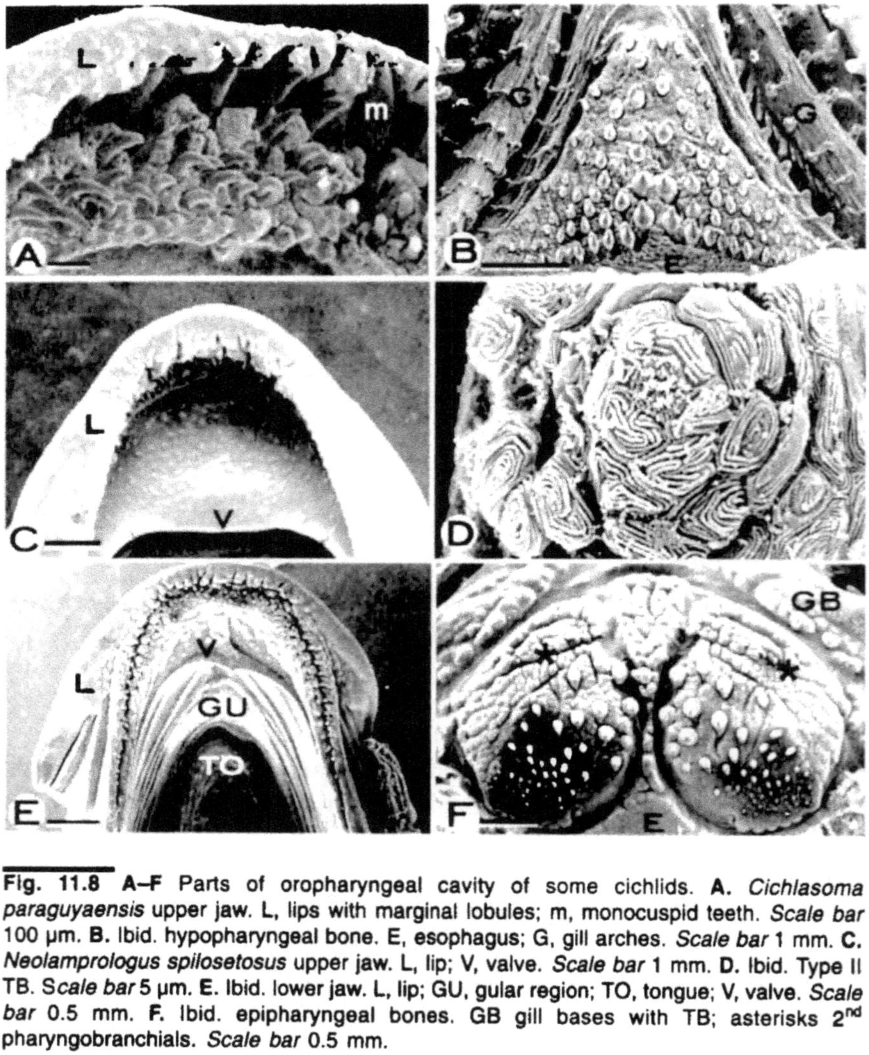

Fig. 11.8 A–F Parts of oropharyngeal cavity of some cichlids. **A.** *Cichlasoma paraguyaensis* upper jaw. L, lips with marginal lobules; m, monocuspid teeth. *Scale bar* 100 µm. **B.** Ibid. hypopharyngeal bone. E, esophagus; G, gill arches. *Scale bar* 1 mm. **C.** *Neolamprologus spilosetosus* upper jaw. L, lip; V, valve. *Scale bar* 1 mm. **D.** Ibid. Type II TB. *Scale bar* 5 µm. **E.** Ibid. lower jaw. L, lip; GU, gular region; TO, tongue; V, valve. *Scale bar* 0.5 mm. **F.** Ibid. epipharyngeal bones. GB gill bases with TB; asterisks 2nd pharyngobranchials. *Scale bar* 0.5 mm.

some sparse TBs. The EBs have 9 rows of teeth each, with strong teeth in the central rows, between which are lobules with Type III TBs. The HB has 22 rows of teeth, the two central rows of which are thick (Fig. 11.8B). In this species, too, the distribution of TBs in the oral cavity is almost uniform (Table 11.2).

Neolamprologus spilosetosus (60–80 mm TL) is an African substrate-brooder. The lips are marginally lobulated (Fig. 11.8C, E); the jaws have a front row of regular bicuspid teeth and sparsely distributed irregular tricuspid teeth. In 80 mm TL fish, interspersed among the teeth on the upper jaw are about 1,050 lobules with about 1,200 papillae bearing Type I and II TBs (Fig. 11.8D). The gular and palate epithelia are folded transversally and longitudinally, bearing 200 Type II and III TBs. The EBs have 14 rows of teeth each, 3–4 teeth in the central one; between each two rows is a row of 18 lobules, each with several TBs, totaling about 1,500 Type III TBs. The ossicles of the 2nd pharyngobranchials bear lobules and about 900 TBs; the lobules situated between the EB also bear Type III TBs (Fig. 11.8F). The teeth on the lower jaw are interspersed with 600 lobules bearing about 800 Type I TBs. The tongue has 20 dispersed papillae with TBs on its apex. The HB has 24 rows of teeth, 5 teeth in the central row, and interspaced lobules with TBs (Table 11.2). Over 70 % of the oropharyngeal TBs are situated in the pharyngeal region.

4. DISCUSSION

Taste buds have been described in numerous publications, with emphasis given mainly to their distribution on the barbels and other cutaneous processes and less to those within the oropharyngeal cavity (Atema, 1971, Hara et al., 1993; references in Hansen and Reutter, 2004). Most studies have dealt with individual species of fish and only a few have compared the distribution and ultrastructure of TBs in several closely related species or morphs (Tsura and Omori, 1976; Livingston, 1987; Boudriot and Reutter, 2001). To the best of my knowledge the only work that compares these organs among a larger group of species from two ecologically close families of fishes is that by Fishelson and Delarea (2004a) on blennies and gobies. The present study, dealing with TBs in a group of cichlid species, is the first to describe the soft parts of the oropharyngeal cavity of these fish and introduces the structure and number of TBs as a possible species-specific marker reflecting adaptive developments in the different ecotypes.

The onset of TB formation was observed 2–5 days following fertilization (Fishelson, 1966; Hansen et al., 2002). This early start of TB development parallels the early development of other vitally important organs in cichlid fish, e.g. thymus and chloride cells (Fishelson, 1995a, b; Fishelson and Bresler, 2002). Consequently, toward the onset of external feeding the larvae are already provided with the organs most important for survival. With maturation of the TBs, their form and cell composition resemble those observed in most species of teleost fishes, namely: each TB resembles a bud composed of a group of cells embedded in the epithelium, resting on the basal membrane, and with receptor microvilli rising above the skin surface (for references, see Hansen and Reutter, 2004, and Reutter and Hansen, 2005, this volume). Of the three types of TBs; Types I and II TBs, rest on high or low papillae, respectively; while Type III TBs level with or below the surrounding epithelium. The present study reveals that in cichlids Type I and II TBs are dominant on the lips and jaws, sites innervated by the facial- and glossopharyngeal nerves (VII. and IX. cranial nerves), while Type III TBs dominate the more posterior situated EB and HB bones, sites innervated predominantly by the vagus nerve (X. cranial nerve) (Spieser, 1970, Reutter and Witt, 1993; Hansen and Reutter, 2004). As these latter sites are encountered immediately before food enters the esophagus, it is possible that the sensory nerve fibers of the vagus provide the final check of the consumed diet. On comparing the distribution of TBs in the oropharyngeal cavity of the fish studied, it becomes evident that in juvenile fish the TBs are denser in the pharyngeal region than in the more anterior part. With growth this relationship changes and more TBs develop along the jaws and on the valves. Possibly this asymmetric development marks the transition from a more predatory diet to a more vegetarian one. In some fish, e.g. *Labeotropheus trewavasae* and *Cichlosoma cyanoguttatum*, of the total number of TBs in the oropharyngeal cavity, 50% are situated in the anterior part (jaws and breathing valves) and 50% in the pharyngeal region; in other species, e.g. *Tilapia zillii* and *Dimidiochromis compressiceps*, 80% and 82%, respectively of the TBs are located along the jaws and valves, and the rest in the pharyngeal region; in other species contrarily, e.g. *Neolamprologus spilosetosus* and *Aulonocara nyassae*, the situation is reversed: only 29% and 38% respectively occur in the frontal part of the mouth. The number of TB per mm^2 in the fish studied calculated for the surface of the entire mouth cavity, ranged between 12–48 TB mm^{-2}. For example, in adult *Oreochromis aureus* there are 12 TB mm^{-2}; in *Pseudotropheus fuelleburnii* 11 TB mm^{-2}, *Astatotilapia flaviijosephii* 20 TB mm^{-2}; *Cichlasoma cyanoguttatum* 35 TB mm^{-2}, and *Tilapia zillii* 48 TB mm^{-2}. However, if we calculate the density for sites with the most numerous TBs, for example on the jaws or pharyngeal bones, then these numbers increase strongly in several of the studied cichlids, to more than 440 TB mm^{-2}, as in *Dimidiochromis compressiceps*, compared to 170 TB mm^{-2} in *Tinca tinca* (Żuwała and Jakubowski, 1993), and 30 TB mm^{-2} in salmonids (Hara et al., 1993). Gomahr et al. (1992) noted 300 TB mm^{-2} as the highest number for the skin surface of cyprinids. It is obvious that there is a great variation of oral TBs in various fish which, as stated by Hansen and Reutter (2004), seems to be species specific in adults.

With fish growth the number of TBs in the oropharyngeal cavity increases. For example, in *Astatotilapia flaviijosefii* of 30 mm TL the upper jaw bears 140 TBs; in 50 mm TL fish 600 TBs, and in the largest, 120 mm TL fish, 1,400 TBs (compare: Finger et al., 1991). Parallel to this it was also noted that the dimensions of the TB receptor area increase with increasing fish length, from 2.5 µm in smaller fish to an average 5.0 µm in adults. In the largest fish of *Tilapia zillii* and *Oreochromis aureus*, TB receptor areas of 8.5–9.0 µm in diameter were often observed, larger than those described in most teleosts. With growth and increase in dimensions of the TB-bearing oropharyngeal epithelia therefore, there is a parallel genesis of the sensory organs and arrangement of neuronal fibers. It is reasonable to assume that this genesis also serves to replace TBs that have been worn out or even damaged during ingestion of food.

The considerable differences in the total number of TB/fish (e.g. around 18,000 in *Tilapia zillii* to the lowest detected 3,500 TBs per fish in *Labeotropheus trewavasae*) are very interesting and seem to be only partly related to the difference in length of these fish (290 mm TL and 100 mm TL respectively). If we consider that an average TB receptor area is 5 µm in diameter and has a receptive area of 19.6 μm^2, then the total sensory surface area of TBs in the mouth of a fully grown *Tilapia zillii* would be about 350 μm^2, and in *Labeotropheus trewavasae* about 68 μm^2.

Table 11.3 provides data on the localization of TBs on the jaws and pharyngeal bones of several of the cichlids studied. As shown, the number of TBs on the upper and lower jaw seems to be almost identical in adult fish. Similarly, the number of TBs on the EBs and HB in some species are equal while in others, e.g. *Tilapia zillii* and *Pseudotropheus fuelleburnii*, the

Table 11.3 Number of TB per mm^{-2} on the jaws and pharyngeal bones (n = 4 – 6 adult fish/species)

Species	*UJ*	*LJ*	*EB*	*HB*
Astatotilapia flaviijosefii	56 (± 18)	55 (± 12)	100 (± 10)	240 (± 28)
Cichlasoma cyanoguttatum	38 (± 14)	30 (± 8)	145 (± 20)	145 (± 15)
Pseudotropheus fuelleburnii	66 (± 16)	70 (± 10)	106 (± 12)	30 (± 8)
Dimidiochromis compressiceps	320 (± 32)	440 (± 24)	87 (± 6)	126 (± 18)
Oreochromis aureus	20 (± 4)	18 (± 4)	36 (± 8)	12 (± 3)
Tilapia zillii	92 (± 8)	37 (± 7)	220 (± 16)	119 (± 12)

UJ = upper jaw; LJ = lower jaw; EB = epipharyngeal bones; HB = hypopharyngeal bone

EBs have 2–3 times more TBs than the HB (Table 11.3). At present we do not know the reasons for these differences nor how far they characterize a specific phenotype. Compared with the above-mentioned species and their pharyngeal jaws, those of the studied *Sarotherodon galilaeus* seem to be unique: the pharyngeal bones are covered so densely with a brush-like form of teeth that almost no room is left for TBs. This species is planktonivorous and it is likely that food recognition at the pharyngeal level plays a very secondary role.

Taking all these data concerned with TB distribution and their number into account, the question arises as to whether it is possible that the primary chemical recognition of food in different species occurs in different parts of the mouth. Further, how might this be connected to the type of food consumed by the various fish? From the diet descriptions of some species we know that *Tilapia zillii* and *Dimidiochromis compressiceps* are predatory fishes, feeding on invertebrates and small fish, while the other species mentioned are more herbivorous. However, current knowledge of the diets of various species is insufficient and does not allow conclusive comments. Nonetheless, the pattern and distribution of TBs on various parts of the oropharyngeal cavity would appear to provide a tool for phylogenetic and adaptive considerations, similar to fish squamation, as suggested by Lippitsch (1998).

Regarding dentition of the various species described in this study, three basic types of teeth are recognized: monocuspid, bicuspid, and tricuspid. Most of the studied cichlids have bicuspid teeth in the front row on the jaws, matching the scheme provided by Liem and Osse (1975). However, in *Cichlasoma cyanoguttatum*, *Aulonocara nyassae*, and *Dimidiochromis compressiceps* the frontal teeth are monocuspid, and in *Tilapia zillii*, *Pseudotropheus fuellebornii*, and *Labeotropheus trewavasae* they are tricuspid. As observed during morphogenesis of the oropharyngeal cavity of the cichlids studied (Fishelson, 2005), the primordial shape of teeth on the jaws is monocuspid and possibly represents the prototype of cichlid dentition. Only at a more advanced stage of morphogenesis did the flanking cusps evolve on either side of the median cusp. Tricuspid teeth therefore appear to be the basic type, with the later regression of one cusp generating the bicuspid form that enables a denser line of cutting edges. On the EBs and HB the teeth in most of the studied species are bicuspid, except for *Tilapia zillii* in which they are tri- and quadricuspid. However, contrary to the tricuspid teeth of the frontal jaws in which the flanking cusps are positioned on either side of the central, larger cusp, on the pharyngeal bones in *Tilapia zillii* the three or four cusps form a single file, with the largest most posterior and the smallest in front. This particular organization of the cusps has not been observed in the other cichlids studied. As noted by Liem (1978, 1979) and observed in the present study, on the teeth of the EBs the higher cusp is curved forward while on the HB backward, toward the esophagus. The horizontal movements of the HB against the more static EBs produce an efficient device for mastication or laceration of the ingested food.

The differences observed in shape and organization of the teeth are possibly partly induced by a "versatile functional design" (Liem, 1978) and partly based on trophic radiation (e.g. Goldschmidt and Witte, 1992). Liem and Kaufman (1984) described two morphs of *Cichlasoma minckleyi* based on the shape of the pharyngeal teeth: one is of a molar-shape morph and the other papilliform, both segregated along the food gradient. Other authors, too, mention the high adaptability of teeth-morph to the character of the diet. In the present study a comparison of *Astatotilapia flaviijosefii* of various ages and from various localities revealed that they all present the scheme of teeth typical for this species. It is possible that among the cichlids, however, there also occur more varied and less varied genotypes.

During the last two decades the study of ecomorphology has grown in importance, as a branch of organism ecology. The latter seeks to reveal the physiological and morphological qualities of various phenotypes whose genetic inheritance is directly exposed to the specific demands of their ecological niche. The present study provides evidence that not only the hard tissues provide a basis for comparisons. The distribution pattern of TBs in the oropharyngeal cavity of various species offers an additional instrument to study ecological adaptation and evolution. Cichlid fish, occurring in flocks of hundreds of species and morphs in common lakes, provide an excellent model for such comparisons.

Acknowledgements

I am grateful to Y. Delarea of the Electron Microscopy Unit for the SEM work, I. Brickner for help in histology, and M. Alexandroni and A. Shoob for photography. Thanks to M. Nicolescu for help with Figures and their arrangement (Tübingen). Thanks also to N. Sharon for help in fish maintenance and Naomi Paz for editing the manuscript. Handling of the fish complied with the law in Israel. This study was partly supported by the Tobias Landau Foundation.

References

Atema, J. 1971. Structure and functions of the sense of taste in the catfish (*Ictalurus natalis*). *Brain Behav. Evol.* 4: 273–294.

Barel, C.D.N. 1983.Towards a constructional morphology of cichlid fishes (Teleostei, Perciformes). *Neth J. Zool.* 33: 357–424.

Beidler, L.M. 1971. *Handbook of Sensory Physiology.* Vol. IV/2. *Chemical Senses. Taste.* Springer-Verlag, Berlin.

Boudriot, F. and K. Reutter . 2000. Ultrastructure of the taste buds in the blind cavefish *Astyanax jordani* ("*Anoptichthys*") and the sighted river fish *Astyanax mexicanus* (Teleostei, Characinidae). *J. Comp. Neurol.* 434: 428–444.

Caprio, J. 1984. Olfaction and taste in fish. In: *Comparative Physiology of Sensory Organs.* L. Bolis, R.D. Kenyes and S.H.P. Maddrell (Eds). Cambridge University Press, Cambridge, pp. 257–284.

Doty, R.L. (Ed.) 2003. *Handbook of Olfaction and Gustation.* 2nd Ed. Marcel Dekker, New York.

Ezeasor, D.N. 1982. Distribution and ultrastructure of taste buds in the oropharyngeal cavity of the rainbow trout, *Salmo gairdneri* Richardson *J. Fish Biol.* 20: 53–68.

Finger, T.E., W.L. Silver and D. Restrepo. 2000. *The Neurobiology of Taste and Smell.* Wiley-Liss, New York.

Finger, T.E., S.K. Drake, K. Kotrschal, M. Womble and K.C. Dockstader. 1991. Postlarval growth of the peripheral gustatory system in the channel catfish *Ictalurus punctatus. J. Comp. Neurol.* 314: 55–66.

Fishelson, L. 1966. Untersuchungen zur vergleichenden Entwicklungsgeschichte der Gattung *Tilapia* (Cichlidae, Teleostei). *Zool. Jb. Anat.* 83: 571–656.

Fishelson, L. 1983. Social behavior of adult tilapia fish in nature and in captivity. In: *Proc. 1st Intl. Symp. Tilapia in Aquaculture*, L. Fishelson and Z. Yaron (Eds). Tel Aviv University press, Tel Aviv, pp. 40–48.

Fishelson, L. 1995a. Ontogenesis of cytological structures around the yolk-sac during embryologic and larval development of some cichlid fishes. *J. Fish Biol.* 47: 439–449.

Fishelson, L. 1995b. Cytological and morphological ontogenesis and involution of the thymus in cichlid fishes (Cichlidae, Teleostei). *J. Morphol.* 223: 175–198.

Fishelson, L. 1997. Comparative ontogenesis and cytomorphology of the nasal organs in some species of cichlid fish (Cichlidae, Teleostei). *J. Zool. Lond.* 243: 281–284.

Fishelson, L. 2002. Flexibility in the reproduction styles of the male St. Peters tilapia, *Sarotherodon galilaeus* (Cichlidae). *Env. Biol. Fish.* 63: 173–182.

Fishelson, L. 2005. Histogenesis of the oropharyngeal cavity taste buds and the relevant nerves and brain centers in substrate-brooding and mouth-brooding cichlid fish (Cichlidae, Teleostei). *Anat. Embryol.* 209: 179-192.

Fishelson, L. and V. Bresler. 2002. Comparative studies on the development and differentiation of chloride cells in Tilapine fish with different reproductive styles. *J. Morphol.* 253: 118–131.

Fishelson, L. and Y. Delarea. 2004a. Taste buds on lips and mouth of some bleniid and gobiid fishes: comparative distribution and morphology. J. Fish Biol. 64: 651–665.

Fishelson, L., Y. Delarea and A. Zverdling. 2004b. Taste buds form and distribution on lips and in the oropharyngeal cavity of cardinal fish species (Apogonidae, Teleostei), with remarks on their dentition. *J. Morph.* 259: 316–327.

Fryer, G. and T. D. Iles. 1972. *The Cichlid Fishes of the Great Lakes of Africa*. Oliver and Boyd, Edinburgh.

Goldschmidt, T. and F. Witte. 1992. Explosive speciation and adaptive radiation of haplochromine cichlids from Lake Victoria: An illustration of the scientific value of lost species flock. *Mitt. Intl. Verein Limnol.* 23: 101–107.

Gomahr, A., M. Palzenberg and K. Kotrschal. 1992. Density and distribution of external taste buds in cyprinids. *Env. Biol. Fish.* 33: 125–134.

Greenwood, P.H. 1991. Speciation. In: *Cichlid Fishes: Behavior, Ecology and Evolution*, M.H.A. Keenleyeside (Ed.). Chapman & Hall, London, New York pp. 86–102.

Greven, H. 2002. Nahrungserwerb bei (Aquarien) Fischen. In: *Verhalten der Aquarienfische*, R. Riehl and H. Greven (Eds). Birgit Schmettkamp Verlag, Bornheim, DFR, Band 2: 37–54.

Grover-Johnson, N. and A.I. Farbman. 1976. Fine structure of taste buds in the barbel of the catfish, *Ictalurus punctatus*. *Cell Tissue Res.* 169: 395–403.

Hansen, A. and K. Reutter. 2004. Chemosensory systems in fish: structural, functional and ecological aspects. In: *The Senses of Fish: Adaptations for the Reception of Natural Stimuli*, G. von der Emde, J. Mogdans and B.G. Kapoor (Eds). Narosa Publishing House, New Delhi, and Kluwer Academic Publishers, Dordrecht, the Netherlands. pp. 55–89.

Hansen, A., K. Reutter and E. Zeiske. 2002. Taste bud development in the zebrafish, *Danio rerio*. *Devel. Dyn.* 223: 483–496.

Hara, T.J. 1992. *Fish Chemoreception*. Chapman & Hall, London, New York.

Hara, T.J. 1993. Chemoreception. In: *The Physiology of Fishes*. D.H. Evans (Ed.). CRC, Boca Raton, pp. 191–218.

Hara, T.J., T. Sveinsson, R.E. Evans and D.A. Klaprat. 1993. Morphological and functional characteristics of the olfactory and gustatory organs of three *Salvelinus* species. *Can. J. Zool.* 71: 413–423.

Hirata, Y. 1966. Fine structure of terminal buds on the barbels of some fishes. *Arch. Hist. Jpn.* 26: 507–523.

Huysseune. A. 1983. Observations on tooth development and implementation in the upper pharyngeal jaws in *Astatotilapia elegans* (Teleostei, Cichlidae). *J. Morphol.* 175: 217–234.

Kapoor, B.G. and T.E. Finger. 2003. Taste and solitary chemoreceptor cells. In: *Catfishes*, G. Arratia, B.G. Kapoor, M. Chardon and R. Diogo (Eds) Science Publishers, Inc. Enfield (NH), USA, and Plymouth, UK. Vol. 2, pp. 753–769.

Kapoor, B.G., H.E. Evans and R.A. Pevzner. 1975. The gustatory system of fish. *Adv. Mar.Biol.* 13: 53–108.

Kaufman, L.S. and K.F. Liem. 1982. Fishes of the suborder Labroidei (Pisces: Perciformes): phylogeny, ecology, and evolutionary significance. *Breviora* 472

Keenleyside, M.H.A. 1979. *Diversity and Adaptation in Fish Behavior*. Springer-Verlag, Berlin.

Keenleyside, M.H.A. 1991. *Cichlid Fishes: Behavior, Ecology and Evolution*. Chapman Hall, London, New York.

Kiyohara, S., S. Yamashita and J. Kitoh. 1980. Distribution of taste buds on the lips and inside the mouth in the Minnow, *Pseudorasbora parva*. *Physiol. Behav.* 24: 1143–1147.

Klett, V. and A. Meyer. 2002. What, if anything, is *Tilapia?* Mitochondrial ND2 phylogeny of Tilapines and the evolution of parental care system in the African cichlid fishes *Molec. Biol. Evol.* 19: 865–883.

Liem, K.F. 1978. Modulatory multiplicity in the functional repertoire of the feeding mechanisms in cichlid fishes. *J. Morphol.* 158: 323–360.

Liem, K.F. 1979. Modulatory multiplicity in the functional repertoire of the feeding mechanisms in cichlid fishes, as exemplified by the invertebrate pickers of Lake Tanganyika. *J. Zool. Lond.* 189: 93–125.

Liem, K.F. 1980. Adaptive significance of intra- and interspecific differences in the feeding repertoires of cichlid fishes. *Amer. Zool.* 20: 295–314.

Liem, K.F. 1991. Functional morphology. In: *Cichlid fishes: Behavior, Ecology, and Evolution*, M.H.A. Keenleyside (Ed.). Chapman & Hall, London, New York, pp. 129–150.

Liem, K.F. and L.S. Kaufman. 1984. Interspecific macroevolution: functional biology of the polymorphic cichlid species *Cichlasoma minckleyi*. In: *Evolution of Fish Species Flocks*, A.A. Echelle and I. Kornfield (Eds). Univ. of Maine, Orono Press, Orono, ME, pp. 203–215.

Liem, K.F. and J.W.M. Osse. 1975. Biological versatility, evolution and food resources exploitation in African cichlid fishes. *Amer. Zool.* 15: 427–454.

Lippitsch, E. 1998. Phylogenetic study of cichlid fishes in Lake Tanganyika: A lepidological approach. *J. Fish Biol.* 53: 752–766.

Livingston, M.E. 1987. Morphological and sensory specializations of five New Zealand flatfish species, in relation to feeding behavior. *J. Fish Biol.* 31: 775–795.

McKay, K.R.L. 1984. Behavioral aspects of cichlid reproductive strategies; patterns of territoriality and brood defense in Central American substrate spawners and African mouth brooders. In: *Fish Reproduction: Strategies and Tactics*, G. W. Potts and R.J. Wootton (Eds). Academic Press, London, pp. 245–273.

Meyer, A. 1990. Ecological and evolutionary consequences of trophic polymorphism in *Cichlasoma citrinellum* (Pisces: Cichlidae). *Biol. J. Linnean Soc.* 39: 279–299.

Mrowka, W. 1987. Brood adaptation in a mouth brooding cichlid fish: Experiments and a hypothesis. *Anim. Behav.* 35: 922–933.

Nagl, S., H. Tichy, W.E. Mayer, I.E. Samonte, J. McAndrew and J. Klein. 2001. Classification and phylogenetic relationships of African Tilapiinae fishes inferred from mitochondrial DNA sequences. *Molec. Phylog. Evol.* 20: 361–375.

Reutter, K. 1971. Die Geschmacksknospen des Zwergwelses *Amiurus nebulosus* (Lesueur). Morphologische und histochemische Untersuchungen. *Z. Zellforsch.* 129: 280–308.

Reutter, K. 1973. Typisierung der Geschmacksknospen von Fischen. I. Morphologische und neurohistochemische Untersuchungen an *Xiphophorus helleri* Heckel (Poeciliidae, Cyprinodontiformes, Teleostei). *Z. Zellforsch.* 143: 409–423.

Reutter, K. 1978. Taste Organ in the bullhead (Teleostei). *Adv. Anat. Embryol. Cell Biol.* 55: 1–97.

Reutter, K. 1991. Ultrastructure of taste buds in the Australian lungfish, *Neoceratodus forsteri* (Dipnoi). *Chem. Senses* 16: 404 (abstract).

Reutter, K. and A. Hansen. 2005. Subtypes of light and dark elongated taste bud cells in fish. In: *Fish Chemosenses*, K. Reutter and B.G. Kapoor (Eds). Science Publishers Inc., Enfield, (NH), USA, and Plymouth, UK, pp. 211–230 (this volume).

Reutter, K. and M. Witt. 1993. Morphology of vertebrate taste organs and their nerve supply. In: *Mechanisms of Taste Transduction*, S.A. Simon and S.D. Roper (Eds). CRC Press, Boca Raton, pp. 29–82.

Reutter, K., W. Breipohl and G.J. Bijvank. 1974. Taste bud types in fishes. II. Scanning electron microscopical investigations on *Xiphophorus helleri* Heckel (Poeciliidae, Cyprinodontiformes, Teleostei). *Cell Tissue Res.* 153: 151–164.

Reutter, K., F. Boudriot and M. Witt. 2000. Heterogeneity of fish taste bud ultrastructure as demonstrated in the holosteans *Amia calva* and *Lepisosteus oculatus*. *Phil. Trans. R. Soc., London*, B, 355: 1225–1228.

Ribbink, A.J. 1991. Distribution and ecology of the cichlids of the African Great Lakes. In: *Cichlid Fishes, Behaviour, Ecology and Evolution*, M.H.A. Keenleyside (Ed.). Chapman & Hall, London, New York, pp. 36–85.

Ribbink, A.J., B.A. Marsh, A.C. Marsh, A.C. Ribbink and B.J. Sharp. 1983. A preliminary survey of the cichlid fishes of rocky habitats in Lake Malawi. *S. African J. Zool.* 18: 149–310.

Spieser, O.H. 1970. Anatomische Untersuchungen an den Hirnnerven von *Tilapia* (Cichlidae, Teleostei). PhD Thesis, University of Tuebingen, Germany.

Stiassny, M.L.J. 1991. Phylogenetic interrelationships of the family Cichlidae: An overview. In: *Cichlid Fishes: Behavior, Ecology and Evolution*, M.H.A. Keenleyside (Ed.). Chapman & Hall, London, New York, pp. 1–35.

Stiassny, M.L.J. and A. Meyer. 1999. Cichlids of the rift lakes. *Sci. Amer.* 280: 64–69.

Tsura, Y. and M. Omori. 1976. Morphological characters of the oral organs of several flatfish species and their feeding behavior. *Tohoku J. Agric. Res.* 27: 92–114.

Verheyen, E., W. Salzburger and A. Meyer. 2003. Origin of the superflock of cichlid fishes from Lake Victoria. Science 300: 325–329.

Whitear, M. 1971. Cell specialization and sensory function in fish epidermis. *J. Zool. (Lond.)* 63: 237–264.

Wickler, W. 1965. Signal value of the genital tassel in the male *Tilapia macrochir* Blgr. (Pisces, Cichlidae). *Nature (Lond.)* 208: 595–596.

Zaccone, G., B.G. Kapoor, S. Fasulo and L. Ainis. 2001. Structural, histochemical and functional aspects of the epidermis of fishes. *Adv. Mar. Biol.* 40: 253–348.

Żuwała, K. and M. Jakubowski. 1993. Light and electron (SEM, TEM) microscopy of taste buds in the tench *Tinca tinca* (Pisces: Cyprinidae). *Acta Zool. (Stockholm)* 74: 277–282.

CHAPTER

12

Role of Gustation in Two Populations of Deep-sea Fish: Comparison of Mesopelagic and Demersal Species Based on Volumetric Brain Data

H.-J. Wagner

ABSTRACT

The sensory brain areas of a sample of more than one hundred deep-sea fish species were studied and the relative volumes of the olfactory bulb, optic tectum, octavolateral area, and the gustatory area determined. In the absence of direct observations on the behaviour of this ichthyofauna these data allow deductions about the kinds of sensory modalities used preferentially in the remote deep-sea environment.

Address for Correspondence: H.-J. Wagner, Graduate School of Neural & Behavioural Sciences and Max Planck Research School, Anatomisches Institut, Universität Tübingen, Österbergstr. 3, D-72074 Tübingen, Germany. E-mail: hjwagner@anatu.uni-tuebingen.de

Key Words: Deep-sea fish; Olfactory bulb; Optic tectum; Octavolateral area; Gustatory area.

1. INTRODUCTION

The deep sea comprises about 99.5% of the total volume of biological habitats on earth (Cohen, 1994; Angel, 1997), and contains the most abundant population of vertebrates. At the limits of sunlight penetration, i.e., between 200 and 1,000 m, there is an abundant and diverse community of fish whose life styles have adapted to the constraints of pelagic life. This environment is dominated by rapidly declining levels of sunlight, currents and associated regional changes in salinity, temperature and nutrients (Pinet, 2000). The ichthyofauna of this mesopelagic habitat consists mostly of relatively small specimens (several cm up to about 35 cm TL) and is surprisingly speciose (509 different species, Merrett and Haedrich, 1997). Among the most typical representatives are hatchet-fish (*Sternoptyx*), viperfish (*Chauliodus*), the 'swimming mouths' (*Eurypharynx*), eels, the well-known anglerfish (*Ceratias*), and *Cyclothone*—the most abundant vertebrates on earth.

By contrast, the floor of the deep sea is home to a community of fish that have adapted to the special conditions of high hydrostatic pressure, low temperature, total absence of sunlight and a sparse food supply. Fishes of the abyss can be divided into a benthic population that inhabits the bottom of the continental slopes, rises, and abyssal plains and secondly, a benthopelagic population roaming the water layers close to the bottom (from 1,000 m to 6,000 m, Marshall and Merrett, 1977; Pinet, 2000). On a gross morphological level, this demersal fish fauna differs remarkably from the mesopelagic community. First of all, most demersal fish are considerably larger than those living at shallower depths in the water column. Further, their coloration, jaw structure, musculature and fin morphology suggest different 'life styles' and adaptations to diverse ecological niches. The pelagic species include many actively swimming fish but the bottom-living population has more passive species, some of which have adopted a sit-and-wait strategy (Merrett, 1987). Food resources in the abyss rely less on local productivity, as in the case of volcanic vent communities, and more on the remains of phyto- and zooplankton and larger organisms such as crustaceans, fish, and mammals (whales) (Pinet, 2000). Among the 84 demersal species recognised in the North Atlantic basin (Merrett and Haedrich, 1997), typical forms include grenadiers (*Coryphaenoides*), eels and slickheads (*Alepocephalids*).

Knowledge of these fish comes mainly from three sources: (i) trawls with specially designed fishing gear which have produced catches of mainly dead specimens, demonstrating the diversity of species and morphological specialisations (for review, see Merrett and Haedrich, 1997); (ii) autonomous vehicles deployed on the sea-floor ('landers') and equipped with imaging facilities that allow observations of fish in their native environment and study of their response to bait (Armstrong et al., 1992; Priede et al., 1994; Priede and Bagley, 2000), and (iii) a few manned submersibles (Pinet, 2000; Priede and Bagley, 2000). In summary, direct observations on the behaviour of deep-sea fish are scarce and fragmentary at best. Therefore inferences about the behaviour of these animals have been mostly drawn from dead material.

Collateral to this situation is the fact that very little experimental evidence is available concerning the sensory environment of deep-sea fish. These data are reviewed by Herring (2002) and summarised here. Vision and the nature of optic stimuli have intrigued scientists for more than 100 years since the majority of deep-sea fish have large eyes although solar light does not penetrate even the clearest ocean water deeper than 800–1,000 m. The mesopelagic zone is exposed to a very dim blue-green light, to which the visual pigments of most animals are well tuned (Douglas et al., 1998). Bioluminescence supplements or substitutes for sunlight as a visual cue at mesopelagic and greater depths matching the downwelling sunlight in spectral composition. It is tempting to speculate that the well developed and even highly specialised eyes in the majority of deep-sea fish (as well as in many crustaceans and cephalopods) have evolved to perceive these stimuli. Water flow and pressure resulting from currents or approaching objects (prey/predator; mate) as a mechanosensory stimulus may be as umed to be present and exploited by deep-sea fish in a manner similar to that in other fish. In addition, audition seems to play an important role in communication and orientation (Popper and Fay, 1993); anecdotal reports indicate that some deep-sea grenadiers do indeed produce sounds audible to humans when brought aboard ship, suggesting that they also use audition in their native environment (Marshall, 1971, 1979).

Chemical cues have been used systematically by several research groups in the form of bait (dead fish) and a number of different species were attracted to these stimuli. This would suggest that many species, especially on the bottom of the sea, use these cues to locate food falls and carrion. Olfactory cues have also been implicated in finding female mates for male anglerfish. Incidental observations may also be mentioned, recorded on film or made by observers in submersibles. The role of gustation as the contact (short distance) version of chemical sensation has received little direct attention in deep-sea fish to date. In many surface dwelling and freshwater fish the morphology of taste buds as well as the organisation of central gustatory areas have been well characterised, however (Morita and Finger, 1985; Kanwal and Caprio, 1987; Finger, 1988; Reutter and Witt, 1993, 1999; Hansen and Reutter, 2004).

Brain morphology in fish is highly diverse and shows a higher degree of divergent differentiation than any other group of vertebrates (Nieuwenhuys et al., 1997). Two main factors responsible for brain morphogenesis have been identified: Phylogeny expressed as the systematic position plays a major role, in addition to environmental factors essential to shaping brain structures by way of adaptive and exaptive processes (Northcutt, 1988; Wagner, 2002). Many earlier investigations have indicated that in teleosts the degree of differentiation of sensory systems correlates highly with the size of the sensory areas in the brain (Lissner, 1923; Geiger, 1956). In several fish families, brain structure has been proposed as a predictor for species sensory ecology (Brandstätter and Kotrschal, 1990; Kotrschal and Palzenberger, 1991; Huber et al., 1997; Kotrschal et al., 1998). Recently, similar studies in deep-sea fish were used as indicators of the kind of stimuli these fish would encounter in their habitat and of the senses most used for intra- and interspecific behaviors (Wagner, 2001a, b, 2002).

In this chapter, volumetric data are considered from a sample of 67 species of mesopelagic fish (caught in the Eastern North Atlantic and in the Central North Pacific) and 35 species of deep demersal fish (caught in the Eastern North Atlantic, and in the Porcupine Seabight and Abyssal Plain south-west of Ireland), complete datasets of which have already been published (Wagner, 2001a, b). Species in which the relative volume of the gustatory area in the rhombencephalon (intermediodorsal zone) indicated above-average importance of taste for (feeding) behaviour were identified; these are listed here and compared with species of different sensory orientations.

2. MEASUREMENTS AND CLASSIFICATIONS

Fish heads were fixed in 4% formalin aboard ship (RRS Discovery and FS Sonne). Dissection exposed the sensory organs, cranial nerves, and brain in the right hemisphere, leaving the left intact for later reference (Figs. 12.1 to 12.8). In some cases, carbocyanine dyes (1,1'-dioctadecyl-3,3,3',3'-tetramethylindocarbocyanine perchlorate [DiI] and 1,1'-dioctadecyl-3,3,3',3'- tetramethylindocarbocyanine, 4-chlorobenzene-sulphonate salt [DiD], both Molecular Probes, Eugene, OR, USA) were applied to the nerves and the projection areas in the brain demonstrated. In the laboratory back home, the dorsal and lateral aspects of the brains were recorded with a digital camera. The length, width and depth of every brain was determined using the measuring tool of the Adobe Photoshop 5 program. Values were scaled by referencing them to a mm-scale included in the micrographs. Quantitative anlaysis basically followed the concepts of Huber et al. (1997) who treated the brain lobes as half-ellipsoids and calculated their volumes from the three cardinal dimensions. To discount for size differences and enable interspecies comparisons, the volumes of the four sensory areas were added and the relative proportions determined. The average value of the relative volumes determined the comparative rank of a given species for each sensory system. This rank was defined with respect to a reference population. Such a population is represented by the environment in which the fish were caught, i.e. mesopelagic (67 species) and demersal (35 species) habitat (Wagner, 2001a, b). In this case, the relative average of each sensory area within the population served as the baseline for the ranking system (Table 12.1). Above-average cases are represented as plus (+) symbols in Table 12.2 and below average cases as minus (–). Species with only a single sensory area above average were considered specialists, species with two areas above average were regarded as dominated by these two senses, and species with three areas above average designated generalists (Wagner, 2001a, b).

3. OBSERVATIONS AND DISCUSSION

3.1 General Considerations

The relative volumes of the four sensory areas show interesting differences between mesopelagic and demersal species (Table 12.1). The visual system takes up more than half the share, yet with a difference of 3.7% the two populations are very similar. The most striking differences are found in the olfactory and the octavolateral systems. In demersal fish, the average olfactory bulb volume is about six times larger than in mesopelagic fish. The importance of the long-distance chemosensory organ is also reflected

Table 12.1 Summary of distribution of sensory specialisations (based on average relative volume of four sensory brain areas) in the total sample of 67 mesopelagic and 35 demersal fish (Wagner, 2001a,b)

	Olfactory bulb		*Optic tectum*		*Octavolateral area*		*Gustatory area*	
	mesopelagic	*demersal*	*mesopelagic*	*demersal*	*mesopelagic*	*demersal*	*mesopelagic*	*demersal*
Mean relative volume (%)	2.76	16.48	61.26	57.56	25.32	11.57	10.08	14.38
Species above mean (%)	34	37	61	54	41	49	40	34
Specialists (%)	1.5	14.2	32.8	20	1.5	0	0	8.6

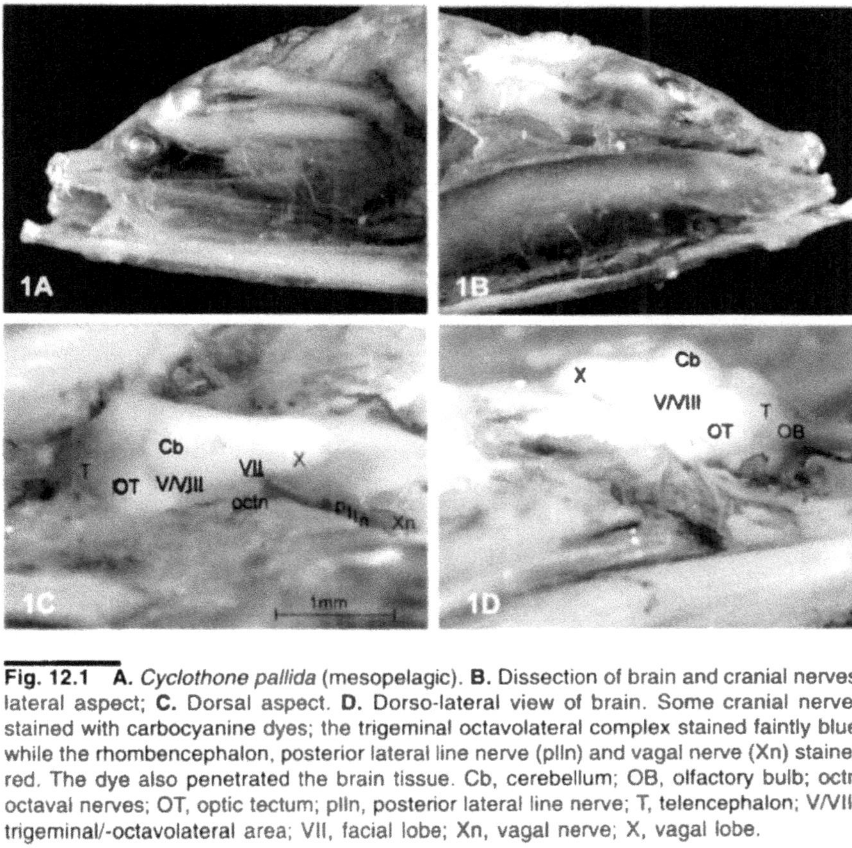

Fig. 12.1 **A.** *Cyclothone pallida* (mesopelagic). **B.** Dissection of brain and cranial nerves, lateral aspect; **C.** Dorsal aspect. **D.** Dorso-lateral view of brain. Some cranial nerves stained with carbocyanine dyes; the trigeminal octavolateral complex stained faintly blue, while the rhombencephalon, posterior lateral line nerve (plln) and vagal nerve (Xn) stained red. The dye also penetrated the brain tissue. Cb, cerebellum; OB, olfactory bulb; octn, octaval nerves; OT, optic tectum; plln, posterior lateral line nerve; T, telencephalon; V/VIII, trigeminal/-octavolateral area; VII, facial lobe; Xn, vagal nerve; X, vagal lobe.

by the number of olfactory specialists in each habitat: There is only a single mesopelagic species as opposed to five in the demersal sample (half the size of the mesopelagic one). Conversely, the octavolateral system seems to be of greater importance in the mesopelagic domain, with an average relative volume about twice as large as in the demersal population. It is mostly asociated with other above-average sensory areas; the only specialist is mesopelagic. On the other hand, the percentage of species with above-average octavolateral areas is similar in the two groups. As for the gustatory area, the relative volume is larger by about 40% in bottom-living than in pelagic fish (actual difference: 4.3%). While there is no gustatory specialist in the mesopelagic group, there are two (8.6%) among the demersal population.

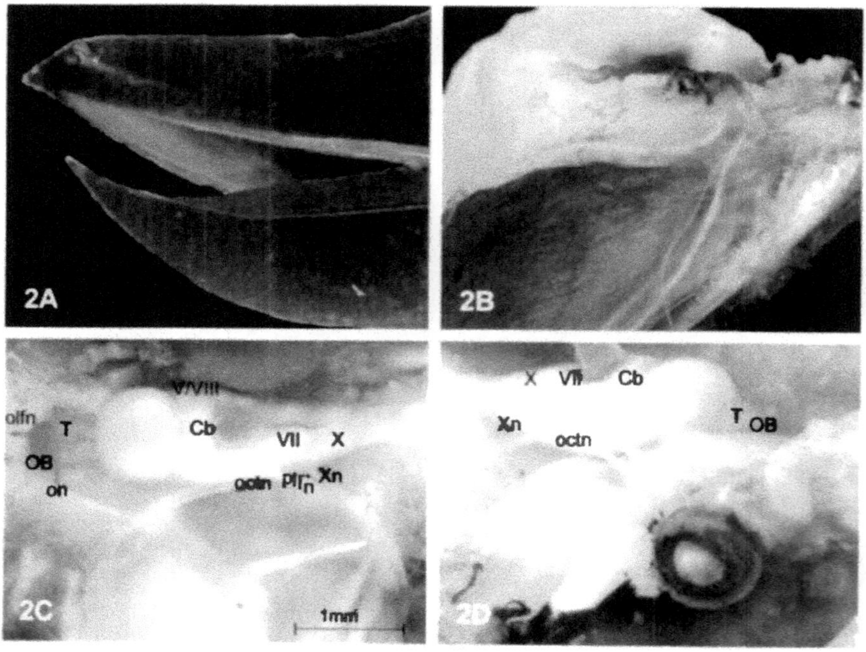

Fig. 12.2 **A.** *Euypharynx pelecanoides* (mesopelagic). **B.** Dissection of brain and cranial nerves, lateral aspect; **C.** Dorsal aspect. **D.** Dorsolateral view of the brain. Some cranial nerves stained with carbocyanine dyes; olfactory nerve (olfn), olfactory bulb and adjacent telencephalon stained faintly blue, while the rhombencephalon, posterior lateral line nerve (plln), and vagal nerve (X) stained red. Cb, cerebellum; OB, olfactory bulb; octn, octaval nerves; olfn, olfactory nerve; on, optic nerve; plln, posterior lateral line nerve; T, telencephalon; V/VIII, trigeminal/octavolateral area; VII, facial lobe; Xn, vagal nerve; X, vagal lobe.

3.2 Mesopelagic Fish

Among the 67 mesopelagic fish analysed, 25 (40%) had gustatory areas above average for this group (Table 12.1). Table 12.2 shows that they belong to numerous and diverse systematic groups, from slickheads (Alepocephalids) to anglerfish (Pediculati), and comprise species with very diverse biology.

The ubiquitous bristlemouth *Cyclothone pallida*, along with other members of its genus, is often regarded as the world's most abundant vertebrate (Herring, 2002). It has small eyes and a subaverage optic tectum (Fig. 12.1; Table 12.2). The three other senses considered here are all above average. Figure 12.1 clearly shows that the rhombencephalon with the gustatory area takes up the largest part. Figure 12.1C and D

Table 12.2 List of species with above-average relative volume of gustatory area. Part A in systematic order, part B in ecologic order.

	Olf. bulb	Telenceph.	Optic tect.	Cerebellum	V / VIII	Gust. area	Depth
Part A							
<u>Clupeiformes</u>							
Alepocephalidae							
Photostylus pycnopterus	G ++		–		+	++	m
<u>Stomiatoidei</u>							
Gonostomatidae							
Cyclothone pallida	G ++		–		++	++	m
Gonostoma bathyphilum	G +++		–		+	+++	m
Gonostoma gracile	G ++		–		+	+	m
Gonostoma elongatum	G +++		–		+	+	m
Gonostoma ebelingi	G +		–		+	+	m
Sternoptychidae							
Argyropelecus hemigymnus	D –		–		+	+	m
Melanostomiidae							
Bathophilus metallicus	D –		+		–	+	m
Thysanactis deutex	D –		–		+	+	m
Pachystomias microdon	D –		–		+	+	m
Malacosteidae							
Aristostomias grimaldi	D –		+		–	+	m
Malacosteus niger	G –		+		+	+	m
Photostomias guerni	D –		+		–	+	m
Idiacanthidae							
Idiacanthus fasciola	D –		–		+	+	m
<u>Iniomi</u>							
Bathysauridae							
Bathysaurus ferox	G – –	– –	+	–	+	+	LS/R

(Table 12.2 contd.)

(Table 12.2 contd.)

Bathysaurus mollis	D – –	– –	–	–	+	+	R/A
Chlorophthalmidae							
Bathypterois dubius	D – –	– –	–	–	+	++	MS/LS
Bathypterois longipes	S – – –	– –	–	– –	–	++	R/A
Bathytyphlops sewelli	D – – –	– –	–	–	++	++	S/R
Neoscopelidae							
Scopelengys tristis	D –		–		+	+	m
Myctophidae							
Lampanyctus ater	D –		–		+	++	m
Lampanyctus intricarius	D –		–		+	+	m
Lyomeridae							
Eurypharynx pelecanoides	D +		–		–	+	m
Anguilliformes							
Cyemidae							
Cyema atrum	G + +		+		–	+	m
Heteromi							
Halosauridae							
Halosauropsis macrochir	G – –	–	+	+	+	+	MS/MR
Notacanthidae							
Polyacanthonotus challengeri	G +	–	–	+	+	+	MS/MR
Gadiformes							
Macrouridae							
Hymenocephalus metallicus	G +		+		–	+	m
Coryphaenoides (Ch.) leptolepis	D –	–	–	+	+	+	LS/R/A
Coryphaenoides mediterraneus	D –	–	–	+	+	+	MS/UR
Coryphaenoides profundicolus	D +	–	–	– –	–	+	A
Coryphaenoides guentheri	G –	–	+	+	+	+	MS/MA

(Table 12.2 contd.)

(Table 12.2 contd.)

Allotriognathi							
Stylephorus chordatus	D –				+	+	m
Berycomorphi							
Melamphaidae							
Poromitra megalops	D –		–		+	++	m
Anoplogasteridae							
Anoplogaster cornuta	D –		–		+	++	m
Ophidiodei							
Ophidiidae							
Bathyonus sp.	S –	–	–	+	–	+	R/A
Pediculati							
Ceratioidei							
Ceratias holboelli	G +		–		+	+	m
Melanocetes johnsoni	D –		–		++	+	m
Part B							
Cyclothone pallida	G ++		–		++	++	m
Gonostoma bathyphilum	G +++		–		+	+++	m
Gonostoma gracile	G ++		–		+	+	m
Gonostoma elongatum	G +++		–		+	+	m
Gonostoma ebelingi	G +		–		+	+	m
Argyropelecus hemigymnus	D –		–		+	+	m
Bathophilus metallicus	D –		+		–	+	m
Thysanactis deutex	D –		–		+	+	m
Pachystomias microdon	D –		–		+	+	m
Aristostomias grimaldi	D –		+		–	+	m
Malacosteus niger	G –		+		+	+	m
Photostomias guerni	D –		+		–	+	m
Idiacanthus fasciola	D –		–		+	+	m

(Table 12.2 contd.)

(Table 12.2 contd.)

Scopelengys tristis	D –		–		+	+	m
Lampanyctus ater	D –		–		+	++	m
Lampanyctus intricarius	D –		–		+	+	m
Eurypharynx pelecanoides	D +		–		–	+	m
Cyema atrum	G ++		+		–	+	m
Photostylus pycnopterus	G ++		–		+	++	m
Stylephorus chordatus	D –				+	+	m
Poromitra megalops	D –		–		+	++	m
Anoplogaster cornuta	D –		–		+	++	m
Ceratias holboelli	G +		–		+	+	m
Hymenocephalus metallicus	G +		+		–	+	m
Melanocetes johnsoni	D –		–		++	+	m
Bathypterois dubius	D – –	– –	–	–	+	++	MS/LS
Halosauropsis macrochir	G – –	–	+	+	+	+	MS/MR
Polyacanthonotus challengeri	G +	–	–	+	+	+	MS/MR
Coryphaenoides mediterraneus	D –	–	–	+	+	+	MS/UR
Coryphaenoides guentheri	G –	–	+	+	+	+	MS/MA
Bathysaurus ferox	G – –	– –	+	–	+	+	LS/R
Coryphaenoides (Ch.) leptolepis	D –	–	–	+	+	+	LS/R/A
Bathytyphlops sewelli	D – – –	– –	–	–	++	++	S/R
Bathysaurus mollis	D – –	– –	–	–	+	+	R/A
Bathypterois longipes	S – – –	– –	–	– –	–	++	R/A
Bathyonus sp.	S –	–	–	+	–	+	R/A
Coryphaenoides profundicolus	D +	–	·	– –	–	+	A

S, specialist: one area above average; D, "dominated" species: two areas above average; G, generalist; – less than relative average; – – less than half rel. mean; – – – very small, negligible. + more than relative average; ++ more than twice rel. mean; +++ Depth information; demersal: S, continental slope; R, continental rise; A, abyssal; U, upper; M, mid; L, lower (Merrett et al., 1991a, b); m, mesopelagic.

indicate that the posterior part of the rhombencephalon, corresponding to the vagal lobe is better developed that the more anterior facial lobe. Bristlemouths are active zooplanktivorous feeders and ingest mostly copepods (Gartner et al., 1997). The morphology of the gustatory area would suggest that *Cyclothone* are rather unselective in what they catch and would use taste information from the oropharyngeal cavity to determine food items for intake or rejection. Like the other Gonostomiids among the present sample, *Cyclothone* is known for its marked sexual dimorphism with males showing conspicuously enlarged olfactory systems (Whitehead et al., 1984; Marshall, 1967). Unfortunately, sex determination was not performed on the specimens studied here; therefore it is not possible to correlate our volumetric findings with the sex of the specimen. However, previous data would suggest that our fish were not mature males but either females or sexually immature specimens. Furthermore, long-distance chemosense olfaction has been implicated with pheromone tracking and mate finding by male fish (Jumper and Baird, 1991; Baird and Jumper, 1995).

While only one species of hatchetfish has an above-average gustatory area (*Sternoptyx hemigymnus*), it is notable that in the melanostomids and malacosteids, the three species known to have specialised visual systems with both red suborbital photophores and far red sensitivity of visual pigments (*Pachystomias microdon*, *Aristostomias grimaldi* and *Malacosteus niger*, Douglas et al., 1998) all possess the aforesaid gustatory regions, with particularly well-developed vagal lobes (not shown). This would suggest that selectivity in food choice might possibly rely as much on taste cues as on visual stimuli.

Pelican eels are characterised by their enormous mouth cavity and small eyes (Fig. 12.2). They have an elongated rhombencephalon with distinct facial and vagal lobes. In *Eurypharynx pelecanoides* the facial lobe is larger than the vagal. This size difference is also seen in the afferent nerves, both stained red with DiI in Figure 12.2B and C, with the posterolateral line nerve considerably thicker than the more posterior vagal nerve. Similarly, the trigeminal and anterior octavolateral nerves supplying the skin of the head as well as the oropharyngeal mucosa are well developed (stained blue with DiD). Contrarily, the optic nerves are very thin (Figs. 12.2B-D). These observations may lead one to speculate that the pelican eel has taste buds along its upper and lower jaws (projecting to the facial lobe) which would enable it to check the oncoming potential food before opening its mouth and allowing it to be swallowed.

E. pelecanoides has been characterised as a broad spectrum generalist feeder whose diet includes caridean shrimps, fishes, and copepods (Gartner et al., 1997).

The fangtooth *Anoplogaster cornuta* is also carnivorous, feeding on crustaceans and fish (Whitehead et al., 1984). Its brain exhibits a cerebellum that is exceptionally large compared to other mesopelagic fish. Well-developed lateral line organs are visible on the head (Fig. 12.3A); accordingly, the nerves associated with the octavolateralis complex (stained blue with DiD in Figs. 12.3 B-F) are robust in size. In the rhombencephalon the vagal lobe is markedly larger than the facial lobe, suggesting that the fangtooth examines prey while holding it in the oropharyngeal cavity, with escape prevented by the large teeth. Judging from the relative volume of the sensory areas, the lateral line system and gustation seem to be more important than olfaction and vision which are both below average.

Angerfish such as *Melanocetes johnsoni* (Fig. 12.4) are ambush predators luring prey with a luminous esca looped anteriorly over the mouth (Gartner et al., 1997). Morpholgy of the head as well as inspection of the brain suggested that neither vision nor olfaction appear to be suffciently developed to play a major role in prey capturing. Instead, there is a system of lateral line canals on the head which subserve mechanosensory information in the near range indicating the approach and presence of organisms attracted by the lure, or mates or enemies respectively. The associated cranial nerves, trigeminal and octavolateral, are prominent and visible running underneath the skin of the upper jaw (Fig. 12.4C). In the large rhombencephalon, facial and vagal lobes are well developed and about equal in size. As for the afferent nerves, it is difficult to identify facial fibres among the strong V-VIII complex; the vagal fibres entering the rhombencephalon are not especially thick. From these observations, one may conclude that *Melanocetes* seem to devote much of the afferent systems to locating approaching objects and examining their chemical properties, either on contact with jaws or the mouth cavity (facial system), or using deeper regions of the oropharyngobranchial cavity (vagal system).

3.3 Demersal Fish

Among the 35 demersal fish analysed, 12 (34%) had gustatory areas above average for this population (Table 12.1). The most important were Iniomi,

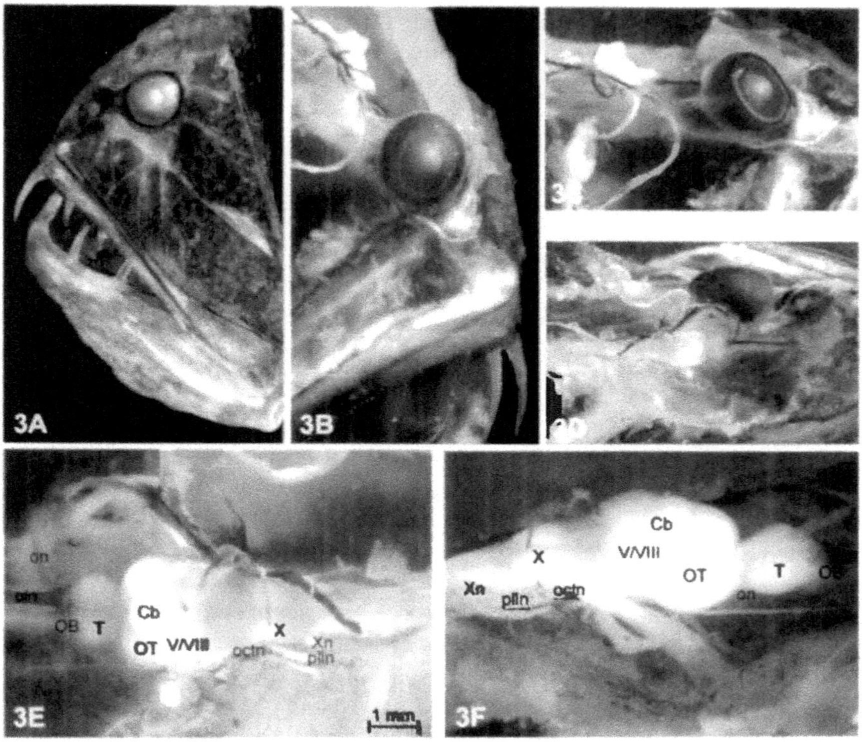

Fig. 12.3 **A.** *Anoplogaster cornuta* (mesopelagic); **B, C.** Dissection of brain and cranial nerves, lateral aspect; **D, E.** dorsal aspect; **F.** dorsolateral view of brain. Some cranial nerves on the left side stained with carbocyanine dyes; olfactory nerve (oln) and adjacent olfactory bulb stained faintly blue, the optic nerve (on) red, the trigeminal/facial nerves, most of the octaval nerve (octn) including the posterior lateral line nerve (plln) blue, and the vagal nerve (Xn) red. Cb, cerebellum; OB, olfactory bulb; OT, optic tectum; T, telencephalon; V/VIII, trigeminal/-octavolateral area; X, vagal lobe.

with genera such as bathysaurids and chlorophthalmids and, besides the halosaurids and notacanthids, the gadiform grenadiers (macrourids). Among them are two gustatory specialists, six 'dominated' species, and four generalists, indicating widely divergent life styles.

Two species of lizardfish (*Bathysaurus sp.*) were observed motionless on the bottom snatching passing fish in a sit-and-wait strategy. Large specimens are not likely to rise more than half a metre from the bottom (Sedberry and Musick, 1978). From the differentiation of their brains, they would be expected to use mainly lateral line input and possibly vision (just above average in *B. ferox* and almost reaching average values in

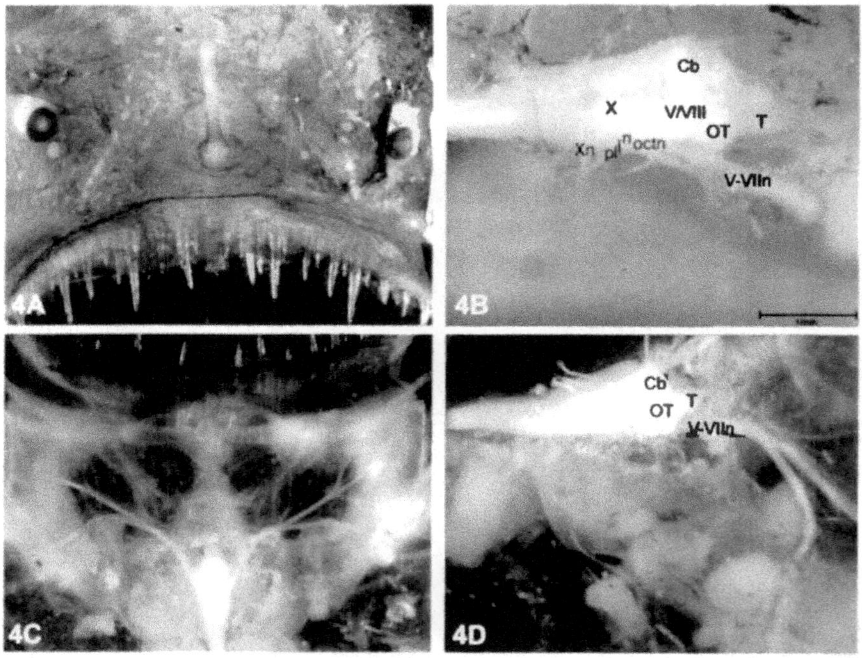

Fig. 12.4 **A.** *Melanocetes johnsoni* (mesopelagic); **B.** Dissection of brain and cranial nerves, dorsolateral aspect; **C.** ventral aspect; **D.** lateral view of brain. Cb, cerebellum; octn, octaval nerves; OT, optic tectum; plln, posterior lateral line nerve; T, telencephalon; V/VIII, trigeminal/-octavolateral area; V-VIIn, trigeminal/facial nerve; Xn, vagal nerve; X, vagal lobe.

B. mollis) to locate their targets (Fig. 12.5C, D). Gustatory control relies on an enlarged vagal lobe (Fig. 12.5D) and may help to decide what to ingest and what to reject. Interestingly, analysis of nematode parasites has demonstrated that *B. ferox* is 'specialised' to bite off only the tails of passing macrourid fish prey (Campbell et al., 1980).

Tripod fish such as *Bathypterois dubius* also belong to the sit-and-wait predators; perched on their elongated pectoral and caudal fins they face into the current. They are considerably smaller than lizardfish, so they feed on epibenthic micronekton (Crabtree et al., 1991). Their small eyes are matched by below-average optic tecta, while their prominent lateral line organs (Marshall and Staiger, 1975) correspond to an above-average octavolateral area in *B. dubius* and even twice the average in *B. sewelli*. In *B. dubius*, three prominent nerves originate in the mandibular and maxillary areas as well as the trunk, including a thick branch from the

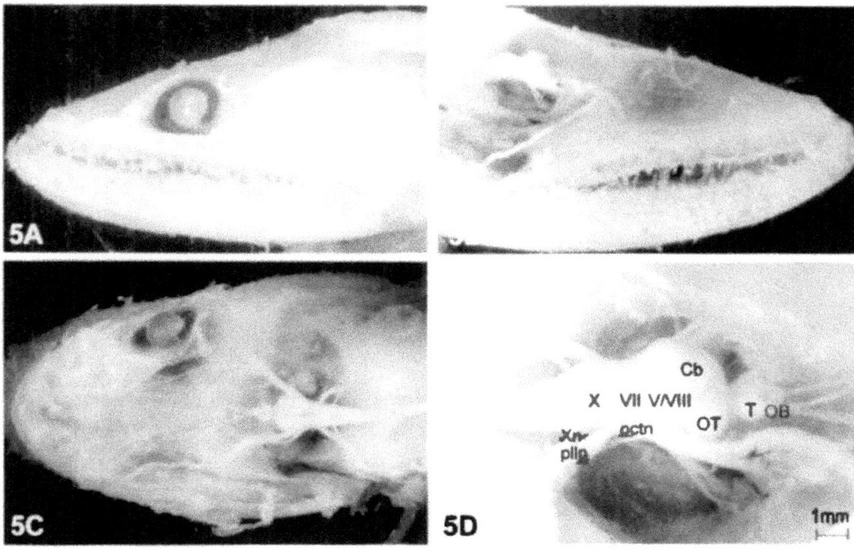

Fig. 12.5 **A.** *Bathysaurus mollis* (demersal); **B.** Dissection of brain and cranial nerves, lateral aspect; **C.** dorsal aspect; **D.** dorsolateral view of the brain. Cb, cerebellum; OB, olfactory bulb; octn, octaval nerves; OT, optic tectum; plln, posterior lateral line nerve; T, telencephalon; V/VIII, trigeminal/-octavolateral area; VII, facial lobe; Xn, vagal nerve; X, vagal lobe.

elongated pectoral ray, and enter the brain stem anterior and posterior to the vestibular nerves, identifying them as trigeminal, facial, and anterior plus posterior lateral nerves (Fig. 12.6 B-D). The vagal/ glossopharyngeal nerves are less conspicuous (Fig. 12.6 F). Dorsal to the emergence of the vagal nerve, a well-developed vagal lobe is clearly seen (Fig. 12.6F), while the facial lobe region shows no particular enlargement. It has been suggested that the elongated pectoral fin rays serve as sensory organs, examining the near-bottom water layer for changes in pressure (currents) and chemical composition (Marshall, 1957). The thick posterior lateral line nerve innervating these fin rays would indicate the presence of numerous mechanosensory neuromasts for this task. On the other hand, it was not possible to trace fibres of the facial nerve to the elongated fin rays. Therefore it is not clear whether taste buds or equivalents (solitary chemosensory cells; Hansen and Reutter, 2004; Hansen, 2005, this volume) are present in the skin covering the fin rays. In sea robins (*Prionotus carolinus*), modified pectoral fin rays contact the substrate and have been shown to be chemosensory despite the absence of taste buds or olfactory receptors (Whitear, 1971): They are innervated by spinal nerves

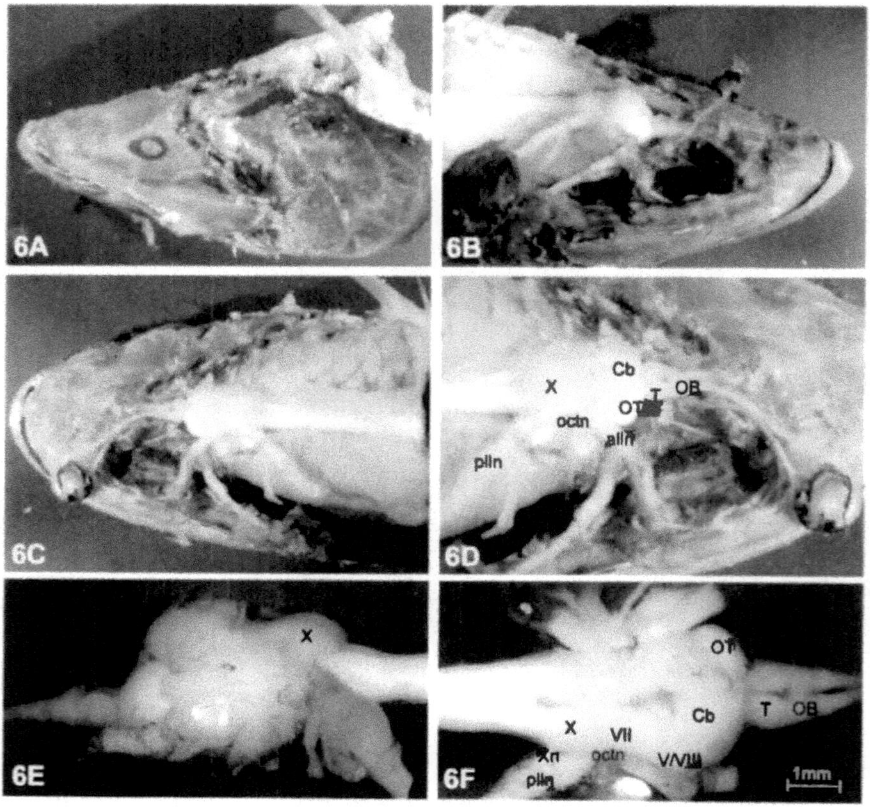

Fig. 12.6 **A.** *Bathypterois dubius* (demersal); **B.** Dissection of brain and cranial nerves, dorsolateral aspect; **C.** dorsal aspect; **D.** dorsal view of brain; **E.** Lateral view of isolated brain; **F.** Dorsal view of the isolated brain. Alln, anterior lateral line nerve; Cb, cerebellum; OB, olfactory bulb; octn, octaval nerves; OT, optic tectum; plln, posterior lateral line nerve; T, telencephalon; V/VIII, trigeminal/-octavolateral area; VII, facial lobe; Xn, vagal nerve; X, vagal lobe.

terminating in accessory spinal lobes (Finger, 1982). It is tempting to speculate that the pectoral fins in tripod fish show a comparable organisation.

Halosaurs have also been observed poised close to the sea-floor and facing into the current (Grassle et al., 1975); they feed on crustaceans, molluscs and echinoderms. When swimming, they hold their long pectorals upwards and forwards, and a sensory function has been suggested for them (Whitehead et al., 1984). The brain morphology of *Halosauropsis macrochir* characterises this species as a generalist with the optic tectum, the octavolateral, and the gustatory area above average relative volume (Fig. 12.7B-D). Both the facial and the vagal lobes are well

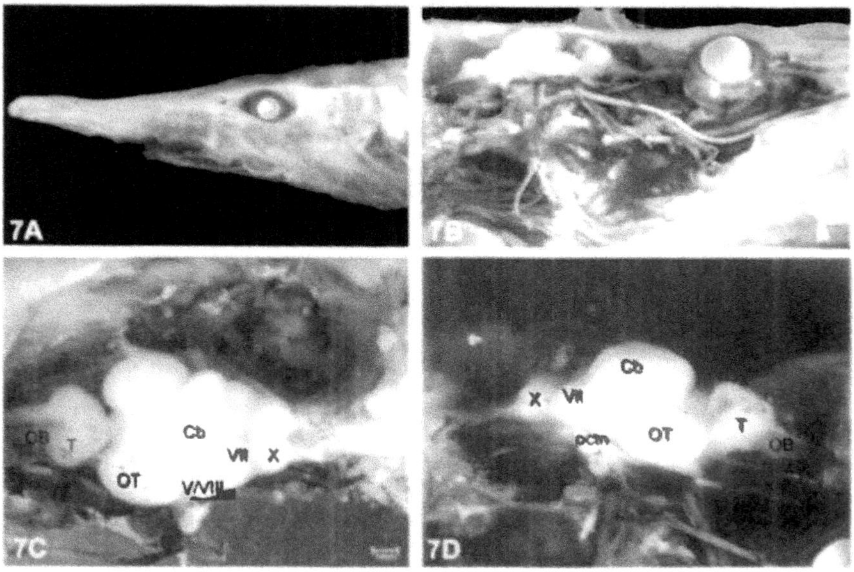

Fig. 12.7 **A.** *Halosauropsis macrochir* (demersal); **B.** Dissection of brain and cranial nerves, lateral aspect; **C.** dorsal aspect; **D.** dorsolateral view of brain. Some cranial nerves stained with carbocyanine dyes; left olfactory nerve and adjacent olfactory bulb stained blue, optic nerve red, trigeminal/facial nerves, most of the octaval nerve including the posterior lateral line nerve are alternately blue and red; vagal nerve stained red. Cb, cerebellum; OB, olfactory bulb; T, telencephalon; V/VIII, trigeminal/-octavolateral area; VII, facial lobe; octn, octaval nerves; X, vagal lobe.

developed and about equal in size. Two prominent nerves are located ventral of the vagal lobe, indicating strong vagal and glossopharyngeal gustatory connections to the oropharyngeal cavity (Fig. 12.7 C). The large facial lobe is associated with prominent afferent nerves from the big rostrum (trigeminal and facial), suggesting that the snout may be used for probing the sediment and therefore contains numerous mechanosensitive and chemosensory receptor organs.

The macrourid grenadiers comprise more than 300 species and are the most speciose deep-demersal family; most of them live in the boundary layer above the bottom and are characterised as benthopelagic. Species richness is greatest in low latitudes on the slope. In the present study, *Hymenocephalus metallicus* and *Coryphaenoides mediterraneus* were found in areas of the upper slope, between 800 and 2,000 m. On the other hand, C. *profundiculus* occupied the deepest end of the habitat, from 2,000 m downwards to the abyssal plain. Many slope-dwellers, among them *H. metallicus*, are bioluminescent ventrally (Herring, 1987).

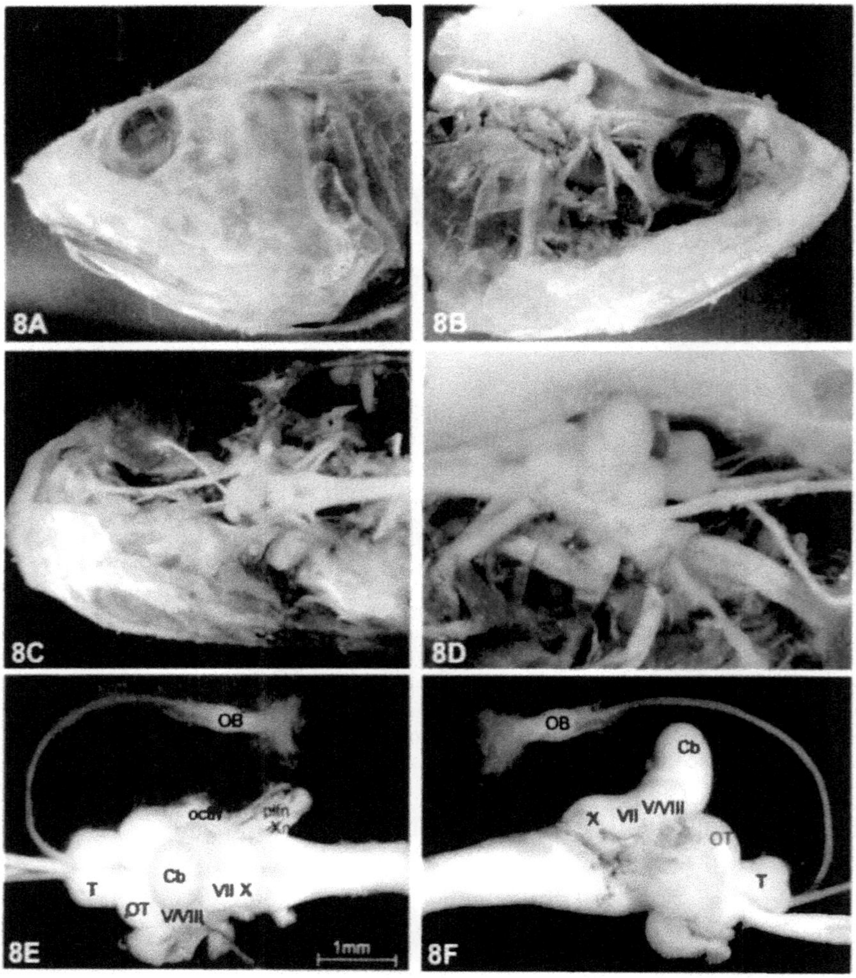

Fig. 12.8 **A.** *Coryphaenoides mediterraneus* (demersal) **B.** Dissection of brain and cranial nerves, dorso-lateral aspect; **C.** dorsal aspect; **D.** dorsolateral view of the brain; **E.** Dorsal view of isolated brain; **F.** Lateral view of isolated brain. Cb, cerebellum; OB, olfactory bulb; octn, octaval nerves; OT, optic tectum; plln, posterior lateral line nerve; T, telencephalon; V/VIII, trigeminal/-octavolateral area; VII, facial lobe; Xn, vagal nerve; X, vagal lobe.

Feeding strategies vary greatly from active scavengers, carrion eaters, and euryphagous species. In terms of differentiation of the sensory brain, the gadiform grenadiers also form a markedly heterogeneous group, including specialists in vision or olfaction, species dominated by various combinations of senses (e.g. trigeminal/octavolateral and gustatory areas; olfaction and vision), and generalists (Wagner 2001b). The benthic generalist *C. guentheri* from the continental rise feeds on small prey that often school off the bottom (Gartner et al., 1997) and is never attracted to bait placed on the sea-floor (Priede et al., 1990, 1994). Its sensory brain shows below-average olfactory bulbs, and an above-average optic tectum, suggesting that visual cues play a more prominent role in locating prey, possibly linked to bioluminescent signals emitted by the targets; octavolateral and gustatory areas are above average. By contrast, *C. profundiculus* is a mobile forager that seems to rely predominantly on chemical cues, both olfactory and gustatory; interestingly, it did not visit baited cameras (Priede et al., 1990, 1994). *C. leptolepis* and *mediterraneus* occupy successive habitats on the continental rise and slope; they have brains with above-average octavolateral and gustatory areas. Figure 12.8 shows that although all four sensory areas are well developed the facial and vagal lobes are more prominent than in most other grenadiers. Afferent information is conveyed via a thick bundle of trigeminal, facial, and octavolateral nerves (Fig. 12.8 D-F). In addition, there are strong posterior lateral line and vagal nerves associated with the rhombencephalon. The strong maxillary component of the trigeminal along with the facial nerves may carry afferents from the chin barbel which contains numerous myelinated nerve fibres (unpubl. data). Video observations in a related species (*C. armatus*) have shown that the barbel is used to guide the fish along the scaffold of a lander vehicle (Priede and co-authors, pers. Comm.); it is not yet clear whether mechanosensory or gustatory information (or both) serve as a cue for this type of behaviour.

4. CONCLUSIONS: ROLE OF TASTE IN THE DEEP SEA

The gustatory system in fish has been best characterised in cyprinids and ictalurids. While the ultrastructure of the peripheral taste buds has been extensively described by Reutter (for comparative summaries, see Reutter and Witt, 1993, 1999; Reutter et al., 2000; Hansen and Reutter, 2004) the central parts have been studied by Finger (1983, 1988). Morphologically, taste buds rest atop a dermal papilla and form an ovoid or pyriform aggregate of specialised epithelial cells, otherwise integrated into the epidermal epithelium. The population of elongated taste bud cells and basal cells shows varying degrees of electron density, secretory granules, and afferent as well as (a few) efferent synaptic (Reutter and Witt, 2005, this volume) connections. Reutter et al. (2000) concluded that there is no common morphological pattern that would define only one type of taste bud in fish. Taste buds may be distributed throughout the body surface but are often concentrated on the head, including the upper and lower lips, and the anterior part of the mouth. In catfish, the barbels carry an especially dense population of taste buds (Atema, 1971; Finger et al., 1991; Caprio et al., 1993). Centripetal projections from taste buds in all of these locations are carried by the facial nerve. Additional taste buds are located in the posterior part of the oropharyngeal cavity, including the gill rakers; their afferent fibres are part of the glossopharyngeal and vagal nerves. The central rhombencephalic nuclei are part of the nucleus of the solitary tract (Meek and Nieuwenhuys, 1997) and associated with a complex system of ascending connections (De Graaf, 1989). Enlargements characterised as facial and vagal lobes show a highly differentiated internal organisation. In the cyprinid vagal lobe, 16 layers have been distinguished (Meek and Nieuwenhuys, 1997) containing a large sensory portion of alternating cell-rich and cell-poor layers, more superficially and efferent fibres, as well as motor neurons for the palatal organ in deeper areas close to the IVth ventricle.

There is a single report on the gustatory system in deep-sea fish, wherein the surface morphology of taste buds on the tongues of three species was studied with the SEM and compared with surface-living teleosts (Meyer-Rochow, 1981). According to a receptor index proposed in this paper, two of the deep-sea species are 'poor tasters', one a 'good taster'. One of the 'poor tasters' is *Sternoptyx diaphana*, whose sensory brain was recently studied; it was established as a visual specialist with all other sensory modalities below average (Wagner, 2001a). So, in this case, analysis of the peripheral taste organ and central representation accord. In spite of these findings one needs to bear in mind that there is no direct evidence that *S. diaphana* reacts less to taste stimuli than species in which volumetric brain analysis has predicted greater sensitivity.

The potential as well as the dangers of the approach for making deductions regarding sensory behaviour of deep-sea fishes from differentiations of the sensory brain areas, were demonstrated in another recent case. Development of *Coryphaenoides armatus*, a macrourid grenadier not contained in the present list because its gustatory area had been found to be below average (Wagner, 2001b), was studied because baited cameras showed that only specimens larger than 0.4–0.5 m were attracted, although smaller specimens had been ascertained by trawls in the same area. The relative volumes of the octavolateral and gustatory areas did not change during growth; by contrast the relative size of the optic tectum decreased until the fish reached a total length of about 0.4 m, while the relative volume of the olfactory bulb increased correspondingly (Wagner, 2003). This finding of a shift in sensory orientation in smaller specimens suggests that fish require a threshold size of the olfactory bulb before they respond to the plume of odor produced by the bait carried with the lander.

These observations indicate that the present judgements based on the relative size of the facial and vagal lobes in the context of other sensory brain regions represent snapshots of the two populations which might be influenced by the developmental state of the individual specimens. It is thus important to note that usually only adult specimens have been included. However, as has been shown for the olfactory system, sexual dimorphism may also influence the volume of this area, depending on the state of maturity (e.g. *Cyclothone microdon*, Marshall, 1967). This aspect was not considered in the analyses forming the basis of this paper.

In summary, based on volumetric analysis of sensory brain areas, we identified 25 of a total of 67 mesopelagic fish and 12 of a total of 35 demersal species with a gustatory area above average. While the relative proportion of the population is similar in both groups, the greater role of gustation in demersal fish is suggested by the fact that the average relative volume is considerably larger than in mesopelagic fish. This is associated on a more general level by an increased importance of the other main chemosensory system, i.e. olfaction in the bottom-living population. In the demersal habitat, the heterogeneous topography of the sea floor and the boundary layer leads to an accumulation of potential food which seems to make chemosensory systems more important for successful survival than in the mesopelagic environment. When considering the combination of an above-average gustatory region with other sensory areas, some interesting, family-related patterns emerge: In many gonostomids, it is associated with an above-average olfactory system, while in several (red sensitive) dragonfish it is combined with a well-developed and even specialised visual system. The sensory modality gustation is most often found associated with, is the mechanosensory trigeminal/octavolateral system. Typical representatives of this strategy are the sit-and-wait tripod fish, some anglerfish, and the fangtooth.

Acknowledgements

This project was financially supported by the DFG (Wa 348/22), BEO (S-107) and indirectly by the British NERC (GR3/12789). I am indebted to

E. Flüh (Kiel), J.C. Partridge (Bristol), I. G. Priede and M. Collins (both Aberdeen) for organising and co-ordinating the work aboard ship as principal scientific officers, and to the masters and crews of Forschungsschiff (FS) Sonne and Royal research ship (RRS) Discovery for highly competent nautical work.

References

Angel, M.V. 1997. What is the deep sea? In: W.S. Hoar, D.J. Randall and A.P. Farrell (Eds). Fish Physiology, vol. 16. *Deep Sea Fishes*. Academic Press, London. pp. 1–41.

Armstrong, J.D., P.M. Bagley and I.G. Priede. 1992. Photographic and acoustic tracking observations of the behaviour of the grenadier *Coryphaenoides (Nematonurus) armatus*, the eel *Synaphobranchus bathybius*, and other abyssal demersal fish in the North Atlantic Ocean. *Mar. Biol.* 112: 535–544.

Atema, J. 1971. Structures and functions of the sense of taste in the catfish (*Ictalurus natalis*). Brain Behav. Evol. 4: 273–294.

Baird, R.C. and G.Y. Jumper. 1993. Olfactory organs in the deep-sea hatchetfish *Sternoptydiaphana* (Stomiformes, Sternoptychidae). *Bull. Mar. Sci.* 53: 1163–1167.

Brandstätter, R. and K. Kotrschal. 1990. Brain growth patterns in four European cyprinid fish species (Cyprinidae, Teleostei): roach (*Rutilus rutilus*), bream (*Abramis brama*), common carp (*Cyprinus carpio*), and sabre carp (*Pelecus cultratus*). *Brain Behav. Evol.* 35: 195–211.

Campbell, R.A., R.L. Haedrich and T.A. Munroe. 1980. Parasitism and ecological relationships among deep-sea benthic fishes. *Mar. Biol.* 57: 301–313.

Caprio, J., J.G. Brand, J.H. Teeter, T. Valentinčič, D.L. Kalinoski, J. Kohbara, T. Kumazawa and S. Wegert. 1993. The taste system of the channel catfish: from biophysics to behavior. TINS 16: 192–197.

Cohen, J.E. 1994. Marine and continental food webs: Three paradoxes? *Phil. Trans. R. Soc. (Lond.)* B 343: 57–69.

Crabtree, R.E., H.J. Carter, and J.A. Musick. 1991. The comparative feeding ecology of temperate and tropical deep-sea fishes from the western North Atlantic. *Deep-Sea Res.* 38: 1277–1298.

De Graaf, P. 1989. Control of respiration in carp (*Cyprinus carpio* L.). Mechanoreceptor input and respiratory rhythm. PhD thesis, Univ Groningen, Netherlands.

Douglas, R.H., J.C. Partridge and N.J. Marshall. 1998. The eyes of deep-sea fish, I: Lens pigmentation, tapeta and visual pigments. *Progr. Retinal Eye Res.* 17: 597–636.

Finger, T.E. 1982. Somatotopy in the representation of the pectoral fin and free fin rays in the spinal cord of the sea robin, *Prionotus carolinus*. *Biol. Bull.* 163: 154–161.

Finger, T.E. The gustatory system in teleost fish. In: *Fish Neurobiology*. I. *Brain Stem and Sense Organs*, R.G. Northcutt and R.E. Davis (Eds). Univ. Michigan Press, Ann Arbor, pp. 285–309.

Finger, T.E. 1988. Sensorimotor mapping and oropharyngeal reflexes in goldfish, *Carassius auratus*. *Brain. Behav. Evol.* 31: 17–24.

Finger, T.E., S.K. Drake, K. Kotrschal, M. Womble and K.C. Dockstader. 1991. Postlarval growth of the gustatory system in the channel catfish, *Ictalurus punctatus*. *J. Comp. Neurol.* 314: 55–66.

Gartner, J.V. Jr., R.E. Crabtree and K.J. Sulak. 1997. Feeding at depth. In: *Deep-Sea fishes*, W.S. Hoor, D.J. Randall and A.P. Farrell (Eds). Academic Press, London, 115–193.

Geiger, W. 1956. Quantitative Untersuchungen über das Gehirn der Knochenfische, mit besonderer Berücksichtigung seines relativen Wachstums. *Acta. Anat.* 27: 324–350.

Grassle, J.F., H.L. Sanders, R.R. Hessler, G.T. Rowe and T. McLennan. 1975. Pattern and zonation: a study of the bathyal megafauna using the research submersible Alvin. *Deep-Sea Res.* 22:643–659.

Hansen, A. 2005. The system of solitary chemosensory cells. In: *Fish Chemosenses*, K. Reutter and B.G. Kapoor (Eds). Science Publ. Inc., Enfield, (NH), USA, and Plymouth, UK. pp. 165–174. (this volume).

Hansen, A. and K. Reutter. 2004. Chemosensory systems in fish: Structural, functional and ecological aspects. In: *The Senses of Fish: The Adaptations for the Reception of Natural Stimuli*, G. von der Emde, J. Mogdans and B.G. Kapoor (Eds). Narosa Publishing House, New Delhi, and Kluwer Academic Publishers, Dordrecht, the Netherlands. pp. 55–89.

Herring, P.J. 1987. Systematic distribution of bioluminescence in living organisms. *J. Biolum. Chemolum.* 1: 147–163.

Herring, P.J. 2002. *The Biology of the Deep Ocean.* Oxford University Press, Oxford.

Huber, R., M.J. van Staaden, L.S. Kaufman and K.F. Liem. 1997. Microhabitat use, trophic patterns, and the evolution of brain structure in African cichlids. *Brain Behav. Evol.* 50: 167–182.

Jumper, G.Y. and R.C. Baird. 1991. Location by olfaction—a model and application to the mating problem in the deep-sea hatchetfish *Argyropelecus hemigymnus*. *Amer. Naturalist* 6: 1431–1458.

Kanwal, J.S. and J. Caprio. 1987. Central projections of the glossopharyngeal and vagal nerves in the channel catfish, *Ictalurus punctatus*: clues to differential processing of visceral inputs. *J. Comp. Neurol.* 264: 216–230.

Kapoor, B.G. and T.E. Finger. 2003. Taste and solitary chemoreceptor cells. In: *Catfishes*. G. Arratia, B.G. Kapoor, M. Chardon and R. Diogo (Eds). Science Publishers Inc., Enfield, (NH), USA, Plymouth, UK. vol 2, pp. 753–770.

Kotrschal, K. and M. Palzenberger. 1991. Neuroecology of cyprinids (Cyprinidae, Teleostei): Comparative quantitative histology reveals diverse brain patterns. *Env. Biol. Fish.* 33: 135–152.

Kotrschal, K., M.J. van Staaden and R. Huber. 1998. Fish brains: Evolution and ecological relationships. *J. Fish Biol. Fish.* 8: 1–36.

Lissner, H. 1923. Das Gehirn der Knochenfische. Wiss. Meeresunters. Helgoland NF 14: 125–184.

Marshall, N.B. 1967. The olfactory organs of bathypelagic fishes. *Symp. Zool. Soc. London* 19: 57–70.

Marshall, N.B. 1971. *Exploration in the Life History of Fishes.* Harvard Univ. Press, Cambridge, MA (USA).

Marshall, N.B. 1979. *Developments in Deep-sea Biology.* Blandford Press, Poole, Germany.

Marshall, N.B. 1996. The lateral line system of three deep-sea fish. *J. Fish. Biol.* 49 (suppl. A): 239–258.

Marshall, N.B. and J.C. Staiger. 1975. Aspects of the structure, relationships and biology of the deep-sea fish *Ipnops murrayi* Gunther (family Bathypteroidae). *Bull. Mar. Sci.* 25: 101–111.

Marshall, N.B. and N.R. Merrett. 1977. The existence of a benthopelagic fauna in the deep sea. In: *A Voyage of Discovery* (George Deacon 70th anniversary volume), M.V. Angel (Ed.). *Deep-Sea Res.* 24 (suppl.): 483–497.

Meek, J. and R. Nieuwenhuys. 1997. Holosteans and teleosts. In: *The Central Nervous System of Vertebrates*. R. Nieuwenhuys, H.J. ten Dondelaar and C. Nicholson (Eds). Springer-Verlag, Berlin, pp. 759–937.

Merrett, N.R. 1987. A zone of faunal change in assemblages of abyssal demersal fish in the eastern North Atlantic: a response to seasonality in production? *Biol. Oceanog.* 5: 137–151.

Merrett, N.R. and R.L. Haedrich. 1997. *Deep-sea Demersal Fish and Fisheries.* Chapman & Hall, London, New York.

Merrett, N.R., J.D.M. Goirdonb, M. Stethmann, R.L. Haedrich. 1991a. Deep demersal fish assemblage structure in the Porcupine Seabight (eastern North Atlantic): Slope sampling by three different trawls compared. *J. Mar. Biol. Assoc.* (UK) 71: 329–358.

Merrett, N.R., R.L. Haedrich, J.D.M. Gordon and M. Stethmann. 1991b. Deep demersal fish assemblage structure in the Porcupine Seabight (eastern North Atlantic): Results of single warp trawling at lower slope to abyssal soundings. *J. Mar. Biol. Assoc.* (UK) 71: 359–373.

Meyer-Rochow, V.B. 1981. Fish tongues—surface fine structure and ecological considerations. *Zool. J. Linn. Soc.* 71: 413–426.

Morita, Y. and T.E. Finger. 1985. Reflex connections of the facial and vagal gustatory systems in the brainstem of the bullhead catfish, *Ictalurus nebulosus*. *J. Comp. Neurol.* 231: 547–558.

Nieuwenhuys, R. and E. Pouwels. 1983. The brain stem of actinopterygian fishes. In: *Fish Neurobiology. 1. Brain Stem and Sense Organs,* R.G. Northcutt, R.E.Davis (Eds). Univ. Michigan Press, Ann Arbor, pp. 25–87.

Nieuwenhuys, R., H.J. ten Donkelaar, and C. Nicholson. 1997. *The Central Nervous System of Vertebrates.* Springer-Verlag, Berlin, vol. 2.

Northcutt, R.G. 1983. Evolution of the optic tectum in ray-finned fishes. In: *Fish Neurobiology. 2. Higher Brain Areas and Functions,* R.G. Northcutt and R.E. Davis (Eds). Univ. Michigan Press, Ann Arbor, pp. 1–42.

Northcutt, R.G. 1988. Sensory and other neural traits and the adaptationest program: mackerels of San Marco? In: *Sensory Biology of Aquatic Animals,* J. Atema, R.R. Fay, A.N. Popper and W.N. Tavolga. Springer-Verlag, New York, pp. 869–883.

Pinet, P.R. 2000. *Invitation to Oceanography.* 2nd Ed. Jones and Bartlett, Boston.

Popper, A.N. and R.R. Fay. 1993. Sound detection and processing by fish: Critical review and major research questions. *Brain Behav. Evol.* 41: 14–38.

Priede, I.G. and P.M. Bagley. 2000. *In situ* studies on deep-sea demersal fishes using autonomous unmanned lander platforms. *Oceanog. Mar. Biol. Ann. Rev.* 38: 357–392.

Priede, I.G., K.L. Smith Jr. and J.D. Armstrong. 1990. Foraging behavior of abyssal grenadier fish: Inferences from acoustic tagging and tracking in the Northern Pacific Ocean. *Deep-Sea Res.* 37: 81–101.

Priede, I.G., P.M. Bagley, A. Smith, S. Creasy and N.R. Morrett. 1994. Scavenging deep demersal fishes of the Porcupine Seabight (NE Atlantic Ocean): Observations by baited camarea, trap and trawl. *J. Mar. Biol. Assoc.* (UK) 74: 481–498.

Reutter, K. and M. Witt. 1993. Morphology of vertebrate taste organs and their nerve supply. In: *Mechanisms of Taste Transduction*, S.A. Simon, S.D. Roper (Eds). CRC Press, Boca Raton, pp. 29–82.

Reutter, K. and M. Witt. 1999. Comparative aspects of fish taste bud ultrastructure. In: *Advances in Chemical Signals in vertebrates*, R.E. Johnston, D. Müller-Schwarze, P.W. Sorensen (Eds). Kluwer Academic Publishers, New York, pp. 573–581.

Reutter, K. and M. Witt. 2005. Efferent Synapses in fish taste buds. In: *Fish Chemosenses*, K. Reutter and B.G. Kapoor (Eds). Science Publishers, Inc., Enfield, (NH), USA, and Plymouth, UK. pp. 231–245 (this volume).

Reutter, K., F. Boudriot and M. Witt. 2000. Heterogeneity of fish taste bud ultrastructure as demonstrated in the holosteans *Amia calva* and *Lepisosteus oculatus*. *Phil. Trans. Roy. Soc.* (Lond.) B 355: 1225–1228.

Sedberry, G.R. and J.A. Musick. 1978. Feeding strategies of some demersal fishes of the continental slope and rise off the Mid-Atlantic coast of the USA. *Mar. Biol.* 44: 357–375.

Wagner, H.-J. 2001a. Sensory brain areas in mesopelagic fishes. *Brain Behav. Evol.* 57: 117–133.

Wagner, H.-J. 2001b. Brain areas in abyssal demersal fish. *Brain Behav. Evol.* 57: 301–316.

Wagner, H.-J. 2002. Sensory brain areas in three families of deep-sea fish (slickheads, eels and grenadiers): Comparison of mesopelagic and demersal species. *Mar. Biol.* 141: 807–817.

Wagner, H.-J. 2003. Volumetric analysis of brain areas indicates a shift in sensory orientation during development in the deep-sea grenadier *Coryphaenoides armatus*. *Mar Biol.* 142: 791–797.

Wagner, H.-J., E. Fröhlich, K. Negishi and S.P. Collin. 1998. The eyes of deep-sea fish, II. Functional morphology of the retina. *Progr. Retinal Eye Res.* 17: 637–685.

Whitear, M. 1971. Cell specialization and sensory function in fish epidermis. *J. Zool.* (Lond.) 163: 237–264.

Whitehead, P.J.P., M.-L. Bauchot, J.-C. Hureau, J. Nielsen, and E. Tontonese. 1984. *Fishes of the North-eastern Atlantic and the Mediterranean* UNESCO, Paris.

Wilson, R.R. Jr. and K.L. Smith Jr. 1984. Effect of near-bottom currents on detection of bait by the abyssal grenadier fishes *Coryphaenoides* ssp., recorded *in situ* with a video camera, free vehicle. *Mar. Biol.* 84: 83–91.

CHAPTER

13

Comparison of Taste Preferences and Behavioral Taste Response in the Nine-spined Stickleback *Pungitius pungitius* from the Moscow River and White Sea Basins

Alexander O. Kasumyan and Elena S. Mikhailova

ABSTRACT

Taste preferences and behavioral taste responses of the nine-spined stickleback *Pungitius pungitius* from two geographically isolated populations were compared (Himka creek in the Moscow River basin and the Lake Machinnoe in the White Sea basin; located in the same latitude, but separated by more than 2,000 km). Using the behavioural assay the palatability of four classical taste substances and 21 free amino acids (L-isomers) were assessed.

Address for Correspondence: Alexander O. Kasumyan, Department of Ichthyology, Faculty of Biology, Moscow State University, 119899, Russia. E-mail: alex_kasumyan@mail.ru

Citric acid, cysteine, glutamine and alanine were among the most palatable substances for fish from the two geographical regions. Sodium chloride, sucrose as well as isoleucine, tryptophan, tyrosine, and threonine were indifferent taste substances. A positive correlation ($r_s = 0.49$; $p < 0.05$) was found among the taste preferences of the fish for the 21 free amino acids tested, indicating that taste preferences do not have a high population specificity in fish. Nine-spined sticklebacks from the two different basins also showed similarity in pellet retention time, which decreased with acceptance ratio. The relationship between pellet acceptance ratio and number of repeated grasps at pellets differed in the sticklebacks from the two geographic groups. We hypothesize that the frequency of spitting out the food item may depend on the density of fish population or/and the hydrological conditions in its native water bodies. Comparative analyses of the results obtained for the nine-spined sticklebacks and the 14 fish species previously investigated showed that taste preferences are highly species specific.

Key Words: Fish; Taste preferences; Free amino acids; Classical taste substances; Population specificity; Species specificity; Feeding behaviour.

1. INTRODUCTION

For many years morphological and electrophysiological approaches dominated studies of the gustatory system of fish. This created the situation wherein a detailed knowledge of the structure and function of the gustatory system is not supported by appropriate data on taste preferences of fishes. Among recent reviews concerning the gustatory system of fish (Jakubowski and Whitear, 1990; Marui and Caprio, 1992; Reutter, 1992; Reutter and Witt, 1993; Hara, 1994; Sorensen and Caprio, 1998; Hansen and Reutter, 2004) only the one by Kasumyan and Døving (2003) deals with fish taste preferences.

A decade ago data on the taste preferences were available for a few fish species only (Sutterlin and Sutterlin, 1970; Appelbaum, 1980; Goh and Tamura, 1980; Carr, 1982; Hidaka, 1982; Mackie, 1982; Johnsen and Adams, 1986; Adams et al., 1988; Jones, 1989; Lamb and Finger, 1995). Many of these studies yielded results which may reflect not only the gustatory but olfactory system also, which set important sensory cues for fish feeding behaviour (Atema, 1980; Pavlov and Kasumyan, 1990). Systematic investigations of fish taste preferences were performed during the last 10–15 years after developing an appropriate bioassay (Mearns et al., 1987; Kasumyan and Sidorov, 1993, 1995a).

Utilizing this bioassay, it was established that fish taste preferences are highly species specific regarding composition and range of substances that evoke stimulatory or deterrent responses (Kasumyan and Nikolaeva, 1997, 2002; Kasumyan and Prokopova, 2001). Even closely related fish species belonging to the same genus, e.g., the Russian sturgeon *Acipenser gueldenstaedtii*, Siberian sturgeon *A.baerii*, and Stellate sturgeon *A.stellatus*, show different oral taste preferences to free amino acids (Kasumyan, 1999). However, high population specificity was not found for fish taste perception. Brown trout (*Salmo trutta*) juveniles belonging to three geographically isolated populations showed similar taste responses to the four classical taste substances: out of sodium chloride, calcium chloride, and sucrose, citric acid was the most palatable substance for brown trout living either in the Caspian Sea, Baltic Sea or White Sea (Kasumyan and Sidorov, 1995c). In the present study the taste preferences of two populations of the nine-spined stickleback (*Pungitius pungitius*), originating from two basins that are located geographically far from each other, were compared. Not only were the classical taste substances tested as taste stimuli, but also free amino acids as common components of food organisms.

2. MATERIAL AND METHODS

2.1 Animals

Two groups of one-year-old specimens of the nine-spined stickleback (*Pungitius pungitius*) were used in the experiments. Fish of the first group (6–7 cm total length) were caught by net in Himka creek running through 'Petrovskoe-Glebovo' park in the western part of Moscow and flowing into the Moscow River (September, 2000). Fish of the second group (5–6 cm TL) were caught in a shallow area of the Lake Mashinnoe (50 × 300 × 1.5 m) located at the southern coast of Kandalaksha Gulf of the White Sea, 40 km from the town of Chupa (June, 2002). Freshwater Lake Mashinnoe is located about 1 km from the sea coast and neither creek nor river flows into or out of it. The distance between the two habitats is about 2,000 km. The nine-spined stickleback was the only fish species in these two habitats. The density of the stickleback population is much higher in Himka creek than in Lake Mashinnoe. Water salinity in Lake Mashinnoe and Himka creek is about 20 mg L^{-1} and 90 mg L^{-1} respectively (Maximova, 1967; Technical report ..., 1992).

Fish were transported to the laboratory and maintained in a 100-litre aquarium during the first 1–2 weeks. They were then placed one by one into 4-litre aquaria. Each aquarium was equipped with an aerator. The cover had a small feeding hole in the centre. The water was partly replaced daily with fresh water. Experiments were performed at a water temperature of 12–16°C (fish from Himka creek) and 10–13°C (fish from the Lake Mashinnoe). Live bloodworms (*Chironomidae* larvae) bought in a Moscow pet shop served as food for both groups of fish. The fish were fed *ad libitum* once a day after completion of experiments.

2.2 Test Procedure

After placing the fish in separate aquaria, they were trained to take the food dropped inside, first bloodworms, then agar pellets containing a water extract of bloodworms (175 g L^{-1}). Pellets were dropped into the aquarium one by one, with an interval of 10–15 minutes. After about 3 days of training, a fish positioned itself under the hole of the aquarium's cover and took the pellet 2–5 seconds after it was dropped into the water. Then a pellet containing one of the test substances was offered to the fish. In each trial, several parameters were registered: number of grasps at the pellet, retention time of the pellet or after the first grasp, and total retention time of all grasps; swallowing of the pellet; eventual rejection of the pellet. The moment of pellet swallowing was determined by the termination of characteristic chewing movements of the jaws and the onset of 'normal' opercular movements. The moment of eventual pellet rejection was determined by fish behaviour: the pellet was spat out and the fish went away and did not return to the pellet for at least one minute.

If the fish did not grasp the pellet within one minute, the trial was halted and not registered. Pellets containing different taste substances were offered to the fish in random sequence. Pellets rejected or not taken were removed from the aquarium immediately after each trial.

Earlier experiments with carp (*Cyprinus carpio*) using the same test procedure showed that the pellets acceptance ratio as well as other characteristics of response to the pellet are independent of the sense of smell. Both anosmic and intact carps exhibited the same pattern of behavioural response to the same type of pellet; they also did not differ in level of sensitivity to the substance contained in the pellet (Kasumyan and Morsy, 1996).

2.3 Preparation of Agar Pellets

Pellets were prepared from agar-agar gel (2%, Reanal) and were bright red. A dye solution (Ponceau 4R, 5 μM) was added to the gel together with one of the test substances or the bloodworm extract at the appropriate concentration. The control gel contained only the dye. Just before starting the experiment, each pellet was cut from the cool agar gel disc with a stainless steel tube. The pellet was 4.0 mm long and 1.35 mm in diameter. (For more details, see Kasumyan and Sidorov, 1995; Kasumyan and Morsy, 1996.) Classical taste substances (citric acid, NaCl, $CaCl_2$, and sucrose) and 21 free amino acids (L-isomers; Fluka, NBC, Calbiochem) were used as taste stimuli. The list of substances and their concentrations are given in Tables 13.1 and 13.2.

In total, 7662 trials were undertaken: 3718 trials with 22 sticklebacks from the Moscow River basin and 3944 trials with 17 sticklebacks from the White Sea basin. The Chi-square test (χ^2) was used for estimation of acceptance ratio and Student's *t*-test for all other characteristics of taste response. To assess the relationships between taste responses and taste preferences of different fish species, the Spearman rank correlation coefficient (r_s) was used.

3. RESULTS

3.1 Nine-spined Stickleback from the Moscow River Basin

The classical taste substances citric acid and calcium chloride significantly increased pellet consumption. Palatability of pellets containing citric acid was almost as high as that of pellets containing bloodworm extract. Pellets with calcium chloride were swallowed less often than pellets with citric acid, but 2.9 times more often than control pellets. Consumption of pellets containing sodium chloride or sucrose was low and did not differ significantly from consumption of control pellets. Fish responses to pellets with citric acid or with bloodworm extract differed significantly from responses to control pellets with respect to the other parameters recorded, such as number of grasps and retention time after the first grasp and during the entire trial. Pellets with citric acid or bloodworm extract were rarely spat out by fish and were retained much longer than control pellets. Pellets with sucrose were also retained by fish significantly longer than control pellets. The retention time for pellets with calcium chloride or sodium chloride were nearly the same as for control pellets (Table 13.1).

Table 13.1 Taste response of the nine-spined stickleback (*Pungitius pungitius*) from the Moscow River basin to agar pellets containing bloodworm (*Chironomidae* larvae) water extract, classical taste substances or free amino acids (L-isomers) (M±m). Concentration of the bloodworm extract given in g l^{-1}. In this and the following tables the significance levels (in relation to the control) are indicated as follows: ***, $p<0.001$; **, $p<0.01$ and *, $p<0.05$. The Chi-square test was used for evaluating the acceptance ratio and the Student's t-test was used for all other characteristics of taste responses.

Substances	*Concentration* M (%)	*Acceptance ratio, %*	*Number of grasps*	*Retention time, s*		*Number of trials*
				first grasp	*all grasps*	
Bloodworm water extract and classical taste substances						
Bloodworm water extract	175	79.2±3.3***	2.6±0.2***	8.3±0.3***	12.7±0.5***	154
Citric acid	0.26 (5)	57.6±4.3***	3.4±0.2*	7.1±0.6***	14.5±0.8***	132
Calcium chloride	0.9 (10)	15.2±3.1**	3.8±0.2	2.8±0.2	7.5±0.4	132
Sodium chloride	1.73 (10)	10.6±2.7	3.1±0.2***	2.9±0.3	6.2±0.4	132
Sucrose	0.29 (10)	7.6±2.3	4.6±0.3	3.6±0.2***	9.0±0.6**	132
Control	–	5.3±2.0	4.3±0.2	2.6±0.2	6.7±0.4	132
Free amino acids						
Glutamine	0.1	72.7±3.9***	1.8±0.1***	9.0±0.6***	12.2±0.7***	132
Cysteine	0.1	67.4±4.1***	2.4±0.2**	9.0±0.7***	13.0±0.7***	132
Alanine	0.1	66.7±4.1***	2.5±0.2*	6.7±0.4***	11.1±0.6***	132
Proline	0.1	47.7±4.4***	2.3±0.2**	6.3±0.5***	10.3±0.7***	132
Histidine	0.1	44.7±4.3***	2.4±0.2**	5.7±0.5***	9.7±0.6***	132
Serine	0.1	34.1±4.1***	3.1±0.2	4.8±0.4***	11.5±0.8***	132
Glycine	0.1	32.6±4.1***	2.9±0.2	4.1±0.3***	8.8±0.6**	132
Arginine	0.1	30.3±4.0***	2.9±0.2	4.9±0.4***	9.7±0.7***	132
Norvaline	0.1	30.3±4.0***	2.6±0.2	6.0±0.5***	10.7±0.8***	132
Lysine	0.1	28.8±4.0***	3.5±0.2	5.2±0.5***	10.9±0.8***	132
Phenylalanine	0.1	25.8±3.8***	2.5±0.2*	4.4±0.4***	8.2±0.7*	132
Methionine	0.1	20.5±3.5**	3.0±0.2	3.6±0.4*	9.1±0.9*	132
Asparagine	0.1	17.4±3.3*	3.3±0.2	3.0±0.3	8.5±0.7*	132
Valine	0.1	15.2±3.1	3.2±0.2	3.4±0.3*	8.4±0.7*	132
Threonine	0.1	9.1±2.5	3.1±0.2	2.7±0.2	8.2±1.0	132
Aspartic acid	0.01	37.9±4.2***	3.2±0.3	5.2±0.4***	10.4±0.8***	132
Leucine	0.01	25.0±3.8***	3.2±0.2	4.4±0.4***	9.7±0.8***	132
Glutamic acid	0.01	20.5±3.5**	3.4±0.2	3.4±0.3	8.2±0.6*	132
Isoleucine	0.01	16.7±3.3	2.8±0.2	4.0±0.4**	8.0±0.7	132
Tryptophan	0.01	13.6±3.0	3.3±0.2	3.6±0.3*	8.6±0.3*	132
Tyrosine	0.001	9.8±2.6	3.0±0.2	3.0±0.3	7.0±0.7	132
Control	–	9.1±2.5	3.1±0.2	2.7±0.2	6.3±0.6	132

Table 13.2 Taste response of the nine-spined stickleback (*Pungitius pungitius*) from the White Sea basin to agar pellets containing bloodworm (*Chironomidae* larvae) water extract, classical taste substances or free amino acids (L-isomers) (M ± m). Concentration of the blood worm (*Chironomidae* larvae) water extract given in g l^{-1}.

Substances	*Concentration* M (%)	*Acceptance ratio, %*	*Number of grasps*	*Retention time, s*		*Number of trials*
				first grasp	*all grasps*	
	Bloodworm water extract and classical taste substances					
Citric acid	0.26 (5)	48.2 ± 3.8***	4.9 ± 0.3***	5.0 ± 0.5***	12.9 ± 0.8***	170
Blood worm water extract	175	24.5 ± 4.3***	6.5 ± 0.5***	2.3 ± 0.2**	12.0 ± 0.9***	102
Sodium chloride	1.73 (10)	3.5 ± 1.4	3.5 ± 0.2	1.7 ± 0.1	5.3 ± 0.3	170
Sucrose	0.29 (10)	2.4 ± 1.2	3.0 ± 0.2	1.9 ± 0.1	4.5 ± 0.3	170
Calcium chloride	0.9 (10)	0***	3.7 ± 0.2	1.5 ± 0.1***	4.5 ± 0.3	170
Control	–	5.9 ± 1.8	3.5 ± 0.2	1.9 ± 0.1	5.2 ± 0.3	170
	Free amino acids					
Cysteine	0.1	66.9 ± 4.0***	3.7 ± 0.3***	10.7 ± 0.9***	21.3 ± 1.1***	136
Glutamine	0.1	13.2 ± 2.9***	4.6 ± 0.3***	3.2 ± 0.3***	11.3 ± 0.9***	136
Alanine	0.1	11.8 ± 2.8***	3.3 ± 0.2**	4.3 ± 0.4***	9.9 ± 0.7***	136
Lysine	0.1	10.3 ± 2.6***	3.2 ± 0.1**	2.4 ± 0.3	7.0 ± 0.8***	136
Proline	0.1	5.1 ± 1.9**	3.3 ± 0.2**	2.4 ± 0.3	6.5 ± 0.7***	136
Valine	0.1	3.7 ± 1.6*	3.5 ± 0.2***	2.1 ± 0.1	6.0 ± 0.4***	136
Serine	0.1	3.7 ± 1.6*	3.0 ± 0.2	2.0 ± 0.1	5.1 ± 0.4*	136
Arginine	0.1	2.9 ± 1.5*	4.0 ± 0.2***	1.9 ± 0.1	6.5 ± 0.5***	136
Asparagine	0.1	2.2 ± 1.3	3.7 ± 0.2**	1.6 ± 0.1*	4.9 ± 0.4*	136
Methionine	0.1	0.7 ± 0.7	3.2 ± 0.1**	1.9 ± 0.1	4.8 ± 0.3*	136
Norvaline	0.1	0.7 ± 0.7	2.8 ± 0.1	2.2 ± 0.1	5.0 ± 0.3*	136
Histidine	0.1	0	3.3 ± 0.1**	1.8 ± 0.1	4.7 ± 0.3*	136
Glycine	0.1	0	3.0 ± 0.2	1.9 ± 0.1	4.8 ± 0.3*	136
Threonine	0.1	0	2.8 ± 0.1	1.9 ± 0.1	4.3 ± 0.2	136
Phenylalanine	0.1	0	2.7 ± 0.1	1.7 ± 0.1*	3.5 ± 0.1	136
Aspartic acid	0.01	49.3 ± 4.3***	3.9 ± 0.3***	5.9 ± 0.5***	14.2 ± 0.8***	136
Glutamic acid	0.01	47.8 ± 4.3***	4.2 ± 0.3***	5.2 ± 0.5***	14.6 ± 0.9***	136
Tryptophan	0.01	0.7 ± 0.7	2.4 ± 0.1	1.8 ± 0.1	3.9 ± 0.2	136
Isoleucine	0.01	0	3.2 ± 0.2*	2.1 ± 0.1	5.2 ± 0.3**	136
Leucine	0.01	0	3.1 ± 0.1*	1.8 ± 0.1	4.5 ± 0.2	136
Tyrosine	0.001	0.7 ± 0.7	2.8 ± 0.1	1.8 ± 0.1	4.2 ± 0.2	136
Control	–	0	2.7 ± 0.1	1.9 ± 0.1	4.0 ± 0.2	136

Sixteen of 21 amino acids used for testing had a positive influence on pellet intake. Glutamine, cysteine, and alanine pellets had the highest acceptance, similar to that of pellets with bloodworm extract. Five of the amino acids (valine, threonine, isoleucine, tryptophan, and tyrosine) had no significant effect on pellet palatability. Further, none of the amino acids decreased pellet consumption significantly. Many of the amino acids tested increased pellet retention time and only the most palatable amino acids (glutamine, cysteine, alanine, proline, histidine and phenylalanine) decreased the number of grasps. For the first grasp, the mean value of pellet retention time varied from 3 to 9 seconds and for all grasps in the trial from 6 to 13 seconds (Table 13.1). Both positive and negative relationships were found between the parameters of fish taste response to pellets with free amino acids (Table 13.3).

3.2 Nine-spined Stickleback from the White Sea Basin

Among the classical taste substances, citric acid had a stimulatory effect on pellet consumption. Fish swallowed pellets containing citric acid 1.9 times more often than pellets with bloodworm extract and 8.2 times more often than control pellets. Calcium chloride had a deterrent effect and none of the 170 pellets with calcium chloride were swallowed. Fish retained the citric acid pellets and those with bloodworm extract much longer in their mouths than pellets containing the other classical taste substances. In contrast to fish belonging to the Moscow River population, sticklebacks from the White Sea basin grasped pellets with the most palatable substances more often than pellets with other taste substances (Table 13.2).

Among the 10 free amino acids which had a stimulatory effect, cysteine was the most palatable one. The remaining 11 amino acids had no significant effect on pellet consumption at the concentrations used. Fish did not swallow pellets containing histidine, isoleucine, leucine, glycine, threonine, and phenylalanine (Table 13.2). A positive relationship was found among all the four parameters registered during the trials (Table 13.3).

4. DISCUSSION

More than 2,000 km separate Himka creek (Moscow River basin) and the Lake Machinnoe (White Sea basin) where the sticklebacks used in the experiments were caught. However, our results clearly indicate that the

Table 13.3 Values of Spearman rank correlation coefficient between different characteristics of taste response to agar pellets containing free amino acids in the nine-spined stickleback *(Pungitius pungitius)* from the Moscow River (MR) and the White Sea (WS) basins. The Student's t-test was used for evaluating the value of Spearman rank correlation coefficient. The significance levels are indicated as follows: ***, $p<0.001$; **, $p<0.01$ and *, $p<0.05$.

Characteristics of taste response	*Number of grasps*		*Retention time*			
			first grasp		*all grasps*	
	MR	*WS*	*MR*	*WS*	*MR*	*WS*
Acceptance ratio	-0.60**	0.70**	0.93***	0.77***	0.86***	0.87***
Number of grasps			–0.62**	0.53*	–0.32	0.83***
Retention time during first grasp					0.87***	0.88***

taste preferences of nine-spined sticklebacks from two geographically isolated basins are rather similar (Fig. 13.1). Citric acid, cysteine, glutamine and alanine are among the most palatable substances for sticklebacks from both the White Sea and Moscow River basins. In both fish populations, sodium chloride, sucrose as well as isoleucine, tryptophan, tyrosine and threonine proved indifferent taste substances. A positive correlation ($r_s = 0.49$; $p < 0.05$) was found among the taste preferences of the fishes for the 21 free amino acids tested.

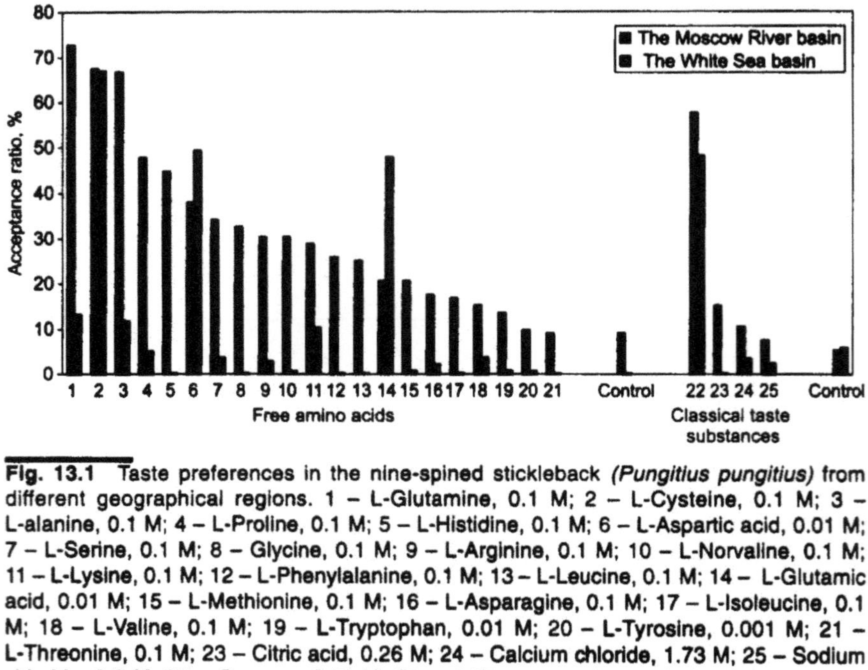

Fig. 13.1 Taste preferences in the nine-spined stickleback *(Pungitius pungitius)* from different geographical regions. 1 – L-Glutamine, 0.1 M; 2 – L-Cysteine, 0.1 M; 3 – L-alanine, 0.1 M; 4 – L-Proline, 0.1 M; 5 – L-Histidine, 0.1 M; 6 – L-Aspartic acid, 0.01 M; 7 – L-Serine, 0.1 M; 8 – Glycine, 0.1 M; 9 – L-Arginine, 0.1 M; 10 – L-Norvaline, 0.1 M; 11 – L-Lysine, 0.1 M; 12 – L-Phenylalanine, 0.1 M; 13 – L-Leucine, 0.1 M; 14 – L-Glutamic acid, 0.01 M; 15 – L-Methionine, 0.1 M; 16 – L-Asparagine, 0.1 M; 17 – L-Isoleucine, 0.1 M; 18 – L-Valine, 0.1 M; 19 – L-Tryptophan, 0.01 M; 20 – L-Tyrosine, 0.001 M; 21 – L-Threonine, 0.1 M; 23 – Citric acid, 0.26 M; 24 – Calcium chloride, 1.73 M; 25 – Sodium chloride, 0.9 M; 26 – Sucrose, 0.29 M; 22 and 27 – respective control pellets.

However, the taste preferences of the fishes from the two groups compared did not coincide completely. Of the 21 amino acids, 16 were palatable for fish from the Moscow River basin and only 10 were palatable for fish from the White Sea basin. None of the amino acids evoked aversive responses in the compared groups of fish. Histidine, glycine, norvaline, phenylalanine, leucine, methionine and asparagine were taste stimulants for sticklebacks from the Moscow River basin and were indifferent taste substances for sticklebacks from the White Sea. One amino acid, valine, was palatable for fish from the White Sea basin but had no effect on fishes from the Moscow River basin. It should be emphasized that the listed amino acids except for histidine, belong to substances of weak palatability for fish of both groups. Calcium chloride is also a good example for taste differences. Pellets containing calcium chloride were palatable for fish from the Moscow River basin but evoked aversive responses in fish from the White Sea basin. It is possible that calcium concentration in the surrounding water might be the main reason for the taste differences: concentration of calcium in Himka creek is almost 30 times lower than in the Lake Machinnoe, 2.2 and 59.1 mg L^{-1} respectively (Maximova, 1967; Technical report ..., 1992). (It is necessary to emphasize that the difference between the concentration of calcium in Himka creek and in calcium chloride-pellets was still relatively high, more than in 500 times.) However, the suggestion is not supported by the results obtained for the three-spined stickleback *Gasterosteus aculeatus* morpha *trachurus* (Kasumyan and Mikhailova, unpubl.). The acceptance ratio for pellets containing sodium chloride was the same for three-spined sticklebacks in sea water and after transferring the fish into fresh water, in spite of the fact that the sodium chloride concentration in the surrounding water was changed more than 10,000 times.

The results presented in this chapter indicate that within one species of fish, taste preferences do not have a high population specificity. Similar data were obtained earlier when the taste preferences of brown trout from the Caspian Sea, Baltic Sea and White Sea basins were compared (Kasumyan and Sidorov, 1995a, c, 1997). In brown trout, citric acid and L-cysteine were the most palatable substances for fish from all three populations. L-isoleucine, L-glutamine, L-valine, L-glycine provoked the strongest aversion taste responses while sodium chloride, calcium chloride, sucrose, and many of the free amino acids were indifferent taste substances.

In the present study taste preferences were similar for single chemical substances, but for more complex taste stimuli such as bloodworm extract, palatability was higher for sticklebacks from the Moscow River basin than for those from the White Sea basin. In a study on grass carp *Ctenopharyngodon idella*, consumption of pellets flavoured with bloodworm or plant extracts differed between the two fish groups, one of which was reared on bloodworms, the other on duckweed *(Lemna minor)* and Romaine lettuce *(Lactuca sativa)* leaves (Kasumyan and Morsy, 1997). However, grass carp of both groups had similar taste preferences with respect to the 4 classical taste substances and the 21 amino acids tested. These results show that long-term rearing on a selected diet does not lead to a shift in taste preferences to simple taste stimuli.

Nine-spined sticklebacks from the two different basins were similar with regard to the time the fish needed for testing the pellets. Retention time for pellets with the most palatable substances was approximately the same for fish of both groups. Total retention time of a pellet exceeded 10 s and retention time for the first grasp varied from 5 to 10 s. It should also be noted that the retention time was much shorter for pellets with indifferent taste substances. It was about the same or even less than the handling time (3 s) needed by the three-spined stickleback *Gasterosteus aculeatus* to decide whether to eat or spit out its natural prey, the freshwater isopod *Asellus aquaticus* (Gill and Hart, 1994). In the nine-spined stickleback, the time of keeping the pellet in the mouth decreased with acceptance ratio. A high positive correlation between these parameters of taste response existed for both groups of nine-spined sticklebacks as well as for many other fish species: tench *(Tinca tinca)*, common carp *(Cyprinus carpio)*, goldfish *(Carassius auratus)*, roach *(Rutilus rutilus)*, and guppy *(Poecilia reticulata)* (Kasumyan and Morsy, 1996; Kasumyan and Nikolaeva, 1997, 2002). These findings indicate that prey handling time is strongly determined not only by the prey width/mouth width ratio, as shown for the three-spined stickleback (Gill and Hart, 1994), but also by the prey's palatability.

It is noteworthy that nine-spined sticklebacks from the two groups showed a reversed relationship between pellet acceptance ratio and number of repeated grasps at pellets. Fish from the Moscow River population made fewer attempts to regrasp the more palatable pellets while fish from the White Sea population increased the number of regrasps with increase in pellet palatability. In total, the mean number of pellet grasps was higher for fish from the White Sea basin than for fish from the Moscow River basin (ranging from 1.8–3.5 and 2.4–4.6 respectively for pellets containing free amino acids). Repeated grasping and spitting out of food items is a typical behavioural pattern inherent in various fish species, including the three-spined sticklebacks (Gill and Hart, 1994, 1996a,b; Batty and Hoyt, 1995; Wanzenböck, 1996; McLaughlin et al., 2000). There might be more than 10 prey manipulations depending on the prey/predator size ratio (Hart and Gill, 1992, 1994). Prey manipulation is essential to reorientate the prey before swallowing. It also depends on fish experience in prey hunting (Hart and Gill, 1992; Ibrahim and Huntingford, 1992). Sticklebacks are known to learn to respond to new prey types and may rapidly modify their feeding behaviour (Beukema, 1968; Milinski and Lowenstein, 1980; Croy and Hughes, 1991).

Competition was described for sticklebacks foraging together in the same patch. No fish abandoned a prey immediately even if a competitor had caught it. The unsuccessful competitor stayed close by to intercept the prey which might be spat out by the other fish. Prey abandoning time increased when the initial stimulation of seeing the prey was reinforced by periodic availability (Gill and Hart, 1996a). We hypothesize that the frequency of spitting out a food item may depend on density of fish population. Thus the number of repeated grasps of a pellet, especially for pellets containing highly palatable substances, was lower for sticklebacks from Himka creek due to the high density of fish in this population vis-à-vis the sparse population from the Lake Mashinnoe. It should be emphasized that retention time for pellets with the most palatable substances was longer than the so-called prey abandoning time (approx. 7 s), during which the unsuccessful competitor remained close to the successful one even after the prey was already swallowed (Gill and Hart, 1996a).

Another explanation for this discrepancy may be associated with differences in the hydrological conditions of the Lake Machinnoe and Himka creek. Usually fishes from the river such as brown trout spat out food items relatively fewer times than bottom feeders which inhabit slowly moving waters (common carp, tench). A lower number of repeated grasps at the food item increases the efficiency of river fish foraging when the prey can be readily swept away by the water current (Kasumyan and Sidorov, 1995a; Kasumyan, 1997). It is known that foraging efficiency of different morphs of the same fish species may differ. For example, the shallow-water morph of the three-spined stickleback requires fewer bites to capture and consume more and larger prey than open-water conspecifics (Robinson, 2000).

During the last decade, significant progress has been made in understanding taste preferences of fish. Several fish species have been tested and the results comparable because the taste preferences for these fish species were determined by the same method as used in the present study. Comparative analyses of the results obtained for nine-spined stickleback and for 14 other fish species revealed that oral taste preferences are highly species specific. Differences among fish species are apparent upon comparing the width of spectra for different types of taste substances: stimulant, deterrent and indifferent (Table 13.4).

Differences in taste preferences were confirmed by correlation analysis of amino acid palatability for 15 fish species (Table 13.5; Kasumyan and Sidorov, 1993, 1995a, b, 2001; Kasumyan and Morsy, 1996; Kasumyan and Nikolaeva, 1997, 2002; Kasumyan, 1999; Kasumyan and Prokopova, 2001; Kasumyan and Marusov, 2003). A significant correlation was found for a small number of pairs of fish species and, in most cases, a positive relationship established between species distant in systematic position, ecology and feeding type. For example, a positive correlation was found between the nine-spined stickleback and the common carp or tench – two benthivorous cyprinid fish inhabiting a more southern area which only rarely occur sympatrically with the nine-spined stickleback. Thus correlation analysis clearly demonstrated that taste preferences are highly species specific. It also showed an important and undeniably leading role of gustatory reception in feeding selectivity in fishes, and in their capacity to consume food items appropriate and specific to them.

In 1992, the nine-spined stickleback from the same stretch of Himka creek was also used for similar experiments in which only the classical taste substances were applied (Kasumyan and Nikolaeva, 2002). Within the 8-year period from 1992 to 2000, 3–4 stickleback generations were replaced. However, essential changes in their taste preferences were not observed during that time: both citric acid and bloodworm extract proved highly stimulatory. Effects of sodium chloride, calcium chloride and sucrose were close to control level (Fig. 13.2). These data underscore stability of fish taste preferences over time.

This study has thus shown that taste preferences in fish are species specific and have no strong population specificity. Fish from geographically isolated populations spent the same time testing the taste properties of food items and deciding whether to swallow or reject them. Characteristically fish behavioural taste response is variable and may be realized in different ways between conspecifics that inhabit still or running waters, or between conspecifics from water bodies distinguished by the fish's population density.

Table 13.4 The number of amino acids that act as stimulants, deterrents or indifferent taste stimuli in 21 fish species. The number of amino acids (L-isomers) tested was 21 for all species except the Siberian sturgeon (19 amino acids tested).

Species	*Stimulants*	*Indifferent*	*Deterrents*	*Source*
Brown trout, *Salmo trutta caspius*	6	10	5	Kasumyan and Sidorov, 1995a
Frolich char, *Salvelinus alpinus erythrinus*	4	16	1	Kasumyan and Sidorov, 1995b
Lake char, *Salvelinus namaycush*	4	14	3	Kasumyan and Sidorov, 2001
Chum salmon, *Oncorhynchus keta*	12	7	2	Kasumyan and Sidorov, 1993
Guppy, *Poecilia reticulata*	5	16	0	Kasumyan and Nikolaeva, 1997
Siberian sturgeon, *Acipenser baerii*	7	11	1	Kasumyan, 1999
Russian sturgeon, *Acipenser gueldenstaedtii*	6	15	0	Kasumyan, 1999
Stellate sturgeon, *Acipenser stellatus*	3	18	0	Kasumyan, 1999
Common carp, *Cyprinus carpio*	6	8	7	Kasumyan and Morsy, 1996
Goldfish, *Carassius auratus*	8	10	3	Kasumyan and Nikolaeva, 2002
Roach, *Rutilus rutilus*	8	13	0	Kasumyan and Nikolaeva, 2002
European minnow, *Phoxinus phoxinus*	4	14	3	Kasumyan and Marusov, 2003
Tench, *Tinca tinca*	12	9	0	Kasumyan and Prokopova, 2001
Arctic flounder, *Liopsetta glacialis*	0	21	0	Kasumyan and Nikolaeva, 2002
Nine-spined stickleback, *Pungitius pungitius* (from the Moscow River basin)	16	5	0	Present study
Nine-spined stickleback, *Pungitius pungitius* (from the White Sea basin)	10	11	0	Present study

Table 13.5 Values of Spearman rank correlation coefficient for taste preferences to 21 free amino acids (19 for Siberian sturgeon) of different fish species.

Species:	*2*	*3*	*4*	*5*	*6*	*7*	*8*	*9*	*10*	*11*	*12*	*13*	*14*	*15*	*16*
1 – Brown trout, *Salmo trutta caspius*	0.64**	0.42	-0.23	-0.03	-0.11	0.47*	0.15	0.33	–0.06	–0.22	–0.59**	–0.04	–0.30	0.07	0.23
2 – Frolich char, *Salvelinus alpinus erythrinus*	–	0.69**	0.01	0.06	0.03	0.31	0.03	0.41	–0.10	–0.34	–0.27	–0.00	–0.07	0.33	0.16
3 – Lake char, *Salvelinus namaycush*		–	0.06	–0.05	–0.09	0.31	0.11	0.18	–0.25	–0.48*	-0.11	–0.07	–0.11	0.06	0.25
4 – Chum salmon, *Oncorhynchus keta*			–	–0.45*	–0.01	–0.63**	–0.03	–0.14	0.10	–0.05	0.21	–0.29	0.11	–0.10	–0.47*
5 – Guppy, *Poecilia reticulata*				–	–0.17	0.26	–0.13	0.45*	–0.18	–0.16	0.30	0.23	–0.22	0.33	0.50
6 – Siberian sturgeon, *Acipenser baerii*					–	0.20	–0.12	–0.26	0.36	–0.02	–0.23	–0.34	0.35	–0.47*	–0.44*
7 – Russian sturgeon, *Acipenser gueldenstaedtii*						–	0.14	–0.05	0.04	–0.06	–0.37	0.08	–0.07	–0.01	0.34
8 – Stellate sturgeon, *Acipenser stellatus*							–	0.08	0.08	–0.02	–0.12	0.22	0.07	0.30	–0.07
9 – Common carp, *Cyprinus carpio*								–	–0.17	–0.34	0.08	0.28	–0.11	0.54*	0.50*
10 – Goldfish, *Carassius auratus*									–	0.28	–0.06	–0.31	0.06	–0.39	-0.24
11 – Roach, *Rutilus rutilus*										–	0.25	0.06	0.08	–0.09	–0.10
12 – European Minnow, *Phoxinus phoxinus*											–	0.32	0.26	0.28	0.21
13 – Tench, *Tinca tinca*												–	–0.03	0.73**	0.66**
14 – Arctic flounder, *Liopsetta glacialis*													–	0.08	–0.12
15 – Nine-spined stickleback, *Pungitius pungitius* (from the Moscow River basin)														–	0.49*
16 – Nine-spined stickleback, *Pungitius pungitius* (from the White Sea basin)															–

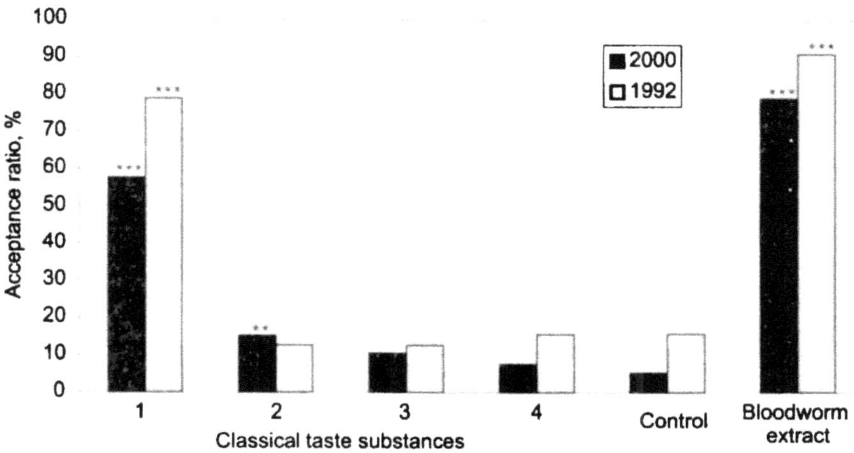

Fig. 13.2 Taste preferences of the nine-spined stickleback *(Pungitius pungitius)* from the creek Himka (the Moscow River basin) for classical taste substances and for blood worm water extract 175 g (wet weight)·L^{-1} (modified from Kasumyan and Nikolaeva, 2002). 1 – Citric acid, 0.26 M; 2 – Calcium chloride, 1.73 M; 3 – Sodium chloride, 0.9 M; 4 – Sucrose, 0.29 M; Control – blank pellets. The significance levels are indicated as follows: ***, $p < 0.001$, ** $p < 0.01$.

Acknowledgements

We thank Anne Hansen for critically reading the manuscript. We are grateful to Eugeny Marusov and Sergey Sidorov for assistance in catching the fish. This research was supported by the Russian Foundation for Basic Research (project 01-04-48460).

References

Adams, M.A., P.B. Johnsen and Z. Hong-Qi. 1988. Chemical enhancement of feeding for the herbivorous fish *Tilapia zillii*. *Aquaculture* 72: 95–107.

Appelbaum, S. 1980. Versuche zur Geschmacksperzeption einiger Süsswasserfische im larvalen und adulten Stadium. *Arch. Fischereiwiss.* 31: 105–114.

Atema, J. 1980. Chemical senses, chemical signals and feeding behaviour in fishes. In: *Fish Behaviour and Its Use in the Capture and Culture of Fishes*, J.E. Bardach, J.J. Magnuson, R.C. May and J.M. Reinhart (Eds). International Center for Living Resources Management, Manila, Philippines, pp. 57–101.

Batty, R.S. and R.D. Hoyt. 1995. The role of sense organs in the feeding behaviour of juvenile sole and plaice. *J. Fish Biol.* 47: 931–939.

Beukema, J.J. 1968. Predation by the three-spined stickleback, *Gasterosteus aculeatus* L.: the influence of hunger and experience. *Behaviour* 31: 1–126.

Carr, W.E.S. 1982. Chemical stimulation of feeding behaviour. In: *Chemoreception in Fishes*, T.J. Hara (Ed.). Elsevier, Amsterdam, pp. 259–273.

Croy, M.I. and R.N. Hughes. 1991. The role of learning and memory in the feeding behaviour of the 15-spined stickleback, *Spinachia spinachia*. *Animal Behav.* 41: 149–160.

Gill, A.B. and P.J.B. Hart. 1994. Feeding behavior and prey choice of the three-spine stickleback: the interacting effects of prey size, fish size and stomach fullness. *Anim. Behav.* 47: 921–932.

Gill, A.B. and P.J.B. Hart. 1996a. Unequal competition between three-spined stickleback, *Gasterosteus aculeatus* L, encountering sequential prey. *Anim. Behav.* 51: 689–698.

Gill, A.B. and P.J.B. Hart. 1996b. How feeding performance and energy intake change with a small increase in the body size of the three-spined stickleback. *J. Fish Biol.* 48: 878–890.

Goh, Y. and T. Tamura. 1980. Effect of amino acids on the feeding behaviour in Red Sea bream. *Comp. Biochem. Physiol.* 66C: 225–229.

Hansen, A. and K. Reutter. 2004. Chemosensory systems in fish: Structural, functional and ecological aspects. In: *The Senses of Fish: Adaptations for the Reception of Natural Stimuli*, G. von der Emde, J. Mogdans and B.G. Kapoor (Eds). Narosa Publishing House, New Delhi, and Kluwer Academic Publishers, Dordrecht, the Netherlands. pp. 55–89.

Hara, T.J. 1994. The diversity of chemical stimulation of fish olfaction and gustation. *Rev. Fish Biol. Fish.* 4: 1–35.

Hart, P.J.B. and A.B. Gill. 1992. Constraints on prey size selection by the three-spined stickleback: energy requirements and the capacity and fullness of the gut. *J. Fish Biol.* 40: 205–218.

Hidaka, I. 1982. Taste receptor stimulation and feeding behavior in the puffer. In: *Chemoreception in Fishes*, T.J. Hara (Ed.). Elsevier, Amsterdam, pp. 243–257.

Ibrahim, A.A. and F.A. Huntingford. 1992. Experience of natural prey and feeding efficiency in threespine sticklebacks, *Gasterosteus aculeatus* L. *J. Fish Biol.* 41: 619–625.

Jakubowski, M. and M. Whitear. 1990. Comparative morphology and cytology of taste buds in teleosts. *Z. mikrosk.-anat. Forsch.* 104: 529–560.

Johnsen, P.B. and M.F. Adams. 1986. Chemical feeding stimulants for the herbivorous fish, *Tilapia zillii*. *Comp. Biochem. Physiol.* 83A: 109–112.

Jones, K.A. 1989. The palatability of amino acids and related compounds to rainbow trout, *Salmo gairdneri Richardson*. *J. Fish Biol.* 34: 149–160.

Kasumyan, A.O. 1997. Gustatory reception and feeding behavior in fish. *J. Ichthyology* 37: 72–86.

Kasumyan, A.O. 1999. Olfaction and taste in sturgeon behaviour. *J. App. Ichthyology* 15: 228–232.

Kasumyan, A.O. and K.B. Døving. 2003. Taste preferences in fishes. *Fish and Fisheries* 4: 289–347.

Kasumyan, A.O. and E.A. Marusov. 2003. Behavioural responses of intact and chronically anosmiated minnows, *Phoxinus phoxinus* (Cyprinidae), to free amino acids. *J. Ichthyology* 43: 528–539.

Kasumyan, A.O. and A.M.H. Morsy. 1996. Taste sensitivity of common carp, *Cyprinus carpio*, to free amino acids and classical taste substances. *J. Ichthyology* 36: 391–403.

Kasumyan, A.O. and A.M.H. Morsy. 1997. Taste preference for classic taste substances in juveniles of the grass carp, *Ctenopharyngodon idella* (Cyprinidae, Pisces), reared on various diets. *Dokl. Biol. Sci.* 357: 284–286.

Kasumyan, A.O. and E.V. Nikolaeva. 1997. Taste preferences of *Poecilia reticulata* (Cyprinidontiformes). *J. Ichthyology* 37: 662–669.

Kasumyan, A.O. and E.V. Nikolaeva. 2002. The comparative analysis of taste preferences in fish with different ecology and feeding. *J. Ichthyology* 42. Suppl. 2. Behavior, distribution and migration of fishes: S203–S214.

Kasumyan, A.O. and O.M. Prokopova. 2001. Taste preferences and the dynamics of behavioral taste response in the tench, *Tinca tinca* (Cyprinidae). *J. Ichthyology* 41: 640–653.

Kasumyan, A.O. and S.S. Sidorov. 1993. Taste sensitivity of chum salmon, *Oncorhynchus keta*, to types of taste stimuli and amino acids. *Sensory systems* 6: 222–225.

Kasumyan, A.O. and S.S. Sidorov. 1995a. Gustatory properties of free amino acids in Caspian trout, *Salmo trutta caspius. J. Ichthyology* 35: 8–20.

Kasumyan, A.O. and S.S. Sidorov. 1995b. The palatability of free amino acids and classical taste substances to frolich char, *Salvelinus alpinus erythrinus* (Georgi). *Nordic J. Freshwater Res.* 71: 320–323.

Kasumyan, A.O. and S.S. Sidorov. 1995c. A comparative analysis of the taste responses of young salmon trout, *Salmo trutta trutta*, from populations of the Baltic and White seas. *Dokl. Biol. Sci.* 343: 289–391.

Kasumyan A.O. and S.S. Sidorov. 1997. Comparison of free amino acids and classical taste substances consumption in brown trout, *Salmo trutta* (Salmonidae, Pisces), juveniles from different sea basin populations. *Chem. Senses* 22: 714–715.

Kasumyan, A.O. and S.S. Sidorov. 2001. Taste preferences in lake char, *Salvelinus namaycush (Salmonidae)*, juveniles. *J. Fish.* Supplement 1: 121–125 (In Russian).

Lamb, C. and T.E. Finger. 1995. Gustatory control of feeding behavior in goldfish. *Physiol. Behav.* 57: 483–488.

Mackie, A.M. 1982. Identification of the gustatory feeding stimulants. In: *Chemoreception in Fishes*, T.J.Hara (Ed.). Elsevier, Amsterdam, pp. 275–291.

Marui, T. and J. Caprio. 1992. Teleost gustation. In: *Fish Chemoreception*, T.J. Hara (Ed.). Chapman & Hall, London, New York. pp.171–198.

Maximova, M.P. 1967. Ionic and organic flow and the correlation of the main ions in the rivers of the Karelian coast of the White Sea. In: *Hydrobiological Investigations of the Karelian Coast of the White Sea* Nauka, Leningrad, pp. 9–14 (In Russian).

McLaughlin, R.L., J.W.A. Grant and D.L.G. Noakes. 2000. Living with failure: the prey capture success of young brook char in streams. *Ecol. Freshwater Fish.* 9: 81–89.

Mearns, K.J., O.F. Ellingsen, K.B. Døving and S. Helmer 1987. Feeding behaviour in adult rainbow trout and Atlantic salmon parr, elicited by chemical fractions and mixtures of compounds identifited in shrimp extract. *Aquaculture* 64: 47–63.

Milinski, M. and C. Lowenstein. 1980. On predator selection against abnormalities of movement: a test of a hypothesis. *Z. Tierpsychol.* 53: 325–340.

Pavlov, D.S. and A.O. Kasumyan. 1990. Sensory principles of the feeding behaviour of fishes. *J. Ichthyology* 30: 77–93.

Reutter, K. 1992. Structure of the peripheral gustatory organ, represented by the siluroid fish *Plotosus lineatus* (Thunberg). In: *Fish Chemoreception*, T.J. Hara (Ed.). Chapman & Hall, London, New York, pp. 60–78.

Reutter, K. and M. Witt. 1993. Morphology of vertebrate taste organs and their nerve supply. In: *Mechanisms of Taste Transduction*, S.A.Simon and S.D.Roper (Eds). CRC Press, Boca Raton, pp.29–82.

Robinson, B.W. 2000. Trade-offs in habitat-specific foraging efficiency and the nascent adaptive divergence of sticklebacks in lakes. *Behaviour* 137: 865–888.

Sorensen, P.W. and J. Caprio. 1998. Chemoreception. In: *The Physiology of Fishes*. D.H.Evans (Ed.). 2nd Edition. CRC Press, Boca Raton, pp. 375–405.

Sutterlin, A.M. and N. Sutterlin. 1970. Taste responses in Atlantic salmon (*Salmo salar*) parr. *J. Fish. Res. Bd. Can.* 27: 1927–1942.

Technical report of the Rublevo tap water station. 1992. Moscow, p. 1 (In Russian).

Wanzenböck, J. 1996. The need for state-dependent foraging models to predict size-selectivity of 0+ zooplanctivorous fish. *Limnologica* 26: 139–143.

Index

PART A EXPRESSIONS

*ff—The same item or expression of further cited in at least four subsequent pages of this chapter.

T

PART B SCIENTIFIC NAMES of NAMED SPECIES

For Product Safety Concerns and Information please contact our EU
representative GPSR@taylorandfrancis.com
Taylor & Francis Verlag GmbH, Kaufingerstraße 24, 80331 München, Germany

www.ingramcontent.com/pod-product-compliance
Ingram Content Group UK Ltd.
Pitfield, Milton Keynes, MK11 3LW, UK
UKHW021503240926
475702UK00035B/232
* 9 7 8 1 5 7 8 0 8 3 1 9 0 *